Sustainable agroecosystems in climate change mitigation

Sustainable agroecosystems in climate change mitigation

edited by:

Maren Oelbermann

Wageningen Academic Publishers

EAN: 9789086862351
e-EAN: 9789086867882
ISBN: 978-90-8686-235-1
e-ISBN: 978-90-8686-788-2
DOI: 10.3920/978-90-8686-788-2

First published, 2014

Wageningen Academic Publishers
P.O. Box 220
6700 AE Wageningen
The Netherlands
www.WageningenAcademic.com
copyright@WageningenAcademic.com

Preface

Agriculture is a relatively recent phenomenon in the overall history of human habitation on this planet. Some of the food-gathering mechanisms used by hunter-gatherer societies were relatively advanced which gave rise to the domestication of animals and plants and contributed to the development of a more complex social structure that promoted co-operation and exchange of knowledge. These activities promoted positive changes in crop yield and animal production leading to the continual improvement in agricultural practices, with the greatest leaps forward in the past century. However, the rise of modern agriculture and its reliance on fossil fuels and fertilizers has also impacted the environment and made a substantial contribution to the increased concentration of greenhouse gases in the atmosphere. As such, one of the challenges under current land management practices is to increase food and soil security to meet projected trends in food production while minimizing greenhouse gas emissions from these agricultural activities.

This book shows that agriculture can also play a large role in mitigating greenhouse gases and help sequester carbon in a variety of different land management practices ranging from rice cultivation to coffee plantations to pasture management for livestock. The chapters within this book suggest that policies and practices integrating microbial technology, modern crop cultivars, conservation practices, increased manure application, organic farming and agroforestry have an enhanced capacity to sequester carbon and reduce the emission of carbon-based greenhouse gases, leading to more robust agroecosystems than conventional agriculture. This book illustrates that empirical models can represent a powerful tool to help assess how mitigation and adaptation strategies can be used to optimize crop yield and predict future changes in soil organic carbon stocks. It also presents new tools by using tomographic methods to measure bulk density with minimal soil disturbance.

The chapters have been written by experienced and internationally recognized scientists in their field. Each chapter helps bridge our current knowledge gaps and recognizes the contribution of sustainable agricultural practices as a way forward in reducing the global carbon and nitrogen footprint. This book is relevant for students, researchers, governmental and non-governmental organisations interested in climate change mitigation, sustainable agriculture, soil science, modern analytical techniques and modelling. It answers the questions: 'How can sustainable agroecosystems help mitigate climate change?' and 'What are the tools to achieve this goal?'

Maren Oelbermann
Editor

Table of contents

Section 1.
Opportunities for carbon sequestration in agroecosystems

Chapter 1.

Effect of soil conservation practices on organic carbon in Vertisols and Luvisols of Northern Italy

S. Brenna[1], A. Rocca[1], M. Sciaccaluga[1] and M. Grandi[2]

[1]ERSAF, Regional Agency for Agriculture and Forests of Lombardy, Via Pola 12, 20124 Milan, Italy; [2]AIGACoS – Italian Association for Conservative Land and Soil Management, Via Emilia Parmense 84, 29122 Piacenza, Italy; stefano.brenna@ersaf.lombardia.it

Abstract

Soils of the Lombardy region in northern Italy contain about 124 million tonnes of soil organic carbon (SOC) in the upper 30 cm. However, the SOC stock varies considerably throughout the region as a result of differences in bioclimatic conditions, soil types and land-use. The lowest average SOC stock (54 t/ha) was found in the croplands of the Po Plain. The current study was part of the AgriCO_2ltura project whose objective was to evaluate the effect of zero tillage and conservation agriculture (CA) on SOC storage. Research sites used in our study were in different pedoclimatic locations to compare the effects of long-term of CA practices to that of traditional agricultural practices (TA) including ploughing on SOC. Results from our study showed that CA in clay soils with vertic features (Vertisols or Vertic Cambisols), formed under a mean annual precipitation of ~650 mm, had the greatest SOC accumulation. Soils under CA had a 46% higher SOC stock compared to TA management practices on the vertic soils. However, using the same CA and TA management practices on a coarse textured Luvisol, with an annual precipitation of ~735 mm, did not lead to a significant difference in SOC stocks.

Keywords: SOC stock, climate change, soil sampling, zero tillage

1.1 Introduction

In terrestrial ecosystems, soil is a major long-term reservoir for organic carbon (C). This reservoir accounts for ~1,500 Gt C, which is double the amount stored in plant biomass or the atmosphere (Tarnocai *et al.*, 2009). Although soil represents the largest terrestrial C sink, this sink is continuously influenced by environmental factors such as global warming (Davidson and Janssens, 2006), land-use change and land management practices (Lal, 2004). For example, land-use and climate change were estimated to contribute to a loss in soil organic C (SOC) at a rate equivalent to 10% of the total fossil fuel emissions produced by Europe (Janssens *et al.*, 2005). Recent soil inventories revealed a decline in SOC including regions in Belgium (Sleutel *et al.*, 2003), Austria (Dersch and Boehm, 2001) and across England and Wales where the annual loss was estimated to be ~13 million tonnes between 1978 and 2003 (Bellamy *et al.*, 2005).

The uncertainty of these estimates is high (Vleeshouwers and Verhagen, 2002) and further measurements through repeated soil inventories are needed (Schrumpf *et al.*, 2011). However, there is consensus that SOC stocks have also declined in many agricultural soils over the past decades because of intensive agroecosystem management practices. Soils in the United States, for example, were estimated to have lost 30-50% of their original SOC (Kucharik *et al.*, 2001) and this decline was mainly attributed to soil cultivation using conventional (mould board plough) techniques (Reicosky, 2003).

Enhancing levels of SOC in agroecosystems can be achieved through the reduction of tillage and a decrease in the SOC mineralization rate or through an increase of organic matter input (Lal, 2003). Soils which have lost a considerable amount of SOC may have a greater potential and capacity to regain it. To achieve this goal, many researchers have proposed the implementation of conservation agriculture (CA) to minimize the negative impacts of traditional tillage-based (TA) management practices. Conservation agriculture follows three main principles: (1) minimal or no soil disturbance by tillage; (2) permanent land cover with crop residues and cover crops; and (3) crop diversity in the form of well-balanced and diverse crop rotations. The combination and interaction of these practices improves soil structure and function, and enhances the capacity of soils to sequester C and increase SOC stocks (Lal *et al.*, 1998). Basch *et al.* (2012) suggested that CA practices may increase the SOC stock by 0.2-0.7 t/ha/year, which may also play a key role in mitigating rising levels of atmospheric CO_2 (Pacala and Socolow, 2004).

There is currently a debate within the literature whether or not CA management practices can actually lead to a significant and measurable level of soil C sequestration (Govaerts *et al.*, 2009). In a review by Govaerts *et al.* (2009), SOC stocks were reported to be greater in CA compared to TA agroecosystem management practices in 40 out of 78 studies. In a compendium of Canadian studies, the mean storage rate of SOC in the CA systems was rather low and occurred only in some pedoclimatic conditions (VandenBygaart *et al.*, 2003). Improved management of cropland resulted in a relative gain in SOC across a wide range of Australian soils (Sanderman *et al.*, 2009). However, in many cases soils showed a significant decline in SOC.

The mechanisms that govern the incorporation of C in soil, after the conversion from TA to CA management practices, are not completely understood and results seem to be strongly influenced by soil type, landscape, climate, and initial SOC background levels. Altering crop rotation or organic matter inputs from livestock manure may also influence SOC stocks. Field measurements of changes in SOC are hampered by the inherently high spatial variability of soil, which is often not accounted for (Schrumpf *et al.*, 2011). Other analyses suggested that soil sampling protocols, limited to a depth of 30 cm or less may have biased the results, overestimating C sequestration in soils (Baker *et al.*, 2007). Ellert and Bettany (1995) also suggested that the importance of changes in soil bulk density (BD) due to the different tillage systems is often neglected when accounting for changes in SOC.

Additionally, CA is a comprehensive system for cultivating crops enhancing the ability of soils to function, while reduced tillage only addresses the physical changes in how the soil is cultivated compared to conventional tillage (Hobbs *et al.*, 2008). Although SOC sequestration has a finite potential and could be considered a riskier long-term strategy for climate mitigation than direct reduction of C emissions, improving agricultural management offers a range of other environmental and economic benefits which may make attempts to improve SOC storage attractive as part of integrated sustainability policies (Follett, 2001).

Currently CA is widely regarded as a technology that can improve soil functionality and is viewed as a pillar of sustainable crop production intensification (SCPI) which was identified by the United Nations Food and Agriculture Organization (UN-FAO) as one approach to ensure food security for a growing population, while conserving the environment and mitigating the effects of climate change (Corsi *et al.*, 2012). At the same time, there is consensus that SOC enhances resilience of agroecosystems and increases sustainability of rural livelihoods, minimizing the negative socioeconomic and environmental consequences of conventional agricultural practices (Stolbovoy *et al.*, 2005). Soil organic C drives the majority of soil ecological functions, including fertility, pH buffering capacity, heavy metal adsorption, water quality, and regulating gas exchange between the lithosphere and atmosphere. Thus, a reliable determination of changes in SOC stocks becomes a prerequisite to assess the role of soils in the global C cycle and the effect of soil management practices on C sequestration.

In Lombardy, Italy ~124 million tonnes of SOC are stored in the upper 30 cm with slight variations across the region depending on bioclimatic conditions, soil type and land-use. Soil organic C stocks are low in the Po Plain, where cropland shows a mean stock of 54 t/ha with the lowest stock ranging from 30-40 t/ha in the western and southern part of this region (Brenna *et al.*, 2010). Thus, increasing SOC content by

0.1% (for example from 2 to 2.1%) in the plough layers of the ~900,000 ha cropland in the Lombardy plain, may lead to an increase in the regional SOC stock by more than 3 million tonnes. This study, was carried out by ERSAF (Regional Agency for Agriculture and Forests of Lombardy) as part of a research and technology transfer project (AgriCO_2ltura) whose objective was to evaluate the effect of zero tillage and conservation agriculture (CA) on SOC storage at two different sites in the Po plain (Figure 1.1). These two sites were characterized by different pedoclimatic conditions and cropping systems. Results were expected to support sustainable soil management practices and the development of C emission offset protocols of interest to agricultural producers in this region of Italy.

1.2 Materials and methods

The study sites were selected to compare agroecosystem management practices between CA using zero tillage and TA practices. The first site was located south of the Po River in the district of Oltrepo Pavese (45°04'N, 09°10'E) at 62.8 m.a.s.l. The soil at this site

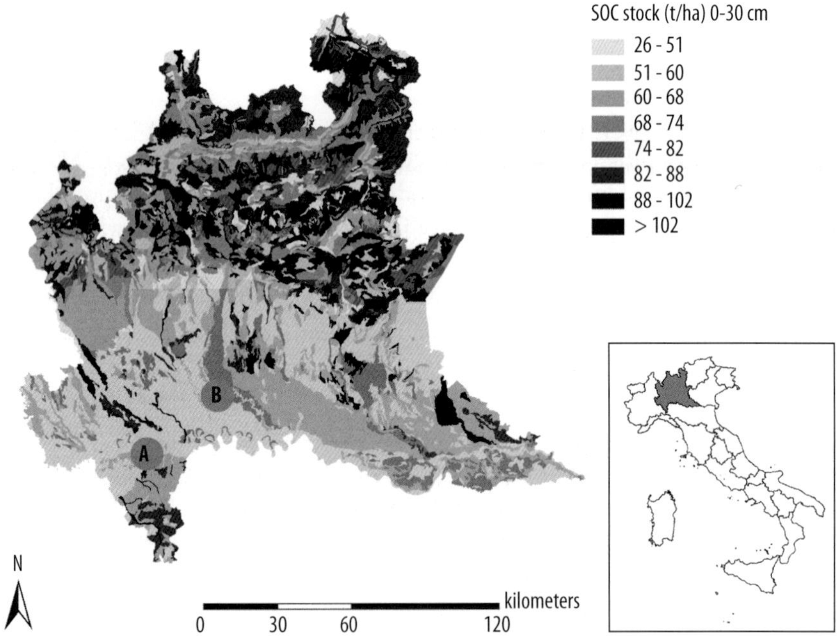

Figure 1.1. Soil organic carbon (SOC) stocks in the region of Lombardy, Italy at a scale of 1:250,000 (Regional Soil Inventory of Lombardy). The sites of the AgriCO_2ltura project, and those used in this study are located in Oltrepo Pavese (A) and Province of Lodi (B).

showed vertic features and was classified as hapli-stagnic vertisol or vertic cambisol (IUSS-WRB, 2007). These soils had a 45 to 55% clay content, a $CaCO_3$ content of 13-14%, and a pH>8. The cropping systems included winter cereals such as wheat (*Triticum aestivum* L.), barley (*Hordeum vulgare* L.), and alfalfa (*Medicago sativa* L.). Summer crops included maize (*Zea mays* L.) mainly produced for grain, soybean (*Glycine max* (L.) Merr.) and sorghum (*Sorghum vulgare* Pers.). The mean annual precipitation was ~650 mm/year and the average annual temperature was ~13.8 °C. Sprinkler irrigation was provided on demand, and urea and ammonium sulphate were used as fertilizers.

The second site was located north of the Po river in the Province of Lodi (45°16'N, 09°32'E) at 73.0 m.a.s.l. The soil at the site was classified as hapli-cutanic luvisol (IUSS-WRB, 2007) with a coarse texture, a clay content of 10-14%, and a pH of 5.5-5.9. Dairy farming predominates in this district with crop rotations of maize, used for silage and/or grain, and wheat, barley and forage crops such as tall fescue (*Festuca arundinacea* (Schreb.) S.J. Darbyshire) and Kentucky bluegrass (*Poa pratensis* L.). The mean annual precipitation at this site was ~735 mm/year, with an average annual temperature of ~13.6 °C. Surface or sprinkler irrigation was used throughout the summer, and cattle manure was used on TA managed soils as a source of fertilizer.

Six test plots, ~5 ha per plot, were selected at Oltrepo Pavese and Lodi. At Oltrepo Pavese, four plots were managed using CA, and two plots were under TA practices. At Lodi, three plots were under CA and three plots were under TA land management practices. Each of the sites were under their respective CA and TA land-use management practices for 10 years. It should be noted that slight differences in agroecosytem management practices between the individual farms, where the test plots were located, may have influenced the results of this study.

Use of the area-frame randomised soil sampling (AFRSS) allowed for the verification of changes in SOC stocks (Stolbovoy *et al.*, 2007). The AFRSS sets out a practical strategy for soil sampling, combining the collection of composite samples with randomised techniques for the geographical positioning of the sampling points. Three monitoring units corresponding to an area of 20×20 m using a regular square grid were used in this study (Figure 1.2). According to the AFRSS, a second set of the same number of monitoring units was located 5 m from the first one (Figure 1.2). This second set of monitoring units was used to provide information on sample reproducibility (R). Reproducibility simulates the error of the average sampling unit that would occur with distance due to the inherent variability of soil characteristics over short distances. Reproducibility is considered an indicator of a minimum detectable change in SOC stock and identifies the boundary condition of the method's applicability. Reproducibility (%) is given for each plot by:

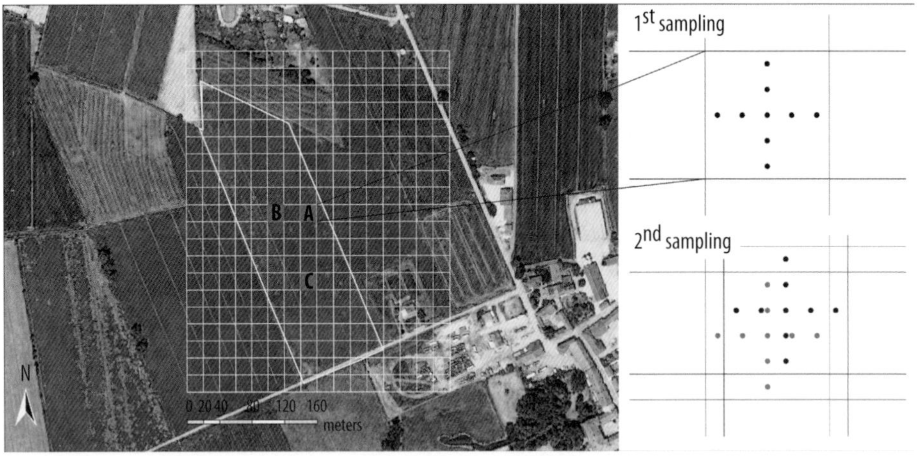

Figure 1.2. Soil sampling schema for one of the plots used in this study using the area-frame randomised soil sampling method to ensure reproducibility and to detect small changes in soil organic carbon. The 1st sampling occurred in A-B-C. The 2nd sampling was carried out in A-B-C, but the sample location was shifted 5 meters. The 2nd sampling took place for the reproducibility (R) evaluation. Both samplings were performed at the same time and after crop harvest. A total of 6 soil samples per plot were collected.

$$R_{plot} = \frac{\Delta S\bar{O}C_{plot}}{S\bar{O}C_{stock1}} \times 100 = \frac{|S\bar{O}C_{stock2} - S\bar{O}C_{stock1}|}{S\bar{O}C_{stock1}} \times 100 \qquad (1)$$

where $S\bar{O}C_{stock1}$ is the mean SOC stock of the soil sampling in the first set of monitoring units and $S\bar{O}C_{stock2}$ is the mean SOC stock coming from the sampling carried out in the second set of monitoring units shifted from the first ones.

Within each monitoring unit, a cross-sampling scheme was used, resulting in nine sub-sampling points. Soil subsamples were collected to a 30 cm depth in autumn 2010 at Oltrepo Pavese and in autumn 2011 at Lodi after the end of the cropping season using a soil auger. Subsamples were bulked together in a single composite sample per monitoring unit. This resulted in six soil composite samples per plot leading to an overall total of 72 samples (6 samples × 6 plots × 2 sites). Soil samples were air dried, sieved at 2 mm and analysed for SOC concentration using the Walkley-Black method.

A similar sampling approach was used to quantify BD. Undisturbed samples, using a cylinder with a minimum volume of 100 cm^3, were extracted from the centre of each monitoring unit. Samples were collected at 10 cm intervals to a total depth of 30 cm in the zero tillage and CA plots and from the middle (at a depth of 10-20 cm) of the plough horizon in TA plots. SOC stock was quantified according to Batjes (1996):

$$\text{SOC stock} = \text{OC} \times \text{BD} \times \text{t} \times (1 - \text{RM}) \times \frac{1}{10} \tag{2}$$

where SOC stock is given in t/ha; OC is the SOC concentration (g/kg); BD is the bulk density (g/cm^3); t is the layer thickness (cm); and RM is the mass proportion of rock fragment content (dimensionless).

Differences in SOC concentration and stock between CA and TA plots were evaluated by analysis of variance (ANOVA; $p<0.05$). The contribution of SOC concentration, soil BD and covariance's between factors to total variance of the SOC stock, CA and TA plots, was evaluated according to the following equation (Schrumpf *et al.*, 2011):

$$\text{Var (SOC stock)} = (\text{OC stock})^2 \times \left(\frac{\sigma^2_{OC}}{OC^2} + \frac{\sigma^2_{BD}}{BD^2} + 2\frac{\sigma_{OC-BD}}{OC \times BD}\right) \tag{3}$$

Where, σ_{OC} and σ_{BD}, respectively, indicated the standard deviation of SOC concentration and BD values, and σ_{OC-BD} was the covariance between the respective factors. Standard deviations and covariance were directly estimated from the measured data sets for individual test plots. In order to quantify the relative contribution of each source of error, including covariance and the total variance, each term of the sum in the equation was multiplied with the square of the respective SOC stock. As some of the covariance terms had a negative sign, we used the sum of absolute values of each term as 100% for the calculation of relative contributions.

1.3 Results

The combined results of the AFRSS sampling method quantified at Oltrepo Pavese are outlined in Table 1.1, whereas Table 1.2 presents the individual collected data. On average the soil under CA (73.9 t/ha) management practices at Oltrepo Pavese showed a higher SOC stock (0-30 cm) compared to that of the TA (50.7 t/ha) management practices. Overall, CA land management practices had a ~45.9% or 23.2 t/ha greater SOC compared to the TA management practices. Assuming that baseline SOC stocks in CA and TA plots were the same 10 years ago, the initiation of CA practices led to a SOC accumulation rate of ~2 t/ha/year. The mean SOC concentration was 17.2 g/kg under CA management practices compared to 12.3 g/kg for TA. The mean soil BD under CA management practices was 1.43 g/cm^3 and 1.39 g/cm^3 for TA.

The ANOVA test confirmed that mean values between CA and TA land management practices were significantly different ($P<0.05$) for SOC concentration and stock. This was also supported by the corresponding frequency histogram, where the dataset for SOC stocks overlapped only marginally (Figure 1.3). Additionally, SOC concentration and stock in the TA plots corresponded well to values of the surrounding soils that

Table 1.1. Combined values of soil bulk density (g/cm^3), soil organic carbon (SOC) concentration (g/kg) and stock (t/ha) to a 30 cm depth under conservation tillage practices (CA) and traditional tillage practices (TA) in the district of Oltrepo Pavese, Italy.

Cultivation practice	Plot no.	Bulk density (g/cm^3)			SOC (g/kg)			SOC stock (t/ha)			R
		Mean	SD	CV%	Mean	SD	CV%	Mean	SD	CV%	
CA	1	1.50	0.03	2.1	18.5	1.5	8.3	83.5	6.9	8.3	2.5
	2	1.38	0.08	5.9	16.6	0.9	5.2	68.5	6.9	10.1	4.7
	3	1.41	0.07	5.2	16.2	1.4	8.4	68.4	5.1	7.4	5.6
	4	1.43	0.09	6.3	17.5	2.5	14.0	75.2	12.0	16.0	14.7
TA	1	1.48	0.07	5.0	10.8	0.4	3.8	48.1	3.3	7.0	8.9
	2	1.30	0.11	8.6	13.7	0.8	6.0	53.2	6.5	12.2	6.4

SD = standard deviation; CV% = coefficient of variation.

were previously investigated in this area as part of the Regional Soil Inventory of Lombardy. This suggested that our values in the TA plots were representative of TA soil management practices in the area of Oltrepo Pavese. Taking all collected data into consideration, the value of R was 7.1%, with a range from 2.5 to 14.7%. This range was strongly influenced by the results of one CA plot; however R is less than the correspondent coefficient of variation in all plots. The combined results achieved at Lodi are instead shown in Table 1.3 and individual values collected in Table 1.4.

The mean SOC concentration quantified under CA management practices was 9.51 g/kg and 10.36 g/kg under TA management practices (Table 1.3 and 1.4). There was no significant difference in soil BD between CA and TA management practices (1.57 g/cm^3). The slightly higher SOC stock under TA (48.7 t/ha) management practices, however, was not significantly different ($P<0.05$) compared to that under CA (44.8 t/ha) management practices. These results were also confirmed by the frequency histograms (Figure 1.4).

At Lodi, the SOC stock was comparable to that provided by the soil inventory for the region of Lombardy. The value of R was 7.1%, with a range from 1.0 to 16.5%, which was lower than the correspondent coefficient of variation. The results of the variance analysis of SOC stocks indicated that SOC concentration had the largest contribution to the total variance, rather than BD and covariance between these two factors (Figure 1.5). The contribution from SOC concentration was higher at Lodi (70.5% for TA;

Table 1.2. Individual values of soil bulk density (BD) (g/cm^3), soil organic carbon (SOC) concentration (g/kg) and stock (t/ha) to a 30 cm depth under conservation tillage practices (CA) and traditional tillage practices (TA) in the district of Oltrepo Pavese, Italy.

Cultivation practice	Plot no.	Monitoring unit	BD (g/cm^3)		SOC (g/kg)		SOC stock (t/ha)	
			1st	2nd	1st	2nd	1st	2nd
CA	1	A	1.55	1.50	16.9	20.7	78.7	93.2
		B	1.50	1.45	19.4	17.3	87.4	75.3
		C	1.50	1.52	19.5	17.3	87.5	78.8
	2	A	1.31	1.28	16.2	15.4	63.7	59.2
		B	1.39	1.37	16.7	16.0	69.7	65.8
		C	1.51	1.41	17.0	17.9	77.1	75.5
	3	A	1.31	1.40	18.3	16.0	72.0	67.4
		B	1.53	1.39	16.1	14.5	74.0	60.4
		C	1.42	1.38	15.3	17.3	65.2	71.6
	4	A	1.46	1.40	14.3	19.7	62.5	82.6
		B	1.55	1.34	19.2	19.6	89.2	79.2
		C	1.32	1.50	14.8	17.6	58.5	79.2
TA	1	A	1.53	1.36	11.0	11.0	50.5	44.8
		B	1.48	1.44	11.0	10.0	48.7	43.2
		C	1.56	1.50	11.0	11.0	51.6	49.5
	2	A	1.36	1.27	14.0	14.0	57.2	53.4
		B	1.43	1.20	14.0	12.0	60.0	43.2
		C	1.14	1.38	14.0	14.0	47.9	57.9

Values under the 1st and 2nd columns represent the AFRSS sampling method where the 2nd column are the values of samples taken at a 5 m distance from that of the 1st sampling (see also Figure 1.2).

79.5% for CA) compared to that in Oltrepo Pavese (44.0% for TA; 61.0% for CA plots). The covariance term was negative for the TA management practices (-16% in Lodi; -33.1% in Oltrepo Pavese) and positive for CA management practices (4.6% in Lodi; 20.2% in Oltrepo Pavese).

1.4 Discussion and conclusions

This study illustrated that SOC stocks may be a reliable indicator to quantify the impact of various soil management practices and their effect on climate change mitigation.

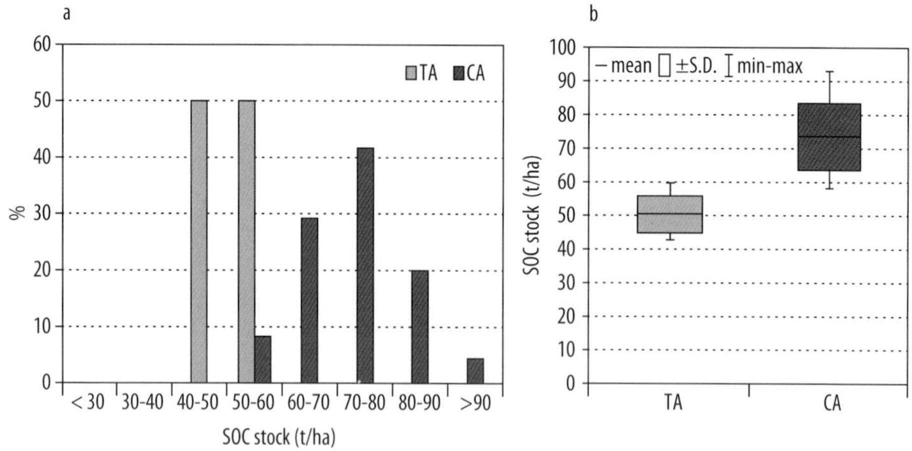

Figure 1.3. (a) Relative frequency histograms, and (b) means, standard deviations and minimum-maximum values of the distribution of soil organic carbon (SOC) stocks under conservation tillage practices (CA) and traditional tillage practices (TA) in Oltrepo Pavese, Italy.

Table 1.3. Combined values of soil bulk density (g/cm^3), soil organic carbon (SOC) concentration (g/kg) and stock (t/ha) to a 30 cm depth under conservation tillage practices (CA) and traditional tillage practices (TA) in the Province of Lodi, Italy.

Cultivation practice	Plot no.	Bulk density (g/cm^3)			SOC (g/kg)			SOC stock (t/ha)			R
		Mean	SD	CV%	Mean	SD	CV%	Mean	SD	CV%	
CA	1	1.61	0.08	4.9	9.5	1.6	16.5	45.9	7.1	15.4	15.4
	2	1.49	0.09	6.1	9.7	1.6	16.8	43.6	8.6	19.8	19.8
	3	1.61	0.05	3.3	9.3	0.2	1.8	45.1	1.2	2.7	2.7
TA	1	1.60	0.12	7.7	9.6	0.8	7.9	46.4	5.9	12.8	11.0
	2	1.49	0.09	5.8	12.4	1.4	10.8	56.0	8.8	15.8	2.2
	3	1.62	0.12	7.3	9.0	0.7	8.0	43.8	4.2	9.6	10.3

Results from our study suggested that cropland, particularly low in SOC, can be managed using CA practices to sequester C (West and Marland, 2002) and increase SOC stocks (Basch *et al.*, 2012). However, results from our study also indicated that SOC sequestration under CA practices provided variable results depending on the site under study. For example, the fine textured Oltrepo Pavese soils showed significant

Table 1.4. Individual values of soil bulk density (BD) (g/cm^3), soil organic carbon (SOC) concentration (g/kg) and stock (t/ha) to a 30 cm depth under conservation tillage practices (CA) and traditional tillage practices (TA) in the Province of Lodi, Italy.

Cultivation practice	Plot no.	Monitoring unit	BD (g/cm^3)		SOC (g/kg)		SOC stock (t/ha)	
			1st	2nd	1st	2nd	1st	2nd
CA	1	A	1.61	1.49	11.6	11.2	56.3	50.3
		B	1.64	1.73	9.0	9.2	44.2	47.7
		C	1.57	1.61	8.0	8.1	37.7	38.8
	2	A	1.45	1.56	10.8	9.6	47.2	45.2
		B	1.49	1.34	10.2	6.5	45.6	26.3
		C	1.51	1.60	11.0	9.9	49.7	47.5
	3	A	1.52	1.66	9.6	9.3	43.8	46.5
		B	1.62	1.57	9.3	9.3	45.2	43.6
		C	1.66	1.63	9.1	9.5	45.2	46.4
TA	1	A	1.65	1.70	10.2	9.6	50.2	49.1
		B	1.63	1.69	10.9	9.1	52.9	45.8
		C	1.58	1.36	9.3	8.8	44.1	36.2
	2	A	1.50	1.52	11.8	11.4	53.1	51.7
		B	1.37	1.41	12.0	11.2	49.4	47.4
		C	1.60	1.56	14.0	14.3	67.3	66.9
	3	A	1.75	1.45	9.0	9.4	47.5	41.1
		B	1.63	1.70	10.2	8.7	49.6	44.5
		C	1.69	1.51	8.1	8.5	41.3	38.6

Values under the 1st and 2nd columns represent the AFRSS sampling method where the 2nd column are the values of samples taken at a 5 m distance from that of the 1st sampling (see also Figure 1.2).

differences in SOC whereas no such difference was observed in the coarse textured Lodi soils. Baker *et al.* (2007) and Govaerts *et al.* (2009) also found great variation in the potential to sequester C in soil under CA management practices. This may be because soil, and more generally pedoclimatic conditions, as well as the history of crop management may have a determinant influence on existing and future SOC stocks (Sleutel *et al.*, 2006).

Results from this study indicated that the vertisols of Oltrepo Pavese responded positively to CA management practices. Together with zero tillage, a wide use of

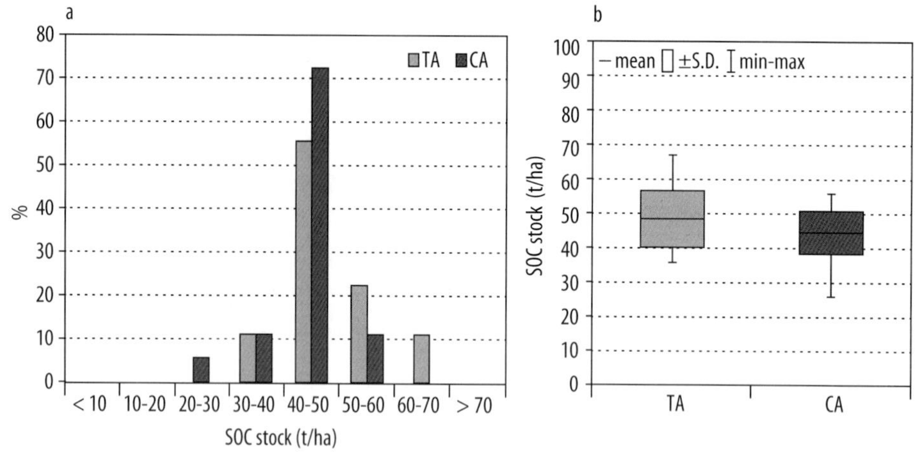

Figure 1.4. (a) Relative frequency histograms, and (b) means, standard deviations and minimum-maximum values of the distribution of soil organic carbon (SOC) stocks under conservation tillage practices (CA) and traditional tillage practices (TA) in Lodi, Italy.

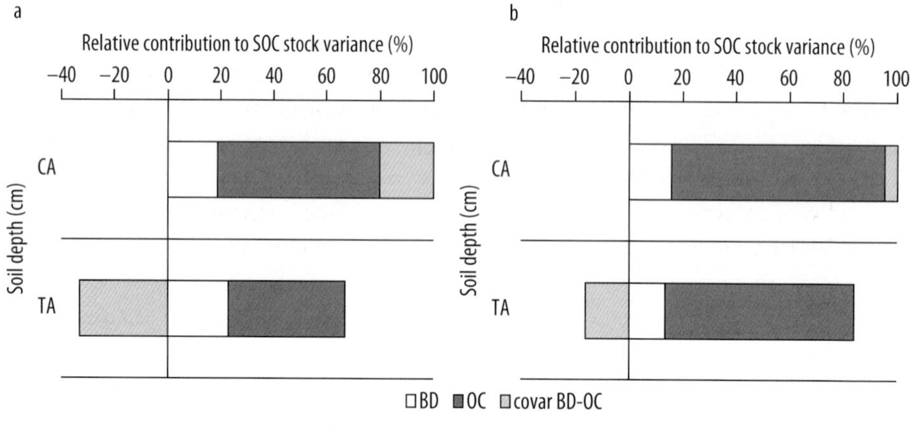

Figure 1.5. Relative contributions of soil bulk density (BD), soil organic C concentration (OC) and covariance between BD and OC at (a) Oltrepo Pavese and (b) Lodi for conservation tillage (CA) and traditional tillage (TA) land management practices.

intercropping and cover crops as well as diversified crop rotations using a variety of crop species characterized CA management practices. When combining these different management practices, as part of CA, an accumulation in SOC occurred compared to TA managed soils. Several studies reported that vertisols have a high potential to accumulate organic C when conservation practices are introduced (Dalal *et al.*, 2011; Follett *et al.*, 2005; Potter, 2010). Additionally, physical protection of SOC in aggregates from biodegradation (Beare *et al.*, 1994) as well as formation of particulate SOC (Chan, 1997), and the formation of C substrates less prone to mineralization when exposed to moisture (Dell'Abate *et al.*, 1998) may also play a key role in the accumulation of SOC.

The study site at Lodi, located in close proximity to Oltrepo Pavese, had different characteristics. For example, coarse soil texture and porous characteristics for gas exchange in a Luvisol may have facilitated microbial activity, resulting in a lower accumulation of SOC (Bajgai *et al.*, 2011). However, it should be pointed out that TA management practices at Lodi also received cattle manure annually for over a decade, which likely contributed to a greater input of organic matter compared to that from crop residues only in the CA plots. It could therefore be argued that, CA plots had a similar SOC stock compared to TA plots, although organic C inputs were lower, the relative balance between C accumulation and loss was positive in the CA plots, even if it was not observable.

Further research on the SOC balance at the farm- and field-scale, as well as a better insight into the accumulation and stabilization SOC processes (Cerli *et al.*, 2012) could improve the knowledge of the potential of CA management practices to sequester organic C into cropland soils in the Po plain of Italy. Moreover, the uncertainty in the accuracy of estimating SOC storage may provide an incentive for monitoring the effects of soil management change on the C cycle in addition to the formulation of targeted C sequestration strategies in the context of global climate change (Panda *et al.*, 2008). In cropland, SOC concentration usually contributes to the variability of SOC stocks more than BD (Goidts *et al.*, 2009) and part of these variance effects may be counterbalanced by the negative covariance of both parameters (Schrumpf *et al.*, 2011). Our study confirmed a greater importance of SOC concentration with respect to the variability of SOC stocks. However, a negative covariance between SOC and BD was observed for TA management practices, while it was positive for CA management practices. This may be due to the different soil disturbances caused by either conservation or conventional tillage management practices.

To date, there is an on-going discussion surrounding the identification of a suitable approach to monitoring long-term changes in SOC (VandenBygaart, 2006). Results from our study contributed to the identification of such an approach through the identification of the type of land management practices required to optimize levels of SOC in Lombardy, Italy.

References

Bajgai, Y., Kristiansen, P., Hulugalle, N. and Mc Henry, M., 2011. A laboratory study of soil carbon dioxide emissions in a vertisol and an alfisol due to incorporating crop residues and simulating tillage. Journal of Organic Systems 6: 1177-4258.

Baker, J.M., Ochsner, T.E., Venterea, R.T. and Griffis T.J., 2007. Tillage and soil carbon sequestration – What do we really know? Agriculture Ecosystems and Environment 118: 1-5.

Basch, G., Kassam, A., Gonzalez-Sanchez, E.J., Streit, B., 2012. Making sustainable agriculture real. CAP 2020. Life+Agricarbon project. Available at: http://www.ecaf.org/docs/ecaf/ca%20and%20cap%20 2020.pdf.

Batjes, N.H., 1996. Total carbon and nitrogen in the soils of the world. European Journal of Soil Science 47: 151-163.

Bellamy, P.H., Loveland, P.J., Bradley, R.I., Lark, R.M. and Kirk, G.J.D., 2005. Carbon losses from all soils across England and Wales 1978-2003. Nature 437: 245-248.

Beare, M.H., Cabrera, M.L., Hendrix, P.F. and Coleman, D.C., 1994. Aggregate protected and unprotected organic matter pools in conventional tillage and no-tillage soils. Soil Science Society of America Journal 58: 787-795.

Brenna, S., Stolbovoy, V., Rocca, A., Sciaccaluga, M., 2010. Potential carbon sequestration of Lombardy soils (Italy). In: Proceedings of the 19th world congress of soil science, soil solutions for a changing world. August 1-6, 2010. Brisbane, Australia, pp. 70-73.

Cerli, C., Celi, L., Kalbitz, K., Guggenberger, G. and Kaiser, K., 2012. Separation of light and heavy organic matter fractions in soil – testing for proper density cut-off and dispersion level. Geoderma 170: 403-416.

Chan, K.Y., 1997. Consequences of changes in particulate organic carbon in vertisols under pasture and cropping. Soil Science Society of America Journal 61: 1376-1382.

Corsi, S., Friedrich, T., Kassam, A., Pisante, M. and Sa, J.D.M., 2012. Soil organic carbon accumulation and greenhouse gas emission reductions from conservation agriculture: a literature review. Integrated crop management Vol. 16. Plant Production and Protection Division, Food and Agriculture Organization of the United Nations, Rome, Italy.

Dalal, R.C., Allen, D.E., Wang, W.J., Reeves, S. and Gibson, I., 2011. Organic carbon and total nitrogen stocks in vertisols following 40 years of no-tillage. Soil and Tillage Research 112: 133-139.

Davidson, E.A. and Janssens, I.A., 2006. Temperature sensitivity of soil carbon decomposition and feedbacks to climate change. Nature 440: 165-173.

Dell'Abate, M.T., Trinchera, A., Benedetti, A. and Dazi, C., 1998. Humic substances as indicators of soil organic matter evolution in vertisols. In: Proceedings of the16th World Congress of Soil Science. August 20-26, 1998. Montpellier, France. Available at: http://natres.psu.ac.th/Link/SoilCongress/bdd/symp12/781-t.pdf.

Dersch, G. and Boehm, K., 2001. Effect of agronomic practices on the soil carbon storage potential in arable farming in Austria. Nutrient Cycling in Agroecosystems 60: 49-55.

Ellert, B.H. and Bettany, J.R., 1995. Calculation of organic matter and nutrients stored in soils under contrasting management regimes. Canadian Journal of Soil Science 75: 529-538.

Follett, R.F., 2001. Soil management concepts and carbon sequestration in cropland soils. Soil and Tillage Research 61: 77-92.

Follett, F.R, Castellanos, J.Z. and Buerger, E.D., 2005. Carbon dynamic and sequestration in an irrigated vertisol in Central Mexico. Soil and Tillage Research 83: 148-158.

Goidts, E., Van Wesemael, B. and Crucifix, M., 2009. Magnitude and sources of uncertainties in soil organic carbon (SOC) stock assessments at various scales. European Journal of Soil Science 60: 723-739.

Govaerts, B., Verhulst, N., Castellanos-Navarrete, A., Sayre, K.D., Dixon, J. and Dendooven, L., 2009. Conservation agriculture and soil carbon sequestration: between myth and farmer reality. Critical Reviews in Plant Science 28: 97-122.

Hobbs, P.R., Sayre, K., Gupta, R., 2008. The role of conservation agriculture in sustainable agriculture. Philosphical Transactions of the Royal Society B 363: 543-555.

IUSS Working Group WRB, 2007. World reference base for soil resources 2006, first update 2007. World Soil Resources Reports No. 103. FAO, Rome, Italy.

Janssens, I.A., Freibauer, A., Schlamadinger, B., Ceulemans, R., Ciais, P., Dolman, A.J., Heimann, M., Nabuurs, G.J., Smith, P., Valentini, R. and Schulze, E.D., 2005. The carbon budget of terrestrial ecosystems at country scale – a European case study. Biogeosciences 2: 15-26.

Kucharik, C.J., Brye, K.R., Norman, J.M., Foley, J.A., Gower, S.T. and Brundy, L.G., 2001. Measurements and modelling of carbon and nitrogen cycling in agroecosystems of southern Wisconsin: potential for SOC sequestration during the next 50 years. Ecosystems 4: 237-258.

Lal, R., 2003. Global potential of soil carbon sequestration to mitigate the greenhouse effect. Critical Reviews in Plant Sciences 22: 157-184.

Lal, R. 2004. Soil carbon sequestration to mitigate climate change. Geoderma 123: 1-22.

Lal, R., Kimble, J.M., Follett, R.F. and Cole, C.V., 1998. The potential of US Croplands to sequester carbon and mitigate the greenhouse effect. Ann Arbor Press, Ann Arbor, MI, USA.

Pacala, S. and Socolow, R., 2004. Stabilization wedges: solving the climate problem for the next 50 years with current technologies. Science 305: 968-972.

Panda, D.K., Singh, R., Kundu, D.K., Chakraborty, H. and Kumar, A., 2008. Improved estimation of soil organic carbon storage uncertainty using first-order Taylor series approximation. Soil Science Society of America Journal 72: 1708-1710.

Potter, K.N., 2010. Building soil carbon content of Texas vertisols. In: Proceedings of the 19th World Congress of Soil Science, Soil Solutions for a Changing World. August 1-6, 2010, Brisbane, Australia. Available at: http://www.iuss.org/19th%20WCSS/Symposium/pdf/0222.pdf.

Reicosky, D.C., 2003. Tillage-induced CO_2 emissions and carbon sequestration: effect of secondary tillage and compaction. In: Garcia-Torres L., Benites J., Martinez-Vilela A. and Holgado-Cabrera A. (eds) Conservation Agriculture. Kluwer Academic Publishers, Dordrecht, the Netherlands, pp. 291-300.

Sanderman, J., Farqhuarson, R. and Baldock, J., 2009. Soil carbon sequestration potential: a review for Australian agriculture. Report prepared for Department of Climate Change and Energy Efficiency. CSIRO Land and Water, Acton, Australia.

Schrumpf, M., Schulze, E.D., Kaiser, K. and Schumacher, J., 2011. How accurately can soil organic carbon stocks and stock changes be quantified by soil inventories? Biogeosciences 8: 1193-1212.

Schulze, E.-D., Luyssaert, S., Ciais, P., Freibauer, A., Janssens, I.A., Soussana, J.F., Grace, J., Levin, I., Thiruchittampalam, B., Heimann, M., Dolman, A.J., Valentini, R., Bousquet, P., Peylin, P., Peters, W., Radenbeck, C., Etiope, G., Vuichard, N., Wattenbach, M., Nabuurs, G.J., Poussi, Z., Nieschulze, J. and Gach, J.H., 2009. Importance of methane and nitrous oxide for Europe's terrestrial greenhouse-gas balance. Nature Geoscience 2: 842-850.

Sleutel, S., De Neve, S., Hofman, G., Boeckx, P., Beheydt, D., Van Cleemput, O., Mestdagh, I., Lootens, P., Carlier, L., Van Campo, N., Verbeeck, H., Vande Walle, I., Samson, R., Lust, N. and Lemeur, R., 2003. Carbon stock changes and carbon sequestration potential of Flemish cropland soils. Global Change Biology 9: 1193-1203.

Sleutel, S., De Neve, S., Singler, B. and Hofman, G., 2006. Organic C levels in intensively managed arable soils – long term regional trends and characterization of fractions. Soil Use and Management 22: 188-196.

Stolbovoy, V., Montanarella, L., Filippi, N., Selvaradjou, S., Panagos, P. and Gallego, J., 2005. Soil sampling protocol to certify the changes of organic carbon stock in mineral soils of European Union. Office for Official Publications of the European Communities, Luxembourg, Luxembourg.

Stolbovoy, V., Montanarella, L., Filippi, N., Jones, A., Gallego, J. and Grassi, G., 2007. Soil sampling protocol to certify the changes of organic carbon stock in mineral soils of European Union. Office for Official Publications of the European Communities, Luxembourg, Luxembourg.

Tarnocai, C., Canadell, J.G., Schuur, E.A.G., Kuhry, P., Mazhitova, G. and Zimov, S., 2009. Soil organic carbon pools in the northern circumpolar permafrost region. Global Biogeochemical Cycles 23: 1029-1040.

VandenBygaart, A.J., Gregorich, E.G. and Angers, D.A., 2003. Influence of agricultural management on soil organic carbon: a compendium and assessment of Canadian studies. Canadian Journal of Soil Science 83: 363-380.

VandenBygaart, A.J., 2006. Monitoring soil organic carbon stock changes in agricultural landscapes: issues and a proposal approach. Canadian Journal of Soil Science 86: 451-463.

Vleeshouwers, L.M. and Verhagen, A., 2002. Carbon emission and sequestration by agricultural land use: a model study for Europe. Global Change Biology 8: 519-530.

West, T.O. and Marland, G., 2002. A synthesis of carbon sequestration, carbon emissions, and the net carbon flux in agriculture. Agriculture Ecosystems and Environment 91: 217-232.

Chapter 2.

Change in pedogenic carbon stocks under different types and duration of agricultural management practices in the central Russian forest steppe

O.S. Khokhlova[1], Y.G. Chendev[2] and T.N. Myakshina[1]

[1]Institute of Physical, Chemical and Biological Problems in Soil Science, Russian Academy of Sciences, Ulitsa Institutskaya 2, Pushchino 142290, Russia; [2]Belgorod State University, Ulitsa Pobedy 85, Belgorod 308015, Russia; olga_004@rambler.ru

Abstract

A large proportion of soil located in semi-arid and arid regions contain high levels of carbonates or soil inorganic C (SIC). This SIC pool may be affected by agricultural activities and thereby influence the global C balance. The purpose of this study was to examine the total (pedogenic) soil C stock in agroecosystems under cultivation for 300 years in the southern forest-steppe ecotone in central Russia. Grey forest soils (Luvisols) under broad-leaved deciduous forest dominated this area prior to cultivation. Our results showed that the pedogenic C stock increased 15-50% due to an accumulation of SIC in the top 200 cm as a result of agricultural activities. A ~10% increase in SOC was also observed in agricultural soils cultivated for more than 200 years. A quantifiable increase in carbonate content in association with the ^{14}C-age of the newly-formed carbonates suggested that changes in the hydrothermal regime of arable lands led to an upward movement of carbonates from the parent material. This upward movement may have occurred via capillary-sized pores in colloidal suspensions. This led to the movement of carbonates without the exchange of young soil CO_2. Our results suggested that the upward movement of carbonates may be a rapid soil transforming process taking place over less than 50 years. The pedogenic C stocks (0-200 cm) in this region's agricultural and forest lands were estimated to be ~11.5 Gt C, of which one third was SIC. We postulated that SIC in agricultural soils was safely deposited and highly protected from dissolution because carbonates were located in the smallest pores and impregnated the soil mass according to micromorphological observations.

Keywords: ^{14}C-dating of carbonates, agricultural management practices, forest soils, soil micromorphology, soil organic and inorganic carbon

DOI 10.3920/978-90-8686-788-2_2, © Wageningen Academic Publishers 2014

2.1 Introduction

When studying changes in the carbon (C) balance under the different types of agroecosystem management practices, typically only soil organic C (SOC) is taken into account. However, the soil inorganic C (SIC) pool is also undergoing significant tranformations in agricultural landscapes in semi-arid and arid regions. Glazovskaya (2009) proposed that the combination of C in soil humus and that from carbonates could be referred to as pedogenic C. Glazovskaya (2009) suggested that the accumulation of carbonate C in sub-arid and arid regions of Eurasia was comparable to that of humus. Therefore, in order to understand the balance of total C or pedogenic C for soils containing both SOC and SIC, the various carbonate forms must also be considered. The main purpose of our research was to evaluate the dynamics of SOC, SIC and pedogenic C stocks in the southern forest-steppe ecotone in central Russia (Belgorod and Voronezh regions) using a 50 to 300 year chronosequence of agricultural activities. We also evaluated the mechanisms of transformation of carbonates in this agricultural chronosequence. In addition, we also evaluated the influence of human activities on the evolution of soils and land cover, specifically related to the grey forest soils, including light and dark grey forest soils (Luvisols) of the central Russian forest-steppe (Egorov *et al.*, 1977). Phaeozems predominated this area before the onset of cultivation.

Agriculture is the longest and most influential factor of soil transformation. However, our knowledge on the evolution of soil under the influence of cultivation is fragmentary, and the transformation of carbonates in agricultural soils along the Russian forest-steppe ecotone remains practically unknown. Recent research suggested that the grey forest soils (Luvisols) may transform to Chernozems under the influence of agricultural activities (Chendev, 2008; Chendev *et al.*, 2011; Kuznetsova *et al.*, 2010). Such changes in the grey forest soils included an increasing depth of the humus layer, an increased number of burrowing holes (krotovinas) which are passageways created by burrowing animals. Krotovinas become backfilled with soil resulting in a textural differentiation with depth as a result of soil mixing by burrowers. These activities increase plant nutrients in the top 100 cm, but also cause alkalinization and the formation of new carbonates. As such, the forest-steppe soils show characteristic qualities of humus and carbonate soil profiles that have undergone pedogenic processes and soil evolution as a result of natural and anthropogenic activities. Therefore, evaluating pedogenic C dynamics as a whole and carbonate specifically, in the soils of this region may help to understand very important aspects of soil evolution in a human-influenced environment and its potential for the long-term sequestration of C.

2.2 Materials and methods

Study sites

The study was conducted on five different sites in the region of Belgorod. Each study site included one chronosequence that consisted of an undisturbed virgin grey forest soil and agricultural soils that were under cultivation for 100, 150, 220 or 240 years (Table 2.1). In the region of Voronezh, research at the Experimental Station of the Voronezh Institute of Maize included an undisturbed dark grey forest soil and three agricultural plots under different management practices (Table 2.1). The first plot included an agricultural soil under maize (*Zea mays* L.) monoculture, the second plot was under bare fallow, and the third plot under crop rotation.

All sites studied were characterized by a similar history of agricultural land-use ranging from cultivation by horse-drawn plough (before 1930) to mechanized farming. As to the application of fertilizers, sites in the Belgorod region received ~4 t/ha manure whereas the sites in Voronezh did not receive any fertilization. These sites were chosen because of their similar characteristics with respect to topography, parent material, and soil type. Native forest refers that the site was under forest cover no less than 300 years (Figure 2.1). For example, at the end of the 17th century the area was covered by forest (Figure 2.1a, grey shading). In 1785, a part of the territory formerly occupied by forest was under cultivation (Figure 2.1b). This suggested that cultivation occurred before 1785 but after the end of the 17th century. In 1864, the area under cultivated land become greater and under forest – smaller than in 1785, and at present only a very small area is under forest (Figure 2.1c and 2.1d). Figure 2.1e shows the boundaries of the forest at different times and locations of the studied soil pits. Pit M 1/09 was located in the forest, pit M 2/09 was located in the area under cultivation for less than 100 years, pit M 3/09 was located in the area under cultivation for ~150 years, and pit Ml 4/09 was located in the area under cultivation for ~240 years.

Detailed description of study sites: Belgorod

The Melikhovo site (chronosequence 1) was located in a typical steppe landscape between the Seversky Donets and Razumnaya Rivers with a level surface. The undisturbed site was composed of an ash (*Fraxinus* spp.) oak (*Quercus* spp.) forest with the dark grey forest soil (Pit M 1/10; Table 2.1). Soils of the agricultural fields were under cultivation for about 100 (Pit M 2/10; Table 2.1) to 150 years (Pit M 3/10, Table 2.1) were classified as dark grey forest soils, and those under cultivation for about 220 years were strongly leached Chernozems (Pit M 4/10; Table 2.1). The parent material for these soils was calcareous loess with a heavy loam texture.

Table 2.1. Site characteristics of the agricultural chronosequences and the undisturbed forest soil in Belgorod and Voronezh.

Name and location of the study site	Latitude and longitude	MAT (°C)	Temperature		MAP (mm)	HTC	Height (m.a.s.l.)	Number of soil pits			
			July (°C)	January (°C)				Forest	Agriculture 100 years	Agriculture 150 years	Agriculture 220-240 years
Melikhovo, Korochansky district of the Belgorod region	36° 47′ E 50° 42′ N	+6.2	+19.5	-8.3	550-600	1.1	216-218	M 1/10	M 2/10	M 3/10	M 4/10
Verkhopenye, Ivnyansky district of the Belgorod region	36°17′ E 50°57′ N	+5.9	+18.8	-8.4	600	1.2	250-253	V 1/10 V 2/10	V 3/10 V 4/10		V 5/10
Kazachya Lisitsa, Graiyvoronsky district of the Belgorod region	35°45′ E 50°32′ N	+6.2	+19.0	-8.2	600	1.2	220-223	K 1/10	K 2/10		K 3/10
Polyana, Shebekinsky district of the Belgorod region	36°52′ E 50°37′ N	+6.4	+19.6	-7.9	550-600	1.1	205-210	P 1/09	P 2/09	P 3/09	
Samarino, Krasnogvardeysky district of Belgorod region	38°25′ E 50°20′ N	+7.0	+20.0	-7.7	475-500	0.9-1.0	202-207	S 1/09 S 10/09 [a]	S 2/09 S 3/09 S 4/09	S 6/10 S 7/10	
Opyt, Khokhols-ky district of the Voronezh region, experimental fields of Maize station	38°58′ E 51°36′ N	+4-5	+22	-8-9	480-550	1.0-1.1	180-185	3 forest/09			1 maize/09 2 fallow/09 4 crop/09

MAT = mean annual temperature; MAP = mean annual precipitation; HTC = hygrothermal coefficient; M = Melikhovo; V = Verkhopenye; K = Kazachya Lisitsa; P = Polyana; S = Samarino.

[a] pit S 10/09 was studied in the field only.

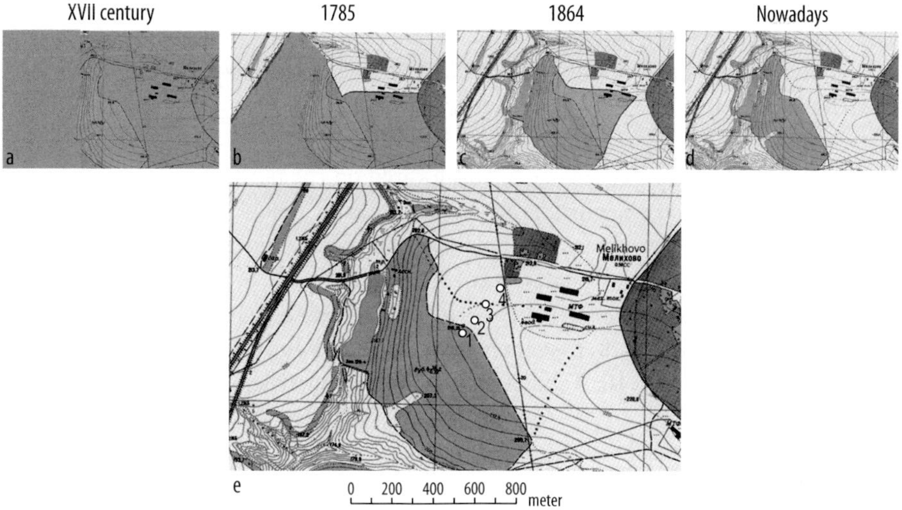

Figure 2.1. An example of change in forest cover at the Melikhovo site (Belgorod region) from the late 17th century to the present day (a-d) and the location of the studied soil pits (e).

The Verhopenye site (chronosequence 2) was located in a typical steppe landscape on a 1-2° slope with a westerly exposure between the Pena and Psyol Rivers. The soil was classified as the podzolized Chernozem under an ash-oak forest (Pits V 1/10 and V 2/10; Table 2.1). The Chernozems transitioned from a podzolized to a leached subtype in agricultural field under cultivation for 100 years (Pits V 3/10 and V 4/10; Table 2.1). Soils under cultivation for ~240 years were classified as typical Chernozems (Pit V 5/10; Table 2.1). The parent material was calcareous loess with a medium loam texture.

The Kazachya Lisitsa site (chronosequence 3) was located within a typical forest-steppe ecosystem between the Vorskla and Lisenok Rivers with a level surface. The soils were classified as the dark grey forest soils under an ash-oak forest (Pit K 1/10; Table 2.1), as well as in agricultural fields under cultivation for ~100 years (Pit K 2/10; Table 2.1), and as a podzolized Chernozem in agricultural fields under cultivation for ~240 years (Pit K 3/10; Table 2.1). The parent material was a loess-like calcareous material with a medium loam texture.

The Polyana site (chronosequence 4) was located within a typical forest-steppe ecosystem. The slope had a gentile gradient of 1-2° with a westerly exposure, and was located between the Seversky Donets and Koren rivers. Soils were classified as a dark grey Forest soils (Pit P 1/10; Table 2.1) under the linden (*Tilia* spp.) – maple (*Acer* spp.) forest, as well as in agricultural fields under cultivation for ~100 years (Pit P 2/10;

Table 2.1), and as the slightly podzolic Chernozems in the agricultural fields under cultivation for ~150 years (Pit P 3/10; Table 2.1). The parent material was a loess-like calcareous material with a light clay texture.

The Samarino site (chronosequence 5) was located on the border between the forest-steppe and steppe ecotone within the central Russian upland. This site has a level surface and was located between the Palatovka and Valuy rivers. Soils were classified as a grey forest soils under ash-oak forest (Pits S 1/10 and S 10/10; Table 2.1) and in agricultural fields under cultivation for ~100 years (Pits S 2/10, S 3/10 and S 4/10; Table 2.1) and as a leached Chernozem in agricultural fields under cultivation for ~150 years (Pits S 6/10 and S 7/10; Table 2.1). The parent material was leached calcareous clay with a fine texture.

Detailed description of study sites: Voronezh

The Experimental Station of Voronezh Institute of Maize was located between the Don and Devitsa Rivers with an absolutely level and uniform surface. The station was located within the southern part of the forest-steppe ecotone. The forests in this region were dominated by oak and linden underlain by the dark grey forest soils. The forest soils had not been disturbed for more than 300 years. Agricultural soils were under cultivation for 300 years and were classified as arable Chernozems. The parent material was a loess-like calcareous loam. At this site, soil was sampled from four different plots including broadleaf forest (Pit 3 forest/09; Table 2.1), and three agricultural sites using different agroecosystem management practices. The specific agricultural sites were under the same management practices for the past 50 years. The first site included a continuous maize monocrop (Pit 2 maize/09; Table 2.1); the second site was a bare fallow (Pit 3 fallow/09; Table 2.1); and the third site was under crop rotation with cereals and vetch-oat mixtures and a fallow every fourth year (Pit 4 crop/09; Table 2.1). These sites were under the same agricultural management practices for the past 50 years. All agricultural plots were ploughed with no fertilizer application. The distance between soil pits in the agricultural fields was less than 5 m. The soil pit in the forest was approximately 500 m from the agricultural sites, but was located within the same geomorphological and lithological positions.

Soil sampling and analysis

Samples were extracted from each of the sites described above in 2009 to 2010. At each site, soil pits were dug to a 200 cm depth. From the upper 100 cm of the profile, samples (n=3) were collected at 10 cm intervals, and at 20 cm intervals at the 100 to 200 cm depth. Soil organic C for all samples was determined using the Turin method

(Arinushkina, 1970). The carbonate C was determined as the amount of CO_2 evolved using sealed vessels with rubber stoppers in which the samples reacted with 10% HCl solution. The CO_2 was analysed using chromatorgraphy, and was subsequently converted to C.

Soil bulk density was sampled at depths of 17, 40, 70, 100 and 130 cm (n=3), which was equivalent to the average depths of the identified soil horizons. To calculate pedogenic C stocks at 10 cm depth increments, values of bulk density were interpolated. A curve of the bulk density distribution throughout the profile was plotted using the measured data at depths of 17, 40, 70, 100 and 130 cm. The distribution curve allowed for the quantification of the soil's bulk density at 10 cm increments. Based on profile distribution of SOC, SIC and the bulk density values for each sampling layer, C stocks within 0-50, 50-100, 100-150, 150-200, 0-100 and 0-200 cm were quantified for all sites.

Undisturbed bulk soil samples or nodules from the Melikhovo, Samarino and Voronezh sites were collected from the profiles and impregnated with resin to prepare thin sections. The thin sections were analysed using a polarising microscope (Carl Zeiss HBO 50, Germany) at the Institute of Physical, Chemical and Biological Problems in Soil Science located in Pushchino, Russia. Fragments of the monoliths from which thin sections were obtained were analysed on a JEOL 6610 LW scanning electron microscope with accelerating voltage of 20 kV (JEOL Ltd., Japan) at the Institute of Geography (Moscow, Russia). Radiocarbon ^{14}C-dating of carbonates was conducted at the Kiev Radiocarbon Laboratory using the scintillation method and a microreactor.

2.3 Results and discussion

Soil organic carbon concentrations

The greatest SOC concentrations occurred in the undisturbed forest soils at a 0-10 and 10-20 cm depth at all sites. The average SOC concentration for all sites was 4% (Figure 2.2). In agricultural soils, the average SOC concentration was 2-3% at a 0-10 and 10-20 cm depth. Soil organic C concentrations from the central section of the soil profile varied in the agricultural soils depending on the length of time the site was under cultivation. For example, SOC concentrations at a 20-100 cm depth in the 100 to 150 year old agricultural soils were similar to that of the undisturbed forest soils. However, in agricultural soils ≥200 years and older, the overall SOC concentration was greater at a 20-100 cm depth and gradually decreased with soil depth (Figure 2.2). This suggested that agricultural soils cultivated for ≥200 years, at a 20-100 cm depth, had an overall greater SOC accumulation than that of the undisturbed forest soils at

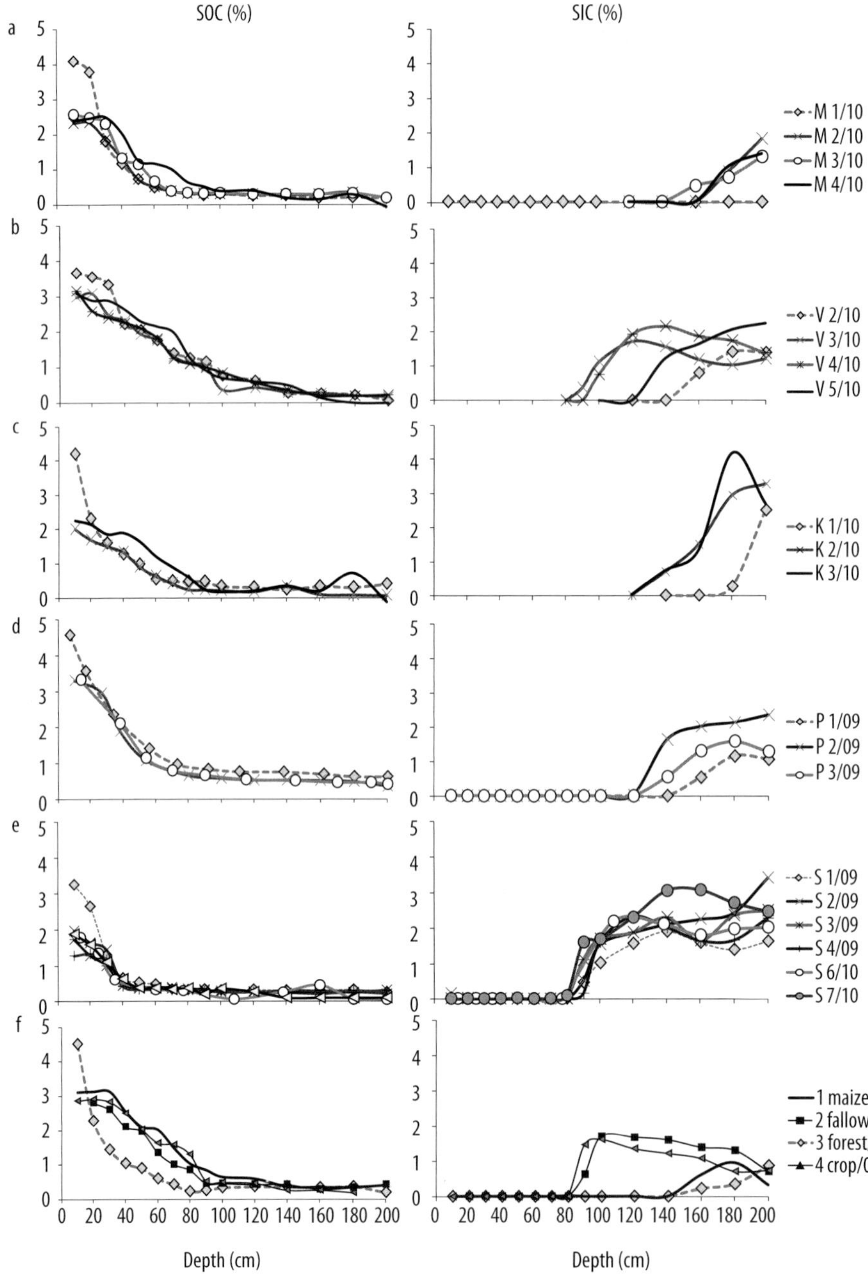

Figure 2.2. Profile distribution of soil organic carbon (SOC) and soil inorganic carbon (SIC) at the various sites and their corresponding description in Table 2.1.

the same depth. Additionally, the distribution of SOC throughout the soil profile in the oldest agricultural soils became smoother and showed a greater resemblance to Chernozem-type soils (Chendev, 2008).

The process of SOC accumulation in subsurface layers was particularly clear in the agricultural soils of the Experimental Station of Voronezh. Before the long-term experiment started in 1959 (Larionova *et al.*, 2012; Stulin, 2007), land at the Experimental Station was under agricultural management for ~300 years. In all three variants of the cultivated soils evaluated in this study, including the bare fallow, SOC in the 20-100 cm layer was higher than that in the forest soil (Figure 2.2e).

Soil inorganic carbon concentrations

SIC concentration varied in the undisturbed forest soils. For example, in some sites there was no SIC to a 200 cm depth (Figure 2.2a) whereas at other sites SIC concentration ranged from 1 to 2% at a depth of 100 to 120 cm and with increasing content with increasing depth (Figure 2.2d). At all sites, the SIC concentration in all the agricultural soils was consistently higher at the same depths compared to that of the forest soils. At some sites, SIC in the agricultural soils was found 20-30 cm higher compared to the forested sites but in the majority of cases the depth of carbonate horizons was the same in agricultural and forest soils (Figures 2.2d and 2.2e). Therefore, our results showed an accumulation of carbonates in all agricultural soils. This was also observed by Yonko and Melnikova (2006) in the region of Voronezh, and by Chendev (2008) in the region of Belgorod. These results suggested that the process of cultivation led to the transformation of soils with properties of a grey Luvisol to that of a Chernozem.

No significant differences in the distribution of SIC within the agricultural soil profiles was observed regardless of the number of years a site was cultivated (Figure 2.2). This suggested that ≤100 years of cultivation of the grey forest soils, carbonate location and/or concentrations rose to a higher level within a soil profile as a result of the changing environmental conditions.

Data from the Voronezh Experimental Station (Figure 2.2f) provided further insight into the mechanisms regulating the changes in SIC accumulation within the soil profiles of the agricultural soils. For example, similar SIC concentrations were observed in the forest and maize monoculture plots, compared to those under crop rotation and bare fallow. Crop rotation and bare fallow plots showed that SIC was located ~60 cm higher within the soil profile compared to the forest and maize monocrop plots. Such difference in the distribution of SIC within the soil profile and between these various land management practices is likely not only related to spatial variability. Instead, according to our hypothesis, the main factor driving these changes may be related to plant cover. For example, the bare fallow has no plant cover throughout the year and crop rotation has no plant cover every fourth year. As a result, the dark surface of the soil without plant cover is strongly heated over the summer months, which creates a temperature gradient between the lower carbonate containing horizons and the surface horizons (Arkhangelskaya, 2012). Such differences in temperature may lead to an upward movement of carbonates. Strong heating does not occur in soils under maize and forest as their surface during the hottest time of the year is protected by vegetation. In addition, roots of plants collect water resulting in restricted movement of water within the soil profile (Kokovina and Lebedeva, 1986). Agricultural activities in the research plots used in our study at the Voronezh Experimental Station have taken place only over the past 50 years; and our results suggested that the observed differences in the distribution and concentration of SIC occurred over a time frame of ≤50 years. This suggested that the upward movement of carbonates due to a change in environmental conditions may be a fast process, requiring less than 50 years. This idea was also supported by a previous study on paleosols buried under kurgans (archaeological earth monuments) in the Russian steppe (Khokhlova, 2008; Khokhlova *et al.*, 2008).

Soil inorganic and organic carbon stocks

The 100 to 150 year old agricultural soils of the five different chronosequences in the Belgorod region showed that the SOC stocks were 5-10% lower compared to the undisturbed forest soils. However, after 220 to 240 years of cultivation, there was an increase in the SOC stock by ~10-15% (Figure 2.3). This increase occurred mainly at the 50-100 cm depth (Figure 2.4a). At the Voronezh Experimental Station where cultivation occurred for 300 years, SOC stocks were ~30% higher at a 50-100 cm compared to the forest soil (Figure 2.3f and Figure 2.4a). It should be noted that such an increase may be an exception rather than the rule because at the Experimental Station appropriate agricultural management practices (e.g. strict time of seeding, fertilizer application) were implemented in addition to this site's level topography which minimized erosion (Stulin, 2007). However, our results showed an overall increase in SOC stocks in all agricultural soils under cultivation for ≥200 years, which may be associated with the process of transforming former grey forest soils to arable Chernozems (Chendev, 2008).

SIC stocks to a 200 cm depth ranged from 0 to 2 t/ha (Melikhovo site; Figure 2.3a) to 250 t/ha (Samarino site; Figure 2.3e). At Samarino, SIC stocks were 150-200 t/ha higher than SOC stocks. In all agricultural soils, SIC stocks increased compared to those under undisturbed forest. In general, the agricultural soils were enriched in carbonates in the lower 100 cm of the soil profile. In rare cases, carbonates accumulated at the

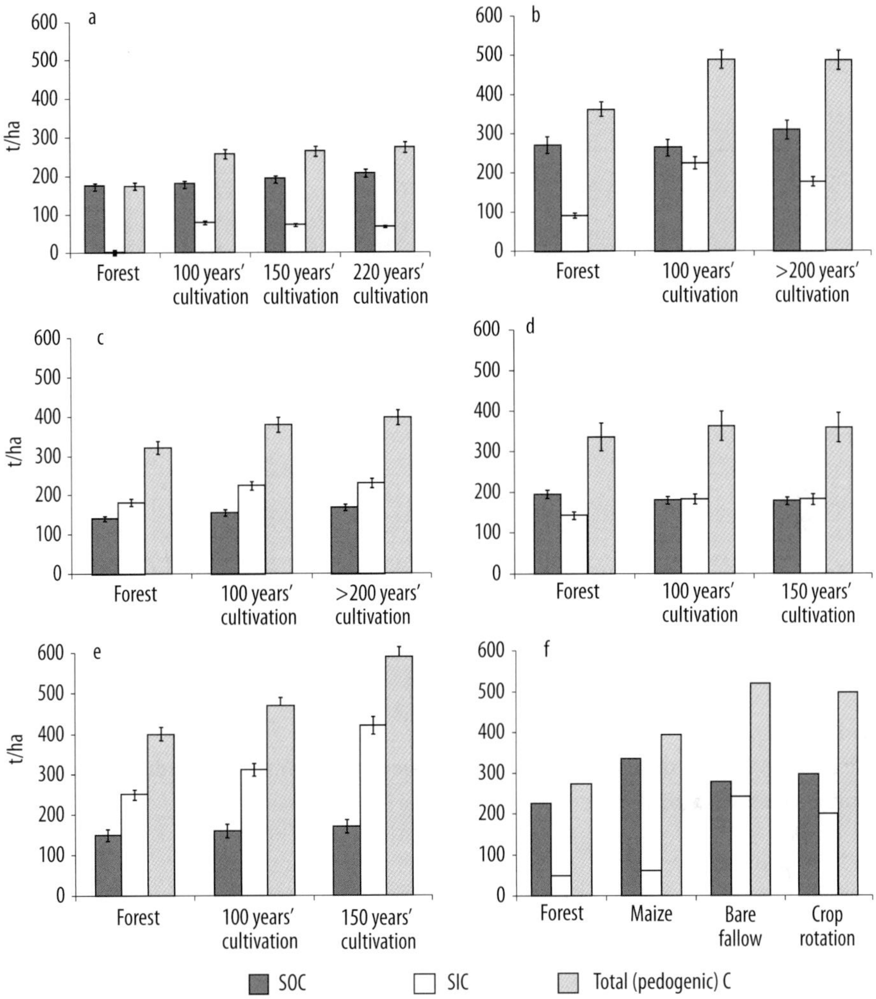

Figure 2.3. Soil organic carbon (SOC), soil inorganic carbon (SIC) and total (pedogenic) C stocks (t/ha) to a 200 cm depth in agricultural chronosequences at the various sites evaluated in Beglorod (see corresponding site descriptions in Table 2.1).

50-100 cm depth of agricultural soils. If SOC stocks rose slightly in the agricultural soils, then SIC stocks increased sharply by 25-35% after 100 years of cultivation. For example, at the Melikhovo site (Figure 2.3a) SIC stocks were 35 to 40 times higher than in the undisturbed forest soils. At all sites, SIC stocks were 50 to 100 t/ha higher in the agricultural soils that were under cultivation for 100 to 150 years compared to

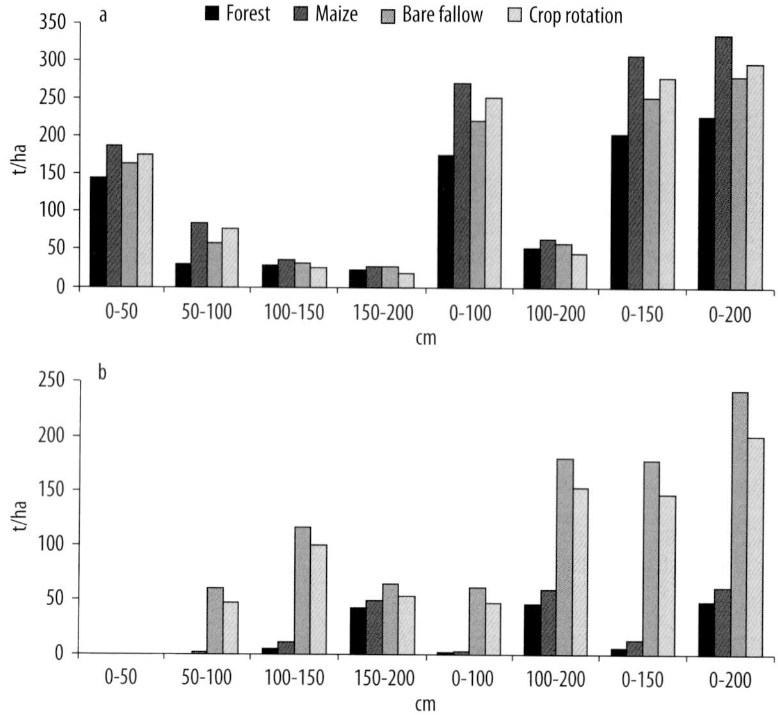

Figure 2.4. (a) Soil organic carbon and (b) soil inorganic carbon stocks (t/ha) at 200 cm depth in agricultural chronosequences evaluated at the Voronezh Experimental Station.

the undisturbed forest soils. Soils under cultivation ≥200, SIC was slightly slower, but still showed a 20-30% increase compared to the undisturbed forest soils.

Thus, in all agricultural soils evaluated in this study, pedogenic C stocks were not reduced compared to the undisturbed forest soils but increased by 15-30% (up to 50%) as a result of SIC accumulation (Akhtyrsev, 1996; Akhtyrsev and Chshetinina, 1969; Chendev, 2008; Fatyanov, 1959; Kharitonychev, 1960). Our study revealed that an increase in the SIC stock occurred after less than 100 years of cultivation and remained stable in soils under cultivation for more than 150 years. The increase in SOC stocks became apparent due to the accumulation of humus in the subsurface layers in the agricultural soils only after 200 to 250 years of cultivation.

At the Voronezh Experimental Station, the largest SOC stocks were observed in the maize monoculture plots, showing the highest SOC accumulation at a 50 to 100 cm depth (Figure 2.4a). It is known that the root system of maize is one of the most massive among agricultural crops grown in central Russia, and dying roots are the most important source of organic matter in agricultural soils (Orlov *et al.,* 1996). Even under the bare fallow there was no significant decrease in SOC stocks after 50 years of cultivation compared to the forest plot (Figure 2.4a). This was likely due to the lack of erosion and the high stability of humus typically found in Chernozems (Orlov *et al.,* 1996; Stulin, 2007). Recently, the agricultural soils at the Experimental Station were classified as a leached Chernozem (Stulin, 2007) or arable Chernozem (Larionova *et al.,* 2012). In addition, soil under bare fallow, showed a maximum enrichment in SIC (Figure 2.4b), and the majority of visible forms of carbonates were located in the closest to surface horizon compared with maize and crop rotation pits. Comparatively, SIC accumulation in soil under the maize monoculture was located at a 100-120 cm depth whereas in the soil under bare fallow it was located at a depth of 60-70 cm. The distance between those two pits (under maize and bare fallow) was about 5 m, and no sign of erosion was observed.

Soil inorganic C stocks to a 200 cm depth showed two distinct patterns. For example, in the forest and maize monoculture at Voronezh, SIC stocks were 50 to 60 t/ha, whereas under the bare fallow and crop rotation, SIC stocks were 240 and 200 t/ha (Figure 2.4b). The latter values were comparable with those of the SOC stocks, where SIC stocks of about 50 t/ha were observed at a depth of 50-100 cm. This confirmed the pattern observed in the profile distribution of SIC in four variants of the experiment (forest, monoculture of maize, crop rotation and bare fallow). In general, the arable Chernozems at the Voronezh Experimental Station had the longest history of cultivation compared to other sites studied in the Belgorod region. Our results showed that conditions at the experimental station varied the most with respect to the pedogenic C stocks in the undisturbed forest. This is due to the most advanced process of transformation of the former grey forest soils to the arable Chernozems. This was supported not only by the accumulation of humus in the subsoil horizons but also by the carbonates within the soil profile to a 200 cm depth of the agricultural soils.

Pedogenic carbon stocks (t/ha) in the forest-steppe ecotone of Russia

An underestimation of the carbonate pool in agricultural soils of forest-steppe ecotone in central Russia may lead to a significant underestimation of pedogenic (total) C stock. This could also result in an incorrect assessment in the direction that the transformation of the pedogenic C pool in agricultural soils taking. For example, instead of a C accumulation, a decrease in C stocks in agricultural soils of this area was postulated (Orlov *et al.,* 1996; Urusevskaya *et al.,* 2000).

Currently SOC stocks in the grey forest soils in this region were estimated to range from 5.6 Gt (Orlov *et al.,* 1996) to 5.8 Gt (Urusevskaya *et al.,* 2000) to a 100 cm depth. This estimation increased to 7.6 Gt (0-200 cm) in undisturbed forest soils or 7.4 Gt in the agricultural soils (Urusevskaya *et al.,* 2000). Our data showed that the SOC stocks increased by ~10% in the agricultural soils that were under cultivation for more than 200 years. Such long-term cultivated soils currently comprise two thirds of all agricultural grey forest soils in the forest-steppe ecotone of Russia (Chendev and Petin, 2006). Taking into account these data, we quantified an increase in SOC by 0.2 Gt or a total of 7.8 Gt to a 200 cm depth.

Results from our study showed that the median values of the SIC stock, to a 200 cm depth, were 150 t/ha in the agricultural soils and 50 t/ha in the undisturbed forest soils. The area of cultivated and non-cultivated grey forest soils in Russia is estimated to be 16.6 and 23.8 million ha, respectively (Urusevskaya *et al.*, 2000), and SIC in all grey forest soils was estimated to be 3.7 Gt. Thus, the total pedogenic C stock in soils within the forest-steppe ecotone of central Russia may be ~11.5 Gt of which one third is SIC.

Soil carbonate micromorphology and sub-micromorphology

Results from the micromorphological analyses within the different chronosequences in our study showed that carbonates were mostly located in large pores and cracks in the undisturbed grey forest soils (Figure 2.5a), and were commonly interweaved with Fe-pedofeatures (see arrow in Figure 2.5a). This indicated a preferential dissolution and leaching of carbonates in soils under undisturbed forest (Gerasimova *et al.,* 1992). In agricultural soils, there were relatively large carbonate pedofeatures in pores (Figure 2.5b) and small carbonate nodules scattered throughout the plasma. In general, the impregnation of the plasma by carbonates in the carbonate horizons of all agricultural soils was more prevalent compared to the undisturbed forested sites (arrows in Figure 2.5c). In some parts of the plasma, typical cracks arranged in a circle were visible in soils with the highest impregnation of carbonates. Such cracks are referred to as desiccation fissures which appear during the drying of a wet carbonate mass (Figure 2.5d). We have already noted in steppe soils in the southern regions of Russia (Orenburg region), that the desiccation fissures occur in the body of carbonate soft nodules. They indicated that carbonate mass has been accumulated all at once and entirely, shrank during drying, and net of cracks formed in dry carbonate mass, has a certain order (Khokhlova, 2008). Such a phenomenon is characteristic of colloidal matter (Kuznetsova and Khokhlova, 2010).

Sub-micromorphological structures of hard carbonate nodules in the lower horizons of the studied soils supported the idea that the process of dissolution of the carbonate

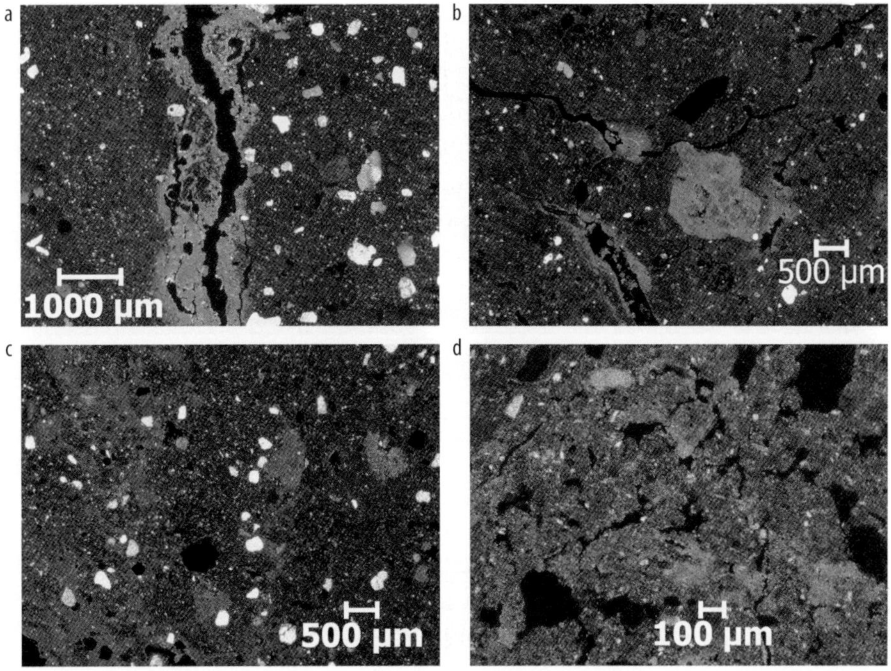

Figure 2.5. Micromorphological structure of carbonate soil horizons in the region of Belgorod (a) carbonate and ferrous accumulations in a long pore-crack (Pit 3forest/09, 190-200 cm); (b) a dense large carbonate pedofeature in plasma and in small pores (Pit S 03/09, 120-130 cm, about 100 years of cultivation); (c) small carbonate nodules are scattered inside strongly carbonate impregnated plasma (Pit 1maize/09, 190-200 cm); and (d) desiccation fissures in the carbonate plasma (Pit 2fallow/09).

nodules prevails in undisturbed forest soils. However, in agricultural soils the process of nodule growth predominates. At the forested site in Melikhovo, the calcite crystals on the cleavage of a hard nodule were fritted (Figure 2.6a) or transformed to a continuous colloform film (Figurc 2.6b); both were covered by etching the pits. In the 100 year old agricultural plot, soil had calcite crystals with multiple planes of growth or small round fresh grains on the surface of formerly perfect calcite crystals (Figure 2.6c). In some cases, the colloform flakes were visible which deposited over the crystals and had a layered structure with numerous growth phases (Figure 2.6d).

Radiocarbon dating of carbonates

Radiocarbon dating of carbonates in soils of the chronosequences at the Experimental Station of Voronezh was carried out for carbonates in the upper horizons (Figure 2.7a) and in the lower horizons (Figure 2.7b). The age of the carbonates in the agricultural

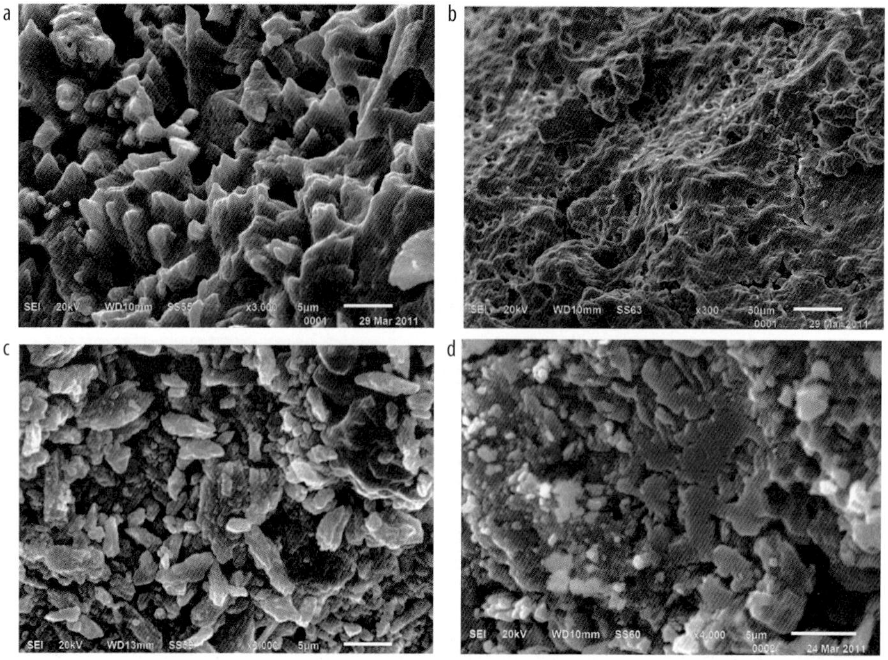

Figure 2.6. Sub-micromorphological structures of hard nodules from soils at Melikhovo in the region of Belgorod (a) the fritted calcite crystals with etching pits on the surface, magnification ×3,000 (Pit MI 1/10); (b) a continuous colloform film in the internal structure of a hard carbonate nodule with etching pits on the surface, ×300 (Pit MI 1/10); (c) calcite crystals with multiple planes of growth, ×3,000; (d) colloform flakes with a layered structure on the surface of calcite crystals, ×4,000; (c) and (d) about 100 years of cultivation, 190-200 cm (Pit MI 2/10).

soils was 2 to 3 times higher compared to the undisturbed forest soil both in the upper and lower soil horizons. An increase in the ^{14}C-dates of carbonates in agricultural soils compared to forest soil was also observed by Khokhlova *et al.* (2013) at the Samarino and Polyana sites. This suggested that in agricultural soils, the carbonate mass moves upward without an exchange with relatively young soil CO_2. Such exchange must occur if carbonates move in true solutions in a dissolved state (Amundson *et al.*, 1994). We believe that the upward movement of carbonates occurred through colloidal solutions in the capillary pores. It means that the carbonates move in the form of whole molecules (ultra-micro-crystals) from the lowermost soil horizons or calcareous parent rocks. The exchange of such molecules with young soil CO_2 is absent. The possibility of such carbonate movement has been discussed previously by Khokhlova (2008).

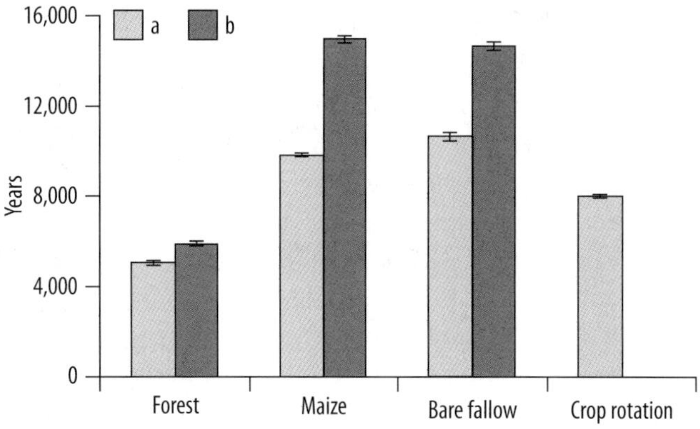

Figure 2.7. Radiocarbon dates of carbonates from (a) the uppermost horizon of their appearance in the studied soil profiles (the uppermost depth of carbonates appearance in the pit under forest is 135-145 cm, 130-150 cm under maize, 80-90 cm under bare fallow and 85-95 cm under crop rotation); and (b) the lower horizon (190-200 cm) in soils of the Voronezh Experimental Station.

Based on micromorphological observations and radiocarbon dating, newly-formed carbonates in agricultural soils may have occurred as a result of pedogenic processes that developed under the changing hydrothermal regimes when the forest was replaced by the agricultural fields. The carbonates moved up from the lowermost soil layers (or calcareous parent material) along capillary-sized pores during the summer. This is especially true when a strong warming and drying of the upper horizons of arable soils without vegetative cover occurs. In addition, the increase in age determined by radiocarbon dating of the agricultural carbonates showed that their old C may be of lithogenic origin. According to our observations from quarries of the Belgorod region, the parent material for recent soils was likely from the late Pleistocene paleosols in which carbonates may have a considerable ^{14}C-age. For instance, the ^{14}C-date for carbonates in the late Pleistocene (Streletskaya) paleosol, in a quarry located in the neighboring Kursk region, is 20,400 yrs. B.P. (Kovda *et al.,* 2009). The Streletskaya paleosol was located just beneath the Holocene soils evaluated in our study.

As carbonates are located in the smallest pores and impregnate the soil mass in the agricultural soils, it is likely to assume that they are safely deposited and highly protected from dissolution. This suggested that the accumulation of carbonates in agricultural soils may lead to the long-term sequestration of SIC in the forest-steppe ecotone of Russia. Results from this study may help to predict SIC accumulation in other semi-arid regions and provide further insight into the potential for the long-term sequestration of C in these soils.

2.4 Conclusions

In the central Russian forest-steppe ecotone, the pedogenic C stock in agricultural soils increased up to 50% due to the accumulation of carbonate C to a 200 cm soil depth. In agricultural soils under cultivation for more than 200 years, little increase in SOC (~10%) in addition to SIC was also observed. This increase in SOC may be due to an accumulation of organic C in the subsurface (80-100 cm) of the soil profile. This is because in this stratum of the soil profile, processes of decomposition and humification dominated due to a dense accumulation of roots from agricultural crops. In the undisturbed dark grey forest and grey forest soils under undisturbed forest cover, SIC stocks varied widely ranging from 0 to 250 t/ha in a 200 cm depth. In the latter case the SIC stocks were on average150-200 t/ha greater compared to SOC stocks. Due to the changing hydrothermal regime in arable lands compared to soils under forest cover, SIC stocks increased as a result of an upward movement of carbonates from the parent material. It is speculated that carbonates move in the capillary-size pores, probably in colloidal suspensions without the exchange of young soil CO_2. This is because in addition to an increase in carbonate C content, the ^{14}C-age of the newly-formed carbonates also increased. These are referred to as pedogenic-lithogenic carbonates because they appear in the profiles of agricultural soils but the source of additional C for them of lithogenic or paleopedogenic origin, which have an old radiocarbon age.

The 100-200 cm depth was especially enriched by carbonates in the agricultural soils. Rarely were carbonates observed at a 50-100 cm depth. The most significant enrichment in carbonates was observed during the first 100 years of cultivation, and subsequent increases in carbonates after 100 years of cultivation were absent. In the long-term field experiment in Voronezh, the different variants of agricultural land management practices demonstrated a clear differentiation in carbonate stocks with respect to the depth of their location within the soil profile. This indicated that the process of capillary uplifting of carbonates may be a rapid soil transforming process, which may occur in less than 50 years. The stocks of pedogenic C to a 200 cm depth in the agricultural and undisturbed forest soils within the forest-steppe ecotone of Russia were estimated to be 11.5 Gt of which one third was SIC. Results from our study suggested that SIC may be highly protected from dissolution in the agricultural soils. This is because carbonates were located in the smallest pores and impregnated in the soil mass according to our micromorphological observations.

Acknowledgements

This work has been supported by RFBR, grants no. 12-05-97512-r-a and no. 12-04-00201-a.

References

Akhtyrsev, B.P., 1996. History of development and antropogenic evolution of the Grey Forest soils within the forest-steppe ecotone. Vestnik of Voronezh' University 2: 11-19.

Akhtyrsev, B.P. and Chshetinina, A.S., 1969. Change of the grey forest soils in the Middle Russian forest-steppe ecotone under agriculrural cultivation. Saransk Pedagogical University Press, Saransk, Russia,164 pp.

Amundson, R., Wang, Y., Chadwick, O., Trumbore, S., McFadden, L., McDonald, E., Wells, S. and DeNiro, M., 1994. Factors and processes governing the ^{14}C content of carbonate in desert soils. Earth and Planetary Science Letters 125: 385-405.

Arinushkina, E.V., 1970. Manual on the chemical analysis of soils. Moscow State University Press, Moscow, Russia, 487 pp.

Arkhangelskaya, T.A, 2012. Temperature regime of complex soil cover. Geos Press, Moscow, Russia, 282 pp.

Chendev, Y.G., 2008. Evolution of forest-steppe soils within the Russian Upland in the Holocene. GEOS Press, Moscow, Russia, 212 pp.

Chendev, Y.G. and Petin, A.N., 2006. Natural and man-caused changes in environments for regions of long-term anthropogenic using – case study in the Belgorod region. Moscow State University Press, Moscow, Russia, 124 pp.

Chendev, Y.G., Aleksandrovskii, A.L., Khokhlova, O.S., Smirnova, L.G., Novykh, L.L. and Dolgikh, A.V., 2011. Anthropogenic evolution of dark gray forest-steppe soils in the southern part of the Central Russian upland. Eurasian Soil Science 44: 1-12.

Egorov, V.V., Fridland, V.M., Ivanova E.N., Rozov, N.N., Nosin, V.A. and Friev, T.A., 1977. Classification and diagnostics of soils in the USSR, 1977. Kolos Press, Moscow, 223 pp.

Fatyanov, A.S., 1959. Experience in analysis of history in development of soils cover in Gorkovskaya oblast. In: Turin, I.V. and Liverovsky, Y.A. (eds.) Soil and geographical investigation and aerophotography using in soil cartography. Academy of sciences of the USSS Press, Moscow, Russia, pp. 3-171.

Gerasimova, M.I., Gubin, S.V. and Shoba, S.A., 1992. Micromorphology of soils of natural zones in the USSR. ONTI PNC RAS Press, Pushchino, Russia, 215 pp.

Glazovskaya, M.A., 2009. Pedolithogenesis and continental cycles of carbon. Librokom, Moscow, Russia, 336 pp.

Kharitonychev, A.T., 1960. Role of human activities in land-use change. Gorkovsky University Press, Gorkiy, Russia, 150 pp.

Khokhlova, O.S., 2008. Carbonate status of steppe soils as an indicator and memory of their spatial and temporal variability. Abstract of dissertation of Doctor of Geographical Sciences, Moscow, Russia, 48 pp.

Khokhlova, O.S., Khokhlov, A.A., Chichagova, O.A., Kuznetsova, A.M. and Oleinik, S.A., 2008. Transformation of carbonate pedofeatures in paleosols buried under kurgans in the North Caucasus region. Eurasian Journal of Soil Science 41: 923-936.

Khokhlova, O.S., Rusakov, A.V., Kuznetsova, A.M., Myakshina, T.N. and Chendev, Yu.G., 2013. Radiocarbon dating of pedogenic carbonates in deep horizons of forest-steppe soils. Eurasian Journal of Soil Science 46: 941-950.

Kokovina, T.P. and Lebedeva, I.I., 1986. Up-to-date hydrothermal regimes, genetic and geographical peculiarites of Chernozems of European part of the USSR. In: Kovda, V.A. and Glazovskaya, M.A. (eds.) Advances of soil science. Soviet soil scientists to the XVIII International Congress of Soil Science, Nauka, Moscow, Russia, pp. 148-153.

Kovda, I., Sycheva, S., Lebedeva, M. and Inozemtzev, S., 2009. Variability of carbonate pedofeatures in a loess-paleosol sequence and their use for paleoreconstructions. Journal of Mountain Science 6: 155-161.

Kuznetsova, A.M. and Khokhlova, O.S., 2010. Morphology of carbonate neoformations in soils of various types. Lithology and Mineral Resources 45: 89-100.

Kuznetsova, A.M., Khokhlova, O.S., Chendev, Y.G. and Aleksandrovskii, A.L., 2010. Evolution of the carbonate state of agriculturalally transformed dark grey forest soils in the central forest-steppe. Eurasian Journal of Soil Science 43: 1527-1534.

Larionova, A.A., Stulin, A.F., Zanina, O.G., Yevdokimov, I.V., Khokhlova, O.S., Buegger, F., Schloter, M. and Kudeyarov, V.N., 2012. Distribution of stable carbon isotopes in an Agro-Chernozem during the transition from C3 vegetation to a corn monoculture. Eurasian Soil Science 45: 768-778.

Orlov, D.S., Biryukova, O.N. and Sukhanova, N.I., 1996. Soil organic matter of Russia. Nauka Press, Moscow, Russia, 256 pp.

Stulin, A.F., 2007. Effect of long-term fertilization in a continuous corn cropping system on the yield of corn and the removal of nutrients on leached chernozem. Agrochemistry 1: 25-30.

Urusevskaya, I.S., Meshalkina, Yu.L. and Khokhlova O.S., 2000. Geographic and genetic features of the humus status of gray forest soils. Eurasian Journal of Soil Science 33: 1213-1225.

Yonko, O.A. and Melnikova, M.G., 2006. The influence of forest plantations on carbonate regime of soils in the 'Kamennaya steppe' wildlife preserve. In: Scheglov, D.I. (ed.) Proceedings of Russian Confederation on Russian Chernozems – their ecological condition and modern soil processes. September 25-28, 2006. Voronezh State University Press, Voronezh, Russia, pp. 84-87.

Chapter 3.

The carbon footprint of coffee production chains in Tolima, Colombia

H.J. Andrade, M.A. Segura, D.S. Canal, M. Feria, J.J. Alvarado, L.M. Marín, D. Pachón and M.J. Gómez

Ecofriendly Production of Tropical Crops (PROECUT), University of Tolima, Apartado:546, Barrio Santa Elena, Ibagué, Colombia; hjandrade@ut.edu.co

Abstract

Agriculture is one of the most important sectors influencing climate change because it can act as net source of greenhouse gases (GHG) or it is able to mitigate global warming. Production systems with woody perennial species, such as coffee (*Coffea arabica* L.) plantations, have shown to mitigate global warming because of their ability to sequester carbon (C) in biomass and soil. In this study, the C footprint of coffee production systems in Líbano, Colombia was assessed by evaluating coffee plantations in monoculture, in agroforestry systems (AFS) with *Cordia alliodora* (Ruiz & Pavón) Oken, and in AFS with plantain (*Musa* sp. var AAB). Carbon sequestration varied between 2.7 and 19.9 tCO_2/ha/y for monoculture and AFS with *C. alliodora*, respectively. All coffee production systems emitted GHG at a rate of 1.4 to 3.5 Mg CO_2e/ha/y; whereas coffee bean processing emitted 7.1 kg CO_2e/kg. Only the agroforestry system with *C. alliodora* had a positive C footprint, showing a net sequestration of 14.2 Mg CO_2e/ha/y in comparison to the AFS with plantain (-2.9 Mg CO_2e/ha/y) and the coffee monoculture (-5.7 Mg CO_2e/ha/y). The inclusion of timber trees, such as *C. alliodora*, in coffee production systems can change a coffee plantation from a C emitter to one of C sequestration. Results from our study suggested that AFS coffee production systems play an important role in mitigating global warming. This provides an incentive not only for coffee producers, but also for the development of policies to adapt AFS for coffee production because they can play an important ecological service in tropical biomes.

Keywords: agroforestry systems, biomass, carbon sequestration, greenhouse gases

3.1 Introduction

It is predicted that climate change will have the greatest impact on developing countries due to their lower capacity to adapt to a changing environment. The United Nations Framework Convention on Climate Change (UNFCCC) proposed that adaptation and mitigation to climate change may be the most effective strategies to combat this global environmental problem. Agriculture is one sector of the economy that can be a major contributor to climate change because of its high contribution to greenhouse gas (GHG) emissions. Johnson *et al.* (2007) estimated that agricultural activities contributed 13.5% of the total global GHG emissions as a result of enteric fermentation, flooding of rice fields, land-use change, and the use of nitrogen (N) fertilizers.

However, land-use systems that incorporate woody perennials, such as agroforestry coffee plantations, are potential technologies that could mitigate climate change through the sequestration of carbon (C) in plant biomass and soil (Albrecht and Kandji, 2003; Andrade *et al.*, 2008; Beer *et al.*, 2003; Brown, 1996; Montagnini and Nair, 2004; Oelbermann *et al.*, 2004; Soto-Pinto *et al.*, 2010). These systems can be included in LULUCF (Land Use, Land Use-Change and Forestry) projects within the Clean Development Mechanisms (CDM) framework (Pearson *et al.*, 2005). Similarly, the tree component in agroforestry systems (AFS) may be an option to mitigate climate change by providing bioenergy through the substitution of fossil fuels and help minimize further deforestation (Verchot *et al.*, 2005).

The possibility to include coffee AFS in voluntary C markets (Soto-Pinto *et al.*, 2010) or in CDM (Van Noordwijk *et al.*, 2005) may be one option to help mitigate climate change. For example, multistrata AFS with coffee were considered as applicable technologies in CDM projects in Indonesia (Van Noordwijk *et al.*, 2005). Coffee production systems have also been promoted as strategies to adapt to climate change because of their higher resilience and stability compared to coffee monoculture. For example, in Chiapas, México, coffee producers diversified their coffee production systems through the inclusion of shade trees as an adaptation strategy to climate change (Frank *et al.*, 2011).

Environmental services generated by such agricultural systems can also generate payment for producers or land managers, and provide consumers with a certified and environmentally sustainable product (Wunder, 2007). In spite of this benefit, the entire C footprint of such production systems must be taken into consideration. Cranston and Hammond (2012) defined the C footprint as the amount of carbon dioxide equivalent (CO_2e) emissions associated with a given activity or community. Post (2006) also defines the C footprint as the total amount of CO_2 and other GHG emitted over the full life cycle of a process or product. In AFS, the C footprint must consider C sequestration in plant biomass and the emission of GHG (Hergoualc'h *et al.*, 2012; Segura and Andrade, 2012). As such, a AFS with a positive C footprint may show a greater capacity to sequester C than the amount of GHG emitted, whereas an AFS with a negative C footprint will have greater GHS emissions compared to the amount of C sequestered (Segura and Andrade, 2012).

Globally, there are ~25 million of coffee producers (Giovannucci and Koekoek, 2003) but certified coffee production such as Fair Trade, and certified organic are minor contributors (<2%). However, certified coffee production benefits 750 thousand families and industries throughout the entire production chain (Giovannucci and Koekoek, 2003). Certified coffee promotes biodiversity protection and improves livelihoods (Philpott *et al.*, 2007). Nevertheless, Segura and Andrade (2012) suggested that in Costa Rica, the shade component rather than certification was the major factor that influenced C sequestration and the C footprint of the coffee plantation. The goal of this study was to estimate the C footprint of the coffee production chain in Líbano, Colombia. The specific objectives of this research were to quantify C sequestration and GHG emissions in coffee production systems at the farm-scale and to evaluate GHG emissions at the processing stage (industrial scale). Results from this study will help identify the most environmentally sound coffee production systems, in terms of climate change, and will help promote the long-term and sustainable production of coffee.

3.2 Materials and methods

Study area

The study was conducted in the municipality of Líbano, which is part of the district of Tolima in Colombia. At the national level, this region is the third largest coffee producer (Federacafé, 2013). Líbano is located in the very humid premontane forest life zone (Holdridge, 1996). The area is 1,565 m above sea level, with a mean annual precipitation of 2235 mm/y and a mean annual temperature of 19.1 °C (IDEAM, 2010). The soils of Líbano are classified as Andosols with good physical characteristics, high fertility, and with moderately steep to steep slopes (25 to 60%). However, some areas have a shallow and coarser soil classified as Umbrisols, with medium fertility and deep soils. These soils are located on steep slopes (>70%), and are highly susceptible to erosion. In this region of Colombia, coffee is produced using a variety of production systems including coffee monoculture, coffee in agroforestry systems with Spanish elm (*Cordia alliodora* (Ruiz & Pavón) Oken), plantain (*Musa* sp. var AAB), and/or rubber (*Hevea brasiliensis* (Willd. ex Adr. Juss.) Muell. Arg.).

Study design

The most dominant coffee production systems (treatments) in Líbano were selected for this study: (1) coffee monoculture; (2) coffee agroforestry systems (AFS) with C. *alliodora*; and (3) coffee AFS with plantain (*Musa* sp. var. AAB). Six coffee producers were selected based on the recommendation by the Municipal Committee of Coffee Producers. Each of the sites was evaluated on the amount of C sequestered in above- and below-ground plant biomass, the development of allometric models for coffee shrubs based on local data, GHG emissions during coffee plantation management (farm-scale), and GHG emitted during the coffee processing phase from whole beans to ground coffee (industrial-scale).

Allometric models for quantifying the total aboveground biomass of coffee shrubs were developed following the approach recommended by Segura and Andrade (2008). A total of 40 coffee bushes with dimensions between 0.6 to 8.0 cm in trunk diameter at a height of 15 cm above ground (D_{15}) and 0.46 to 3.00 m height (ht), were selected. These sizes represented a range of dimensions of coffee plants within the three different production systems evaluated in this study, and within the region of Tolima. An equal number of individuals were selected from *Caturra* and *Castillo* cultivars, which were produced at each of the study sites.

Each bush was measured (D_{15} and ht) and subsequently cut at ground level. If coffee shrubs were previously pruned, the D_{15} was measured at ground level, in addition to ht and pruning height. Coffee shrubs were separated into separate components of trunk, leaves, fruits and roots and weighted for fresh weight. A 200 g subsample of each component was oven dried at 60 °C until a constant weight was obtained and weighed. Pearson's correlation coefficients between total aboveground biomass and D_{15} or ht were quantified and the best-fit allometric biomass equation was developed (Segura and Andrade, 2008). Best-fit allometric models were selected based on those equations with the highest coefficient of determination (R^2), the highest adjusted R^2 (Adj.R^2), the lowest root mean square error (RMSE), the lowest quadratic mean error of prediction (ECMP), and a best fit for a logistic model. These models were also compared to those generated by Segura *et al.* (2006).

$$B = -0.357 + 0.371 \times D_{15} \quad (1)$$

$$\log B = -1.181 + 1.991 \times \log (D_{15}) \quad (2)$$

where B is the total aboveground biomass (kg/plant) and D_{15} is the diameter of the trunk at a height of 15 cm. Root systems were carefully extracted from the soil and washed over a 2 mm sieve to remove soil. Root biomass quantified by weighing fresh material and that of a 200 g oven-dried (60 °C) subsample. From this information the belowground/aboveground biomass ratio (Br/Ba) was quantified.

Subsequently, C storage in coffee plantations was quantified using non-destructive sampling. This data was based on four replicates per treatment. Coffee shrubs and shade trees were sampled and their aboveground biomass was estimated using the previously developed allometric equations. Two rectangular sampling plots (600 m^2) were established and all shade trees with a diameter of trunk at breast height (dbh) higher than 10 cm were measured (dbh and ht). In the north-east corner of each plot, a subplot of 25 m^2 was established to measure all coffee shrubs within this area for their diameter and height (D_{15} and ht).

Biomass of *C. alliodora* (Equation 3) was estimated using allometric models previously developed from Costa Rican cacao (*Theobroma cacao* L.)-*C. alliodora* plantation (Andrade *et al.*, 2008). For coffee shrubs, aboveground biomass was estimated using the biomass equations developed in this study. Belowground biomass of coffee plants was estimated using a (Br/Ba) ratio of 0.15, which was based on the destructive sampling of coffee shrubs in this study. Root biomass of *C. alliodora* was quantified according to Cairns *et al.* (1997) (Equation 4).

$$B = 10^{(-0.51 + 2.08 \log (dbh))} \qquad (Adj.R^2 = 0.92) \tag{3}$$

where B is the total aboveground biomass (kg/plant) and D_{15} is the diameter of the trunk at a breast height (130 cm).

$$Br = e^{(-1.0587 + 0.8836 \ln (Ba))} \tag{4}$$

where Br is belowground biomass (Mg/ha) and Ba is aboveground biomass (Mg/ha). Total biomass was the addition of above- and below-ground components for coffee plants plus the shade trees. Biomass was converted to C sequestration using a C fraction of 0.5 (IPCC, 2006). The mean C sequestration rate was estimated dividing the C storage by the age of the coffee or trees. Carbon was converted to CO_2 using a constant of 3.67.

The emission of GHG from coffee plantations at the farm- and industrial-scale was quantified for each of the six coffee farms in Líbano and for six different bean processing plants located in Líbano, Ibagué, Armenia and Pereira, using semi-structured interviews. At the farm-scale, producers were interviewed on energy inputs required for the production and processing of coffee, in addition to their usage of fossil fuels, N fertilizers and limestone. At the industrial-scale, a review of electricity bills provided insight into the energy required to process whole coffee beans to the finished and ready-to-market product. In summary, the following activities emit GHG in the management of coffee plantations, and coffee bean transportation and processing:

a. The amount fertilizer and its N content used by coffee producers. We used an emissions factor for N application of 0.01 kg N_2O/kg N (IPCC, 2006).
b. The amount and type of limestone added, and how often limestone was applied to the soil. A factor of 0.122 kg C/kg of limestone was considered for the application of calcium and magnesium carbonates (IPCC, 2006).
c. The amount of gasoline used in the management of coffee plantations in addition to the gasoline required for the transportation of coffee beans, workers, and other incidental uses of gasoline required for coffee production at the farm-scale. A fuel emission factor of 2.33 kg CO_2e/l of gasoline (IPCC, 2006) was used.
d. Emission of GHG at the industrial-scale was determined by the amount of energy used during coffee bean processing based on electrical consumption of the production plant. An emission factor of 130 g CO_2e/kWh was used (Camargo *et al.*, 2013).

The C footprint was based on the amount of C sequestered in above- and below-ground plant biomass minus the GHG emitted as part of coffee plantation management, transportation and processing of beans to ground coffee. In our study, a positive (+) C footprint refers to C, whereas a negative (–) C footprint refers to the system as a C emitter.

3.3 Results and discussion

Tools for estimating above- and below-ground biomass in coffee bushes

The best-fit model (P<0.05) to quantify C sequestration for both coffee cultivars was based on the trunk diameter at a 15 cm height and that based on ht as the second independent variable (Table 3.1). Both models had a high R^2, low error and best-fit for the logistic function.

The models developed in this research have high accuracy (high R^2) to estimate aboveground biomass and could be applied to the coffee producing region of Líbano, Colombia. Also in Colombia, Aristizábal (2011) developed biomass models for coffee based on non-destructive sampling. However large errors were associated with this research due to the lack of destructive sampling. Because coffee shrubs were destructively sampled for above- and below-ground biomass, the accuracy in our study is greater than that of Aristizábal (2011). Additionally, it is possible to develop models

Table 3.1. Allometric models for estimating above- and below-ground biomass of the *Caturra* and *Castillo* coffee cultivars in Líbano, Colombia.

Models	R^2	Adj. R^2	ECMP[1]	RSME[1]
$\ln(B) = -1.92 + 0.95 \times \ln(D_{15}) + 1.42 \times \ln(ht)$	0.88	0.82	0.46	0.61
$B = 0.36 - 0.18 \times D_{15} + 0.08 \times D_{15}^2$	0.82	0.81	0.45	0.61

[1] ECMP = mean error of prediction; RSME = root mean square error.

where the dependent and independent variables are logarithmically transformed, providing greater accuracy of the allometric equation (Kanninen and Pérez, 2002; Lott *et al.*, 2000; Segura and Kanninen, 2005; Segura *et al.*, 2006). Additionally, our allometric models were similar to those developed by Segura *et al.* (2006) for coffee shrubs in Nicaraguan agroforestry systems. Given the high accuracy of the currently available allometric equations for coffee shrubs, it may be possible to construct a neotropical model which may be used at other sites to estimate C sequestration and nutrient cycling in coffee.

Carbon sequestration in biomass

Coffee bean yield was not statistically different ($P>0.05$) between the different plantation management practices. For example the bean yield ranged from 1,520 kg/ha/y in the monoculture to 1,178 kg/ha/y in the AFS with *C. alliodora* and 1,153 kg/ha/y in the coffee AFS with plantain (Table 3.2). The amount of C sequestered in the different coffee production systems ranged from 2.6 to 19.4 Mg CO_2e/ha/y (Table 3.2). The coffee AFS with *C. alliodora* (19.4 Mg CO_2e/ha/y) sequestered a significantly greater ($P<0.05$) amount of C compared to the AFS with plantain (2.6 Mg CO_2e/ha/y) and the coffee monoculture (2.7 Mg CO_2e/ha/y) (Table 3.2). In Costa Rica, Hergoualc'h *et al.* (2012) found that a coffee AFS with *Inga densiflora* Benth. (16.9 Mg CO_2e/ha/y) sequestered 2.3 times more C compared to a coffee monoculture (7.3 Mg CO2e/ha/y).

The greatest amount of C was sequestered in the coffee AFS with *C. alliodora* where the tree component contributed 83% of the total amount of C sequestered in biomass. Root systems sequestered 13-17% of the total amount of C sequestered in biomass. Segura and Andrade (2012) estimated rates of C sequestration between 5.9 and 13.8 Mg CO_2e/ha/y in AFS under conventional and certified, including organic and fair trade, coffee plantations in Costa Rica.

Table 3.2. Coffee bean yield (kg/ha/y) and carbon sequestration (Mg CO_2e/ha/y) in biomass of coffee shrubs and woody perennial plants in three differently managed coffee production systems in Líbano, Colombia.[1]

	Monoculture	AFS with *Cordia alliodora*	AFS with plantain
Coffee bean yield (kg/ha/y)	1,520.0±350.4[a]	1,177.5±420.4[a]	1,152.5±410.9[a]
Carbon sequestration (Mg CO_2e/ha/y)			
Coffee bushes			
Aboveground	2.32±0.51[a]	2.93±2.51[a]	2.30±1.13[a]
Belowground	0.35±0.08[a]	0.44±0.38[a]	0.35±0.17[a]
Total	2.67±0.59[a]	3.37±2.89[a]	2.64±1.30[a]
Shade trees			
Aboveground	0.00±0.00[b]	13.08±6.91[a]	0.00±0.00[b]
Belowground	0.00±0.00[b]	2.91±1.19[a]	0.00±0.00[b]
Total	0.00±0.00[b]	15.99±8.09[a]	0.00±0.00[b]
Total			
Aboveground	2.32±0.51[b]	16.02±6.21[a]	2.30±1.13[b]
Belowground	0.35±0.08[b]	3.35±1.08[a]	0.35±0.17[b]
Total	2.67±0.59[b]	19.37±7.26[a]	2.64±1.30[b]
Total (kg CO_2e/kg)	1.75±0.39[b]	16.45±6.17[a]	2.29±1.12[b]

[1] Values are means ± standard error. Values with the same superscript letter, comparing differences between coffee production systems (monoculture vs. agroforestry system (AFS) with *Cordia alliodora* or plantain), are not significantly different at $P<0.05$.

Overall, coffee produced in AFS with *C. alliodora* (16.45 kg CO_2e/kg) was more environmentally sustainable because of their greater potential to sequester C per unit of coffee bean produced compared to the coffee monoculture (1.75 kg CO_2e/kg) and the coffee AFS with plantain (2.29 kg CO_2e/kg). However, our results on net C sequestration were lower compared to those in Costa Rica (34.5 kg CO_2e/kg), but were more similar to coffee produced under certification programs including fair trade, Rainforest Alliance and UTZ Kapeh (3.7 to 14.0 kg CO_2e/kg) (Segura and Andrade 2012).

Emission of greenhouse gas during coffee bean production

Greenhouse gas emissions during coffee production at the farm-scale ranged from 1.43 to 3.46 Mg CO_2e/ha/y (Table 3.3), where the monoculture had a 20 and 46% greater

Table 3.3. Greenhouse gas emissions (Mg CO_2e/ha/y) from coffee production systems in monoculture or in agroforestry systems in Líbano, Colombia.[1]

		Monoculture	AFS with *Cordia alliodora*	AFS with plantain
Greenhouse gas emissions (kg CO_2e/kg)				
Nitrogen fertilizers		1.37±2.01[a]	0.37±0.33[a]	0.63±0.28[a]
Limestone		0.16±0.04[a]	0.12±0.08[a]	0.13±0.09[a]
Fossil fuels	Machinery	0.28±0.47[a]	0.39±0.46[a]	0.42 ± 0.43[a]
	Transportation	0.15±0.19[a]	0.47±0.65[a]	0.52 ± 0.43[a]
Total		1.96±1.78[a]	1.34±0.73[a]	1.69±0.59[a]
Total (Mg CO_2e/ha/y)		3.46±4.01[a]	1.43±0.83[a]	1.79±0.60[a]

[1] Values are means ± standard error. Values with the same superscript letter, comparing differences between coffee production systems (monoculture vs. agroforestry system (AFS) with *Cordia alliodora* or plantain), are not significantly different at $P<0.05$.

emission compared to the AFS with plantain and *C. alliodora*, respectively (Table 3.3). Use of N fertilizers resulted in the greatest contribution to GHG emissions. For example, N fertilizers contributed 45% of the total GHG emissions from coffee plantations. However, N fertilizers in coffee monoculture plantations contributed 70% or 1.37 kg CO_2e/kg of the total GHG emissions (Table 3.3). Segura and Andrade (2012) observed similar results in Costa Rica where the use of N fertilizers contributed 63 to 82% of the total GHG emissions in coffee production systems. However, Noponen *et al.* (2012) found that N from pruning inputs contributed only 7% (Costa Rica) and 42% (Nicaragua) of the C footprint in coffee AFS production systems. This was because N losses via leaching are lower in coffee grown in AFS due to the presence of shade trees. However, when shade trees are legumes such as *I. densiflora*, N_2O emissions were greater than those in monoculture coffee plantations (Hergoualc'h *et al.,* 2012).

Greenhouse gas emissions as a result of fuel consumption in coffee plantation management at the farm-scale were relatively low (24% of total – 0.74 kg CO_2e/kg). Segura and Andrade (2012) found similar results in Costa Rican coffee plantations. Limestone application (8%) was also less relevant in terms of GHG emissions compared to other sources (Table 3.3). Overall, coffee monoculture (3.46 kg CO_2e/kg) had the greatest GHG emissions, which was 93% and 142% higher compared to coffee AFS with plantain or *C. alliodora* respectively (Table 3.3).

Emission of greenhouse gas in coffee bean processing

The coffee processing industries evaluated in this study differed significantly (P<0.05) with respect to the volume of coffee beans processed (3.26 to 13.96 Mg/y) (Table 3.4). As expected, the GHG emission from coffee processing plants was scale-dependent as a result of the amount of coffee processed yearly. For example, coffee processing plants in Armenia and Pereira were more environmentally sustainable due to a lower GHG emission per unit of coffee processed compared to those in Ibagué and Líbano (Table 3.4). On average, grinding 1 kg of coffee in this region emits 7.1 kg CO_2e to atmosphere (Table 3.4). These results suggested that coffee processing may be more environmentally sustainable in large coffee processing plants. This promotes the idea of coffee producer's cooperatives where coffee beans could be processed at a larger scale.

Carbon footprint and the coffee production chain

Significant differences (P<0.05) between components of C sequestration were observed between coffee production systems. Carbon sequestration in above- and below-ground biomass was significantly different. For example, the coffee AFS with *C. alliodora* had the highest C sequestration (Figure 3.1). Contrarily, GHG emissions due to management activities in coffee plantations were not significantly (P>0.05) different between production systems (Figure 3.1). These results contrasted with those reported by Segura and Andrade in Costa Rica (2012), who found GHG emissions between 0.5 and 1.1 kg CO_2e/kg green coffee in different types of coffee AFS.

The AFS with *C. alliodora* was the most environmentally sustainable coffee production system because it presented the greatest potential for C sequestration. This system had a

Table 3.4. Emissions of greenhouse gases (kg CO_2e/kg ground coffee) based on the consumption of electricity required for coffee bean processing in Líbano, Colombia. Standard errors are given in parentheses.

Locality	Volume of coffee processed (Mg/y)	Total emission (kg CO_2e/kg ground coffee)[1]
Líbano	3.26	45.47±3.57[a]
Ibagué	4.00	9.81±5.05[b]
Armenia & Pereira	13.96	2.77±3.57[b]

[1] Values are means ± standard error. Values with the same superscript letter, comparing differences between coffee bean processing plants, are not significantly different at P<0.05.

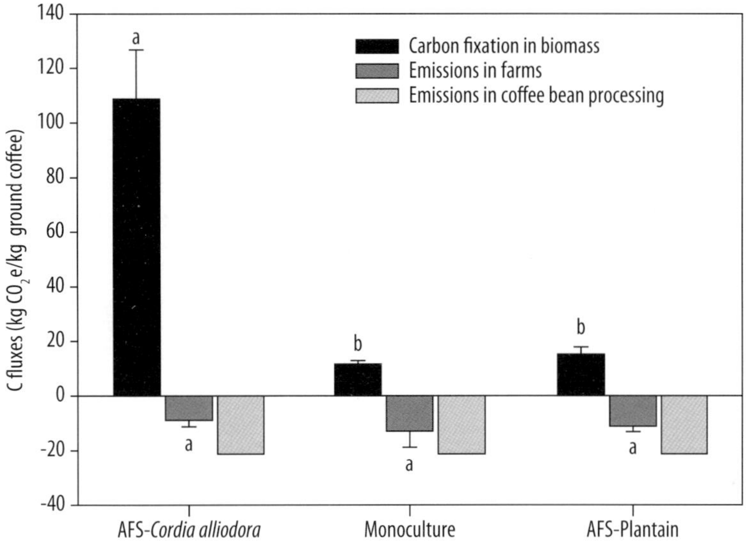

Figure 3.1. Components of the carbon footprint in the coffee production chain in monoculture or agroforestry (AFS) in Líbano, Colombia. Positive values refer to carbon sequestration in biomass, whereas negative values are emissions of greenhouse gases at the farm- and industrial-scale.

positive C balance, where C sequestration in biomass was markedly higher than GHG emissions (14.2 Mg CO_2e/ha/y). This was significantly greater (P<0.05) compared to the coffee AFS with plantain (-2.9 Mg CO_2e/ha/y) and the coffee monoculture (-5.7 Mg CO_2e/ha/y) (Figure 3.2a). Hergoualc'h *et al.* (2012) also found that the GHG balance was ~4 times greater in AFS with *I. densiflora* (14.6 Mg CO_2e/ha/y) compared to that in coffee in monoculture (3.8 Mg CO_2e/ha/y) in Costa Rica.

The carbon footprint produced during the production of coffee was expected to be significantly different (P<0.05) between coffee production systems (Figure 3.2b). The coffee AFS with *C. alliodora* (78.6 kg CO_2e/kg ground coffee) was more environmentally sustainable compared to the coffee AFS with plantain (-17.3 kg CO_2e/kg ground coffee) and coffee monoculture (-22.6 kg CO_2e/kg ground coffee). These results suggested that the tree component plays a major role in enhancing net C sequestration in coffee production systems.

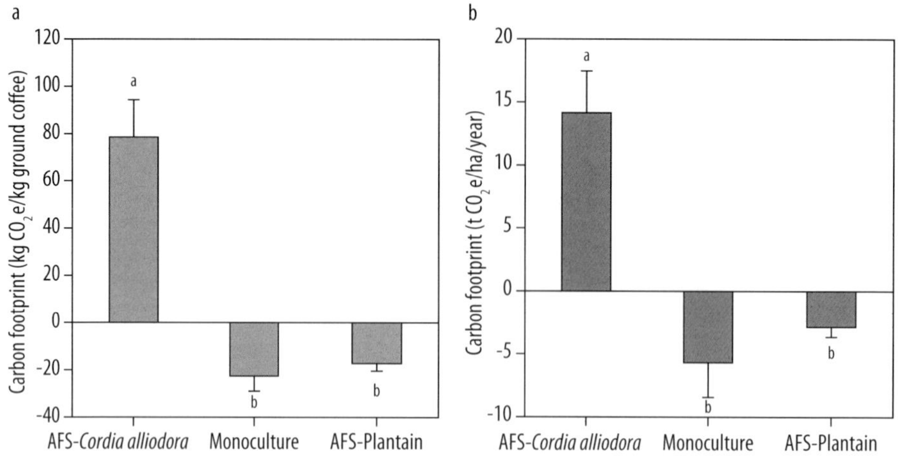

Figure 3.2. The carbon footprint of the most dominating coffee production systems under monoculture or in agroforestry (AFS) in Líbano, Colombia (a) by unit of coffee produced, and (b) per area of coffee plantation.

3.4 Conclusions

The development of allometric biomass models, based on local coffee plantation biomass data, could be successfully implemented to estimate overall C sequestration. Results from our study showed that coffee AFS, especially those with *C. alliodora*, had a greater capacity to sequester C compared to coffee monoculture or coffee AFS with plantain. This was attributed to the tree component in these coffee production systems. Our study suggested that through national policies, coffee production with timber species may be more environmentally sustainable compared to producing coffee in monoculture. As such, the adoption of coffee AFS may be an incentive to initiate a payment for environmental services system which will result in the long-term and sustainable production of coffee in this region of Colombia.

Acknowledgements

We thank all of the participating coffee producers in Líbano, Colombia, and their input of labor and time to help complete this study. We also thank the Municipal Committee of Coffee Producers of Líbano and the Central Committee of Research of the University of Tolima, who financially support this study (project number 80111).

References

Albrecht, A. and Kandji, S.T., 2003. Carbon sequestration in tropical agroforestry systems. Agriculture, Ecosystems and Environment 99: 15-27.

Andrade, H.J., Segura, M., Somarriba, E. and Villalobos, M., 2008. Valoración biofísica y financiera de la fijación de carbono por uso del suelo en fincas cacaoteras indígenas de Talamanca, Costa Rica. Revista Agroforestería en las Américas 46: 45-50.

Aristizábal, J.D., 2011. Desarrollo de modelos de biomasa aérea en sombríos de cafeto (*Coffea arabica* L.) mediante datos simulados. Revista U.D.C.A Actualidad and Divulgación Científica 14: 49-56.

Beer, J., Harvey, C., Ibrahim, M., Harmand, J.M., Somarriba, E. and Jiménez, F., 2003. Servicios ambientales de los sistemas agroforestales. Agroforestería en las Américas 10: 80-87.

Brown, S., 1996. Present and potential roles of forests in the global climate change debate. Unasylva 185: 3-10.

Cairns, M.A., Brown, S., Helmer, E.H. and Baumgardner, G.A., 1997. Root biomass allocation in the world's upland forests. Oecologia 111: 1-11.

Camargo, L.A., Arboleda, M.N. and Cardona, E., 2013. Producción de energía limpia en Colombia, la base para un crecimiento sostenible. Boletin Virtual XM. Compañía Expertos en Mercados, Filial de ISA, Colombia. Available at: http://www.xm.com.co/BoletinXM/Documents/MDLColombia_Feb2013.pdf.

Cranston, G.R. and Hammond, G.P., 2012. Carbon footprints in a bipolar, climate-constrained world. Ecological Indicators 16: 91-99.

Federación Nacional de Cafeteros de Colombia (FEDERACAFÉ), 2013. Tolima café de alta calidad. Available at: http://tolima.federaciondecafeteros.org.

Frank, E., Eakin, H. and López-Carr, D., 2011. Social identity, perception and motivation in adaptation to climate risk in the coffee sector of Chiapas, Mexico. Global Environmental Change 21: 66-76.

Giovannucci, D. and Koekoek, F.J., 2003. The state of sustainable coffee: a study of twelve major markets. Feriva S.A., Cali, Colombia, 198 pp.

Hergoualc'h, K., Blanchart, E., Skiba, U., Henault, C. and Harmand, J. M., 2012. Changes in carbon stock and greenhouse gas balance in a coffee (*Coffea arabica*) monoculture versus an agroforestry system with *Inga densiflora*, in Costa Rica. Agriculture, Ecosystems and Environment 148: 102-110.

Holdridge, L., 1996. Ecología basada en zonas de vida. 4ª reimpresión, Instituto Interamericano de Cooperación para la agricultura IICA, San José, Costa Rica, 216 pp.

Instituto de Hidrología, Meteorología y Estudios Ambientales (IDEAM), 2010. Cambio Climático en Colombia. Available at: http://www.derechoambientalcolombiano.org/Assets/PDFs/ICDA-Generalidades_CC_Ideam.pdf.

Intergovernmental Panel on Climate Change (IPCC), 2006. IPCC guidelines for national greenhouse gas inventories. Intergovernmental Panel on Climate Change National Greenhouse Gas Inventories Programme. Available at: http://www.ipcc-nggip.iges.or.jp/public/2006gl.

Johnson, J., Franzluebbers, A.J., Weyers, S.L. and Reicosky, D.C., 2007. Agricultural opportunities to mitigate greenhouse gas emissions. Environmental pollution 150: 107-124.

Kanninen, M. and Pérez, L.D., 2002. Wood specific gravity and aboveground biomass of *Bombacopsis quinata* plantations in Costa Rica. Forest Ecology and Management 165: 1-9.

Lott, J.E., Howard, S.B., Black, C.R. and Ong, C.K., 2000. Allometric estimation of above-ground biomass and leaf area in managed *Grevillea robusta* agroforestry systems. Agroforestry Systems 49: 1-15.

Montagnini, F. and Nair, P.K.R., 2004. Carbon sequestration: an underexploited environmental benefit of agroforestry systems. Agroforestry Systems 61: 281-295.

Noponen, R., Edwards-Jones, G., Haggar, J.P., Soto, G., Attarzadeh, N. and Healey, J.R., 2012. Greenhouse gas emissions in coffee grown with differing input levels under conventional and organic management. Agriculture, Ecosystems and Environment 151: 6-15.

Oelbermann, M., Voroney, R.P. and Gordon, A.M., 2004. Carbon sequestration in tropical and temperate agroforestry systems: a review with examples from Costa Rica and Southern Canada. Agriculture, Ecosystems and Environment 104: 359-377.

Parliamentary Office for Science and Technology (POST), 2006. Carbon footprint of electricity generation. Available at: http://www.parliament.uk/documents/post/postpn268.pdf.

Pearson, T., Walker, S. and Brown, S., 2005. Sourcebook for land use, land-use change and forestry projects. Biocarbon Fund and Winrock International. Available at: http://www.winrock.org/sites/default/files/publications/attachments/Winrock-BioCarbon_Fund_Sourcebook-compressed.pdf.

Philpott, S., Bichier, P., Rice, R. and Greenberg, R., 2007. Field-testing ecological and economic benefits of coffee certification programs. Conservation Biology 21: 975-985.

Segura, M. and Andrade, H.J. 2008. ¿Cómo construir modelos alométricos de volumen, biomasa o carbono de especies leñosas perennes? Agroforestería de las Américas 46: 89-96.

Segura, M. and Andrade, H.J., 2012. Huella de carbono en cadenas productivas de café (*Coffea arabica*) con diferentes estándares de certificación en Costa Rica. Revista Luna Azul 35: 60-77.

Segura, M. and Kanninen, M., 2005. Allometric models for tree volume and total aboveground biomass in a tropical humid forest in Costa Rica. Biotropica 37: 2-8.

Segura, M., Kanninen, M. and Suárez, D., 2006. Allometric models for estimating aboveground biomass of shade trees and coffee bushes grown together. Agroforestry Systems 68: 143-150.

Soto-Pinto, L.M., Anzueto, J., Mendoza, G. and Jiménez-Ferrer, B., 2010. Carbon sequestration through agroforestry in indigenous communities of Chiapas, Mexico. Agroforestry Systems 78: 39-51.

Van Noordwijk, M., Widayati, A., Lusiana, B., Hairiah, K. and Arifin, B., 2005. What can a clean development mechanism do to enhance trees in the land scape? Experience with rubber, coffee and timber-based agroforestry systems in Indonesia. In: Murdiyarso, D. and Herawati, H. (eds.) Carbon forestry: who will benefit? CIFOR, Bogor, Indonesia, pp. 92-111.

Verchot, L., Mackensen, J., Kandji, S., Van Noordwijk, M., Tomich, T., Ong, C., Albrecht, A., Bantilan, C., Anupama, K. and Palm, C., 2005. Opportunities for linking adaptation and mitigation in agroforestry systems. In: Robledo, C., Kanninen, M. and Pedroni, L. (eds.) Tropical forests and adaptation to climate change – in search of synergies. Center for International Forestry Research (CIFOR), Bogor, Indonesia, 186 pp.

Wunder, S., 2007. The efficiency of payments for environmental services in tropical conservation. Conservation Biology 21: 48-58.

Chapter 4.

Carbon stock in planted woodlots at Kongowe, Kibaha, Tanzania

A.A. Kimaro[1], K.D. Novak[2], R.S. Shemdoe[3] and S.A.O. Chamshama[4]

[1]World Agroforestry Centre (ICRAF), ICRAF Country Programme, P.O. Box 6226, Dar-es-Salaam, Tanzania; [2]University of Saskatchewan, Department of Soil Science, 51 Campus Drive, Saskatoon, SK S7N 5A8, Canada; [3]Institute of Human Settlement Studies, Ardhi University, P.O. Box 35124, Dar-es-Salaam, Tanzania; [4]Faculty of Forestry and Nature Conservation, Sokoine University of Agriculture, P.O. Box 3010, Morogoro, Tanzania; a.kimaro@cgiar.org

Abstract

Forest degradation due to extraction of wood for fuelwood and charcoal production or agricultural expansion is one of the main sources of carbon dioxide (CO_2) emissions in Africa. Woodlots can offset CO_2 emissions while sustainably meeting the wood demand of local communities. However, the amount and dynamics of carbon (C) stored in soils in coastal area woodlots and the contribution of this practice to climate change mitigation is largely unexplored. This study evaluated wood biomass and C stocks in woodlots, forest reserves and farmland with *Acacia crassicarpa* Benth., *Acacia mangium* Wild., and *Acacia auriculiformis* Benth. to determine their ability to sequester soil C and offsetting CO_2 emissions. Woodlots of fast-growing *Acacia* spp. produced wood (30 to 62 Mg C/ha), which may meet the household fuelwood demand for up to 10 to 20 years and offset up to 31 Mg C/ha of C emissions. Moreover, these species stored soil organic carbon (SOC) (58 Mg C/ha), which was similar to levels in the native forests; demonstrating climate change mitigation capacity. The effects of tree species on SOC was confined to the top 40 cm with native forest consistently recording the highest values. This improvement was attributed to litter and root turnover, C leaching and mixing of SOC by microorganisms, and increased recalcitrant form of SOC with depth. Because of the sandy soil texture in the study site, most of SOC stocks (up to 70% in the bulk soil and 50% in fractionated soils) were in the labile fractions and prone to degradation. Of the tested woodlot species, *A. auriculiformis* showed the highest promise to produce wood and improve SOC because of higher survival. Our results showed that woodlots systems utilizing the fast growing tree species hold high promise to mitigate atmospheric CO_2 increase through on-farm wood supply to alleviate harvesting pressure and accumulation of SOC in the soil.

Keywords: agroforestry, carbon sequestration, soil fractions, soil organic carbon, REDD+

4.1 Introduction

The carbon (C) sequestration potential of agroforestry systems (AFS) offers the ability to help offset the 1.6 billion tonnes of C emitted due to deforestation and forest degradation annually (Funder, 2009). As a result, the Kyoto Accords recognized AFS as one of the greenhouse gas (GHG) mitigating strategies under Afforestation-Reforestation Activities (IPCC, 2000). As part of this acknowledgement, the Kyoto Accords implemented the clean development mechanism (CDM). This C trading scheme enables industrialized nations to meet their GHG requirements by investing in sustainable development projects within developing countries that reduce emissions or sequester C in forests – allowing industrialized countries to offset present emission levels, while theoretically providing much-needed financial investments to emerging nations (Sutter and Parreño, 2007). To increase the involvement and contributions of developing countries and small-scale farmer, in particular, additional C trading schemes were negotiated. These include the United Nation's Reducing Emissions from Deforestation and Forest Degradation in Developing Countries (REDD+), and C trading programs focusing on the agriculture, forestry, and other land uses (AFOLU) sectors. In both the REDD+ and AFOLU initiatives, developing countries can be compensated for adopting strategies that reduce the emission of GHG via carbon sequestration, thus funding sustainable rural development and enhancing native resource conservation (Mountino *et al.,* 2005; Negra and Ashton, 2009).

The potential to sequester C in soils, enhance aboveground biomass and reduce non-CO_2 emissions creates an opportunity for the agricultural sector to contribute to climate change mitigation. Although this sector has largely been neglected in previous climate change negotiations, it is increasingly been recognized that comprehensive approaches for accounting C in agriculture and forestry can be a win-win option for integrating climate change mitigation with efforts to enhance food security and sustainable rural development (Minang *et al.*, 2011; Scherr *et al.*, 2012). In the context of REDD+, agroforestry has the potential to reduce forest degradation by supplying timber and fuelwood that would otherwise be sourced from adjacent or distant forests. Thus, agroforestry can contribute to REDD+ either as part of REDD+ under certain forest definitions and / or as part of a strategy for achieving REDD+ in multifunctional landscapes (Minang *et al.*, 2011). Agroforestry can also provide a platform for harnessing synergies between climate mitigation and adaption options because the multi-functionality nature of AFS allows for the diversifications food and income sources while sequestering C.

Although AFS theoretical potential in both C sequestration and climate change mitigation has been solidified (Nair *et al.,* 2004; Oelbermann *et al.,* 2004), system-specific measurements are required for C quantifying purposes. Most studies investigating the sequestration potential of African AFS have focused on intercropping and agricultural-based systems in sub-Saharan Africa (Isaac *et al.,* 2005; Kaonga and Bayliss-Smith, 2009; Kimaro *et al.,* 2011; Takimoto *et al.,* 2008), while largely ignoring the coastal region of eastern Africa. Planted woodlots and traditional woodlots like *Ngitili* are one of the practices being evaluated and promoted for fuelwood supply and as a C management strategy under the piloted REDD+ scheme in Tanzania. Unlike other AFS, which focus largely on agricultural gains, woodlots emphasize the production of fuelwood as a way to alleviate forest degradation while meeting the fuelwood demand for domestic and agro-processing industries like bricks making, tobacco curing, etc. in urban and rural areas (Kimaro *et al.,* 2011; Ramadhani *et al.,* 2002; Sileshi *et al.,* 2007). Despite these advantages, the amount of C stored in trees and soil in this practice and its contribution to C sequestration remains largely unexplored. Soil contains larger amount of C and has a higher capacity to store C compared to vegetation and atmosphere (Brady and Weil, 2004). The retention of soil C depends on many factors including the nature of soil aggregates. Soil C associated with macro-aggregates (>250 μm diameter) and micro aggregate (<250 μm diameter) is considered short-lived compared to the more stable form of soil C associated with the silt-and-clay fractions with diameter less than 50 μm (Saha *et al.,* 2010; Tonucci *et al.,* 2011). Because of the sandy texture, a large proportion of SOC in coastal soils may be in the labile form, hence influencing the C sequestration capacity of tree-based systems. Thus, understanding the C retention potential of coastal soils is necessary because co-benefits such as climate change adaptation benefits through adopting AFS, like woodlots, are mainly linked to improved land productivity where SOC is the major driving factor. Additionally, the efficacy of the woodlot technology in sequestering soil C and offsetting CO_2 emissions in Tanzania and other countries in sub-Saharan Africa is poorly understood. This study, therefore, determined the effects of woodlot tree species on wood biomass supply and wood C and SOC stocks in order to assess the contribution of the woodlot technology in reducing harvesting pressure on native forests, offsetting CO_2 and climate change mitigation. Such information can be useful to REDD and other C trading schemes for evaluating strategies that mitigate atmospheric CO_2 emissions.

4.2 Materials and methods

Study area

The study was conducted at Kongowe, Kibaha, Tanzania (6°32' and 6°43' S and 38°48' and 39°02' E) within the Ruvu North Forest Reserve, located at 80 m.a.s.l. The area experiences a sub-humid type of climate and has irregular rainfall patterns with a mean annual precipitation of 700 to 1000 mm. The minimum and maximum temperature averages at 18 °C and 30 °C, respectively (Kindo *et al.*, 2010). The native vegetation is open savanna miombo (*Brachystegia* spp.) woodland, dominated by *Dalbergia melanoxylon* Guillemin & Perrottet, *Chlorophora excelsa* (Welw.) Benth., *Pterocarpus angolensis* D.C. and *Newtonia buchananii* (Baker) G. Gilbert & Boutique. The grass species present in the area include *Penisetum purpurcum* L., *Panicum maximum* Jacq. and *Panicum trichuclachum* L. (Kindo *et al.*, 2010). Soils are primarily sandy, sandy loam and gravel with considerable texture variation over short distances. The water-holding capacity is generally low resulting from diminished effective rainfall and free-draining, marine sediment soils. Clay soils are present in valley bottoms and near streams (Kindo *et al.*, 2010). Selected physicochemical properties of the study site are presented in Table 4.1. Overall the soils are slightly acidic with pH ranging from 6.1-6.7 and very low levels of total nitrogen (N) and extractable phosphorus (P).

Experimental design and management

This study adopted the previously established provenance trials of *Acacia* spp. from Australia. The purpose of the trial was to evaluate the best provenances for fuelwood production to broaden a range of species planted in the Fuelwood Pilot Project to address the increasing demand for fuelwood and an overall increase in deforestation in the areas surrounding the Ruvu North and South Forest Reserves, including Dar-es-Salaam City. 22 provenances from four Australian *Acacia* species, namely: *Acacia auriculiformis, Acacia crassicarpa, Acacia julifera* sub. *julifera*, and *Acacia mangium*

Table 4.1. Soil chemical properties in planted woodlots at Kongowe, Kibaha, Tanzania.

Soil parameter	Soil depth (cm)					LSD[1]	*P* < F-value
	0-10	10-20	20-40	40-80	80-120		
pH	6.294^{bc}	6.628^{a}	6.665^{a}	6.451^{ba}	6.103^{c}	0.226	<0.0001
Total nitrogen (%)	0.045^{a}	0.031^{b}	0.019^{c}	0.019^{c}	0.021^{c}	0.006	<0.0001
Total inorganic N (mg/kg)	9.925^{a}	5.323^{b}	4.000^{cb}	2.975^{c}	3.612^{cb}	1.794	<0.0001
Extractable P (mg/kg)	4.374^{a}	2.055^{b}	1.308^{b}	1.961^{b}	1.969^{b}	1.349	0.0007
Exchangeable Ca (mg/kg)	43.76^{a}	29.61^{b}	21.82^{c}	16.77^{dc}	14.06^{d}	5.346	<0.0001
Exchangeable Mg (mg/kg)	7.086^{a}	4.575^{b}	4.328^{b}	4.727^{b}	5.384^{b}	1.318	0.0011
Exchangeable K (mg/kg)	5.515^{a}	4.883^{ba}	4.909^{ba}	4.629^{bc}	3.997^{c}	0.803	0.0114

[1] LSD = least significant difference.

(Kindo *et al.,* 2010), were planted in March 2002. The trials were laid out in a randomized complete block design (RCBD) consisting of three replications. Each plot consisted of 25 trees planted at a spacing of 3×3 m, covering a total of 2.5 ha. Two border rows of the same species surrounded each plot to prevent cross-border interference. This study adopted 3 provenances based on the survival rate and productivity at 2 and 4 years after establishment (Kindo *et al.,* 2010) and during the fieldwork when the species were 8 years old. These provenances are *A. auriculiformis* (seed lot number 15697), *A. crassicarpa* (seed lot number 19748) and *A. mangium* (seed lot number 19387). Details of seed sources and early growth performance of all these provenances are described in Kindo *et al.* (2010).

Biomass yield and wood carbon determination

Within each plot, the number of trees surviving in each plot were counted and expressed as a percentage of the planted trees (25 trees per plot). Diameter at breast height (dbh) and tree height (m) were measured using a clinometer and veneer caliper, respectively. The following mixed-species allometric equation for Australian *Acacia* species developed by Kindo (2008) in the study site was used for wood biomass estimation:

$$\text{Biomass (kg/tree)} = 0.126\ \text{dbh}^{1.481}\ (\text{ht}^{0.674}) \qquad (1)$$

Where dbh is the diameter at breast height (1.3 m) in cm, and ht the tree height in m (R = 0.72 and SE = 0.20).

Carbon stock in woody biomass was estimated as 50% of the biomass yield as recommended by the Intergovernmental Panel on Climate Change (IPCC, 2000) and employed in other studies where destructive sampling of trees in farmers' fields is not possible (Kaonga and Bayliss-Smith, 2009; Takimoto *et al.,* 2008). Data on biomass yield and C stock were extrapolated to a hectare basis based on the survival rate of each species.

Soil sampling, fractionation and laboratory analysis

Soil samples were collected from each plot of the selected *Acacia* tree species in August 2010 when the woodlots were eight years old. To determine the efficacy of woodlot technology on soil C accumulation, soil samples were also collected in triplicates from native miombo woodlands within the Ruvu North Forest Reserve and from farmland, which was under continuous maize cultivation. A 1.2 m soil pit was dug in the middle of each test plot. Soil samples were collected at 0-10 cm, 10-20 cm, 20-40 cm, 40-80 cm, and 80-120 cm depths from three sides of the soil pit, bulked and mixed thoroughly for each depth increment, and then sub-sampled to obtain a composite sample for chemical and physical analysis. Prior to analysis, samples were air dried and sieved through a 2 mm sieve. The wet sieving procedure (Saha *et al.,* 2010; Tonucci *et al.,* 2011) was adopted to fractionate the bulk soil samples into macro-size (250 to 2000 250 μm), micro-size (53 to 250 μm) and silt-and-clay fractions (<53 μm). This procedure involved slaking 100 g of the composite soil sample in a 500 ml beaker of deionized water for about 5 min in order to remove all air pockets within the aggregates. The soil-water suspension was poured on top of the 250 μm sieve and agitated manually in both circular and vertical motions for about three minutes. The soil remaining on top of the sieve (250 to 2000 250 μm fractions) was collected into an aluminum baking pan. The wet sieving procedure was repeated for the suspension that passed through the first sieving by using a 53 μm sieve to collect the 53 to 250 μm fractions. Because of the colloidal nature of clay particles, 0.5 M calcium chloride was added in the suspension to flocculate clay particles and speed up the sedimentation process before decanting a clear solution to collect the <53 μm fractions (silt-and-clay fractions).

For soil organic carbon (SOC) determination, the bulk and fractionated samples were ground to fine powder using a ball grinder. Then SOC was analyzed by dry combustion using a LECO CNS analyzer and expressed in g C/kg dry soil. Other soil parameters were conducted on the unfractionated soil samples. Soil texture was analyzed using a soil texturing analyzer and total N, extractable P, and exchangeable cations (potassium, calcium and magnesium) by Atomic Absorption Spectrophotometer (AAS) after the extraction with H_2O_2-H_2SO_4 digestion method, the modified Kelowna procedure, and 1 N ammonium acetate solution, respectively (Carter and Gregorich, 2008). All laboratory chemical analyses were conducted in the Department of Soil Science, at the University of Saskatchewan, Canada.

Data analysis

Prior to the analysis of variance (ANOVA), data of tree and soil parameters were tested for normality and homoscedasticity of residuals using SAS version 9.2. Thereafter, tree species effects on dbh, height, wood yield, and C stock were tested using the generalized linear model procedure in SAS. Similarly soil C stocks in tree species (*Acacia* woodlots), native forests and farmland were analyzed to assess the variation in SOC in the bulk and fractionated soils in these land use types and by sampling depth. ANOVA of basic soil biophysical parameters by sampling depth was also conducted to characterize the study site (Table 4.1). Following ANOVA, significant means were compared using Fishers protected least significant difference (LSD) at 5% level of significance.

4.3 Results

Tree growth and wood biomass production

Survival of *Acacia* spp. after 8 years was the highest (P=0.0007) in *A. auriculiformis* (85%) while *A. crassicarpa* and *A. mangium* showed poor survival (Table 4.2). No statistical differences between species were noted for dbh and height. However, dbh and height of *A. crassicarpa* were the largest, reflecting high growth performance over the 8 years. Comparatively, *A. auriculiforms* grew better in terms of height than dbh increment as noted by highest height (15.5 cm) and lowest dbh (13.4 cm). Likewise, *A. mangium* ranked the second on dbh size but it was the last based on height measurements. Wood biomass production between species was variable. *A. crassicarpa* produced the largest amount of wood per tree (P=0.0110) while *A. auriculiformis* recorded the lowest wood biomass (Table 4.2). This trend was reversed when wood production was extrapolated into hectare, reflecting the effects of survival on wood yield. As expected, C stocks per tree and on per hectare basis for the tested tree species followed a similar trend to wood biomass since C stock was estimated as 50% of the wood biomass (Table 4.2).

Table 4.2. Survival, dbh, height, wood yield, and wood carbon in 8-year-old *Acacia* woodlots at Kongowe, Kibaha, Tanzania

Acacia species	Survival (%)	dbh[1] (cm)	Height (m)	Wood biomass		Wood carbon	
				(kg/tree)	(Mg/ha)	(kg C/tree)	(Mg C/ha)
A. auriculiformis	85.2^{a} [2]	13.4^{a}	15.5^{a}	61.0^{b}	61.6^{a}	30.5^{b}	30.8^{a}
A. crassicarpa	51.9^{b}	16.5^{a}	15.5^{a}	82.9^{a}	40.7^{ba}	41.4^{a}	20.4^{b}
A. mangium	33.3^{b}	15.7^{a}	13.6^{a}	70.4^{ba}	30.2^{b}	35.2^{ba}	15.1^{b}
Probability	0.0007	0.0773	0.6321	0.0110	0.0005	0.011	0.0005
LSD[3]	24.7	3.5	8.6	18.1	15.0	9.1	0.9

[1] dbh = diameter at breast height.

[2] Means within a column marked with the same superscript letter are not statistically different at P>0.05 according to LSD (n=3).

[3] LSD = least significant difference.

Soil organic carbon in the bulk soil

Total amount of SOC in the bulk soil up to 1.2 m soil depth varied significantly (P=0.0369) between *Acacia* spp. woodlots, native forest and the farmland (Figure 4.1). Soil organic C was the highest in the forest (64.1 Mg C/ha) and the lowest in the farmland (42 Mg C/ha). Overall, SOC stored in *Acacia* spp. woodlot soils averaged 58 Mg/ha, which was similar to the levels obtained in the forest soils and 37% higher than SOC stock in the farmland (Figure 4.1). Within the woodlot species, there was little variation in SOC with *A. auriculiformis* recording the highest amount (61.4 Mg C/ha) followed by *A. crassicarpa* and *A. mangium*. Accumulation of SOC in the soil under *A. mangium* (55.6 Mg/ha) was comparable to the SOC level in the farmland, reflecting little C input from this species during the fallow period.

As expected, SOC content in the bulk soil declined with soil depth in *Acacia* spp. woodlots, native forests and farmland (Figure 4.2a). In all sampling depths, native forest and farmland soils recorded the highest and the lowest amounts, respectively. In the top 10 cm, SOC stocks in the native forest (20.4 Mg C/ha) and *A. auriculiformis* woodlot (17.2 Mg C/ha) were significantly (P=0.0211) higher than SOC in the farmland (7.6 Mg C/ha). Similar results were obtained in the 10-20 cm soil depth. Relative to farmland, *A. auriculiformis* (11.3 Mg C/ha) and *A. crassicarpa* (11.8 Mg C/ha)

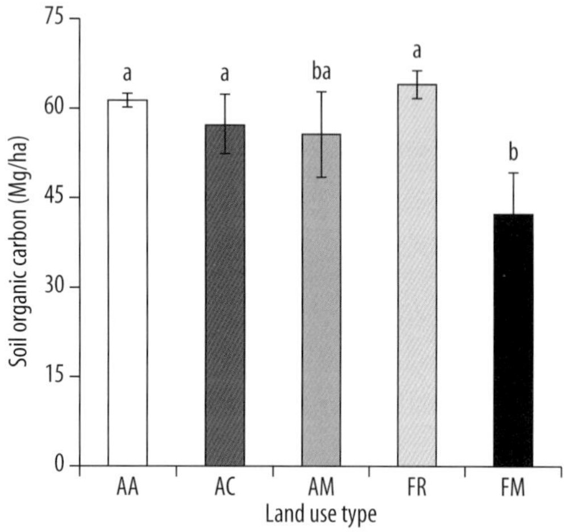

Figure 4.1. Total soil organic carbon stock in the bulk soil up to 1.2 m depth in *Acacia* woodlots (AA = *A. auriculiformis*, AC = *A. crassicarpa*, AM = *A. mangium*), native forest (FR), and farmland (FM) at Kongowe, Kibaha, Tanzania. Vertical bars indicate standard error of means (n=3). Means followed by the same letter are not statistically different at α=0.05.

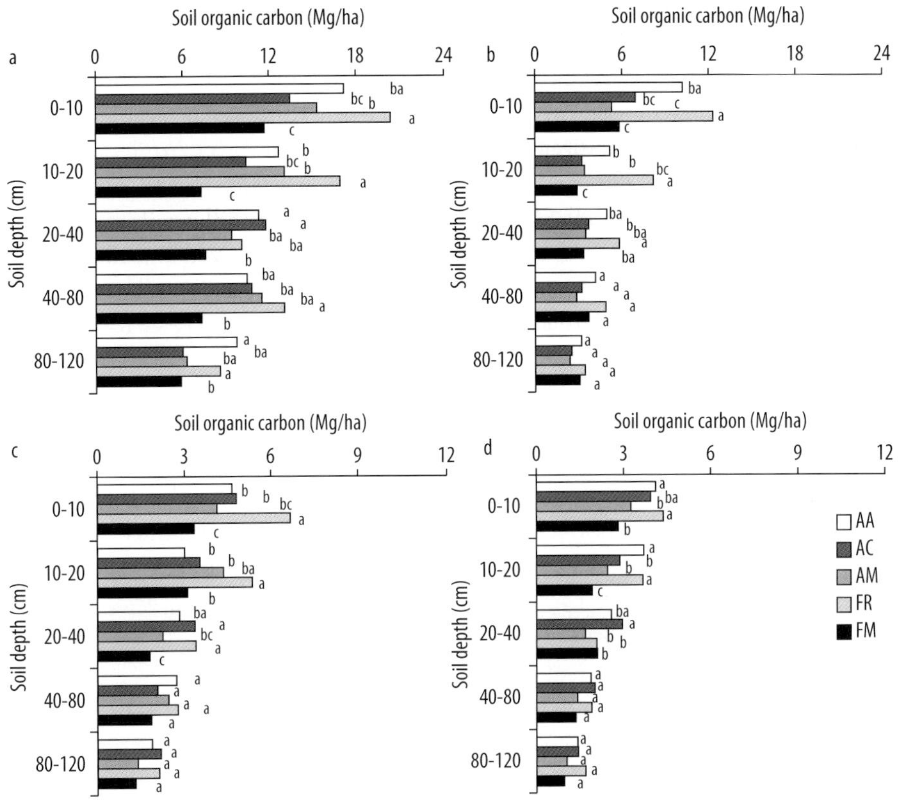

Figure 4.2. Soil organic carbon in (a) the bulk and (b) macro-sized fraction, (c) micro-sized fractions and (d) silt-and-clay fractions in *Acacia* woodlots (AA = *A. auriculiformis*, AM = *A. mangium* and AC = *A. crasicarpa*), native forests (FR) and farmland (FM) at Kongowe, Kibaha, Tanzania. Horizontal bars indicate standard error of means (n=3). Means followed by the same letter within a given soil depth are not statistically different at α=0.05. Note the scale in Figure 4.2c and d is reduced by half to minimize clutter.

woodlots had a significantly (P=0.0102) higher SOC stock at the 20-40 cm soil depth. The amount of SOC in these woodlots, however, was similar to SOC levels in the soils under the native forest and *A. mangium* (9.4-10.1 Mg C/ha) within this sampling depth (Figure 4.2a). Overall, the difference in SOC stocks between tree-based land use systems (*Acacia* woodlots and forests) and the farmland became less apparent at a soil depth greater than 40 cm, although significance differences (P=0.0310) among land use types were observed even at a 80-120 cm depth (Figure 4.2a). The proportions of SOC in the top 20 cm relative to total SOC in the whole soil profile (up to 1.2 m) were 49%, 46% and 51% for *A. auriculiformis, A. crassicarpa* and *A. mangium*, respectively. Corresponding values for forests and farmland were 54% and 48%, indicating that most of the SOC was found in the top soils. The percentage of SOC stored below the 40 cm depth across the tested land uses ranged from 31-33%.

Soil organic carbon in the macro-sized fractions (250 to 2,000 μm)

Overall, SOC content in the macro-sized fractions (250 to 2,000 μm) was two-folds higher in all depths compared to levels obtained in the micro-sized and silt-and-clay fractions (Figures 4.2b-d). Soil organic C in the micro-sized fraction was the highest in the native forest and *A. auriculiformis* in the 0-10 cm (12.3 Mg C/ha and 10.2 Mg C/ha; P=0.0114) and in the 10-20 cm (8.2 Mg C/ha and 5.2 Mg/ha; P=0.0110) sampling depths (Figure 4.2b). There was no significant difference in SOC stocks in *A. crassicarpa*, *A. mangium* and farmland in the top 20 cm. The amount of SOC in these land-uses averaged 6.0 Mg C/ha and 3.2 Mg C/ha at the 0-10 cm and 10-20 cm, respectively; indicating a declining trend with depth (Figure 4.2b). There was no significant difference in SOC stocks observed between the tested land uses below 20 cm depth. The percentage of SOC stored in the top 20 cm in all land uses was 53.7%.

Soil organic carbon in the micro-sized fractions (50 to 250 μm)

Within the top 10 cm, SOC content in the micro-sized fractions (50 to 250 μm) from native forest was the highest (6.7 Mg C/ha), followed by *Acacia* woodlots species (4.1-4.8 Mg C/ha) and farmland (3.3 Mg C /ha) (Figure 4.2c). Soil organic C stocks at the 10-20 cm depth, was also significantly (P=0.0440) higher in the native forest (5.4 Mg C/ha) compared to SOC levels under *A. crassicarpa*, *A. auriculiforms* and farmland (Figure 4.2c). At the 20-40 cm depth, SOC levels were higher (P=0.0384) in the native forest and *A. crassicarpa* (3. 4 Mg C/ha) and the lowest in the farmland soil (1.8 Mg C/ha). Variation in SOC in the micro-sized fractions between land uses was not significantly different below 40 cm. As expected, SOC levels in the micro-sized fractions declined relative to the amount found in the top 10 cm with farmland consistently recording comparatively lower values at all depths (Figure 4.2c).

Soil organic carbon in the silt-and-clay fractions (<50 μm)

The amount of SOC in each soil depth in the silt and clay fractions (<50 μm; Figure 4.2d) was comparatively lower than corresponding values in the macro- and micro-sized fractions (Figures 4.2b and 4.2c). Soil organic C in the top 10 cm ranged from 2.8 to 4.4 Mg C/ha and it was significantly higher (P=0.0355) in the native forest and *A. auriculiformis* and the lowest in the farmland and *A. mangium* (Figure 4.2d). Significant difference in SOC (P=0.0209) was also noted in the 10-20 cm soil depth, with the highest values obtained with soils from the native forest and *A. auriculiformis*

(3.7 Mg C/ha), followed by *A. crassicarpa* and *A. mangium* (2.7 Mg C/ha), and finally farmland (1.9 Mg C/ha). Similar results were also noted in the 20-40 cm depth with the native forest and *A. auriculiformis* having the highest (P=0.0389) SOC values. Like the other soil fractions, there was little variation in SOC among land uses in the 40-80 cm and 80-120 cm sampling depths.

4.4 Discussion

Wood biomass and carbon stocks in *Acacia* woodlots

Australian *Acacia* spp. are exotic tree species to Tanzania, but were introduced in the country in the 1980s to establish fuelwood plantations, woodlots and other niches on farmland and on the homestead (Karachi *et al.*, 1997). Earlier work indicated that high survival and productivity of these species in semiarid Tanzania is associated with local adaptation, high water table and mycorrhizal association (Kwesiga *et al.*, 2003; Nyadzi *et al.*, 2003) and high nutrient use efficiency (Kimaro *et al.*, 2007). Thus, large variations in survival rates noted in this study possibly reflect site specific conditions as average rainfall in the study site is 700 to 1000 mm/year (Kindo *et al.*, 2010), which is similar to the amount of precipitation at other sites where these species are grown in Tanzania (Kimaro *et al.*, 2007; Nyadzi *et al.*, 2003). Kindo *et al.* (2010) indicated that parasite damage (*Loranthus* spp.) and a narrow genetic base of the seed source of *A. mangium* were the main reasons for poor survival (50%) of this species in the 4-year old provenance trial on the same site. The survival of *A. auriculiformis* (93.3%), *A. crassicarpa* (66.3%) at 4 years were similar to the survival rates for these species in this study (85% and 52%), signifying that the impact of these factors on survival was significant even after 8 years of growth.

Tree dbh and height varied little among tree species, which indicated that tested tree species had a similar potential to adapt to local growth conditions. However, *A. crassicarpa* produced the largest (P=0.0110) amount of wood per tree. Given that tested species were exposed to the same low nutrient levels and other site conditions, such high productivity possibly reflect the ability to accumulate more biomass per unit nutrient uptake (Doran *et al.*, 1997; Kimaro *et al.*, 2007). Australian *Acacia* species form symbiotic associations with mycorrhizae and rhizobia that enhance access to immobile soil nutrients including P and atmospheric N_2 (Kwesiga *et al.*, 2003). These mechanisms contribute to efficient use of soil nutrients for biomass production (Kimaro *et al.*, 2007), making the species adaptable to poor soil conditions typical in semi-arid Tanzania and other countries in Sub-Saharan African. *Acacia crassicarpa* is widely known as the most productive species in woodlots and short rotation plantations eastern and western Africa (Karachi *et al.*, 1997; Ramadhani *et al.*, 2002; Sileshi *et al.*,

2007). As mentioned earlier, the highest amount of wood biomass in *A. ariculiformis* was not expected and may be the result of a higher survival rate of this species.

Despite low survival by some tree species, wood yield in this study (30.2 to 61.6 Mg C/ha) was within the range (23.2 to 51.0 Mg C/ha) reported in similar studies of Australian *Acacia* woodlots (Kimaro *et al.*, 2007; Nyadzi *et al.*, 2003; Sileshi *et al.*, 2007). Assuming a household consumption rate of 10 kg/week (Biran *et al.*, 2004), wood yield from this study may satisfy the fuelwood demand of a six member family household for up to 10 to 20 years. These results demonstrated the high potential of the woodlot technology to reduce harvesting pressure in native forests. About 4 to 26% of deforestation in eastern and southern Africa is linked to wood extraction for agro-processing industries such as tabacco curing and brick making, and for domestic use (Sileshi *et al.*, 2007). This study, however, suggested that woodlot technology holds high promise to reduce this harvesting pressure and the associated CO_2 emissions. The amount of C stored in wood (15 to 31 Mg C/ha) was similar to 23.1 Mg/ha, the average C stock in native miombo woodlands in eastern and southern Africa (Hammarstrand and Särnberger, 2013). Although C sequestered in wood biomass in woodlots can be released during wood fuel processing and consumption, it counterbalances CO_2 emissions from clearing local forests for fuelwood supply or from using fossil fuels. In this way, woodlots may help to meet the wood demand of forest depend communities while contributing to climate change mitigation and adaptation. However, such high potential of woodlots and other agroforestry practices may be undermined in the absence of comprehensive landscape approaches for C accounting and financing, which consider contributions from other sources and sinks in the farmlands or outside the forests.

Soil carbon stocks in *Acacia* woodlots

Besides wood production and aboveground C sequestration, woodlots accumulate a substantial amount of C in the soil but information on SOC contents and dynamics in woodlots in sub-Saharan African countries is scant. Our study demonstrated that within a short period (8 years), woodlots of fast growing tree species like Australian *Acacia* spp. stored SOC, which was similar to those in the native forest and higher than in farmland. As expected, lower SOC in farmland (28 vs. 34%) compared to *Acacia* woodlots and native forest possibly reflect cultivation-induced loses (Solomon *et al.*, 2000). Increased aeration and soil aggregate destructions from tillage are some of the processes associated with increased decomposition rates and SOC degradation under continuous cultivation (Brady and Weil 2004), typical in the study site. Conversion of native tropical vegetation to continuous cultivation has also been noted to reduce labile SOC fractions by 50% within 3-5 years (Detwiler, 1986; Solomon *et al.*, 2000).

However, depending on the tree species, integrating trees in farmland and improved land management may mitigate this loss by building and/or maintaining SOC to levels similar to those in the native forests. The build-up of SOC in woodlots is often attributed to C inputs through plant litter and root turnover during the fallow period. In the short-term, root biomass and turnover was reported to be more important for SOC accumulation than litter inputs (Guo and Gifford, 2002). However, litter input probably accounted for species variation in SOC content in this study as the highest values were obtained with *A. auriculiformis*, the woodlot species with the highest survival rate. Also accumulation of litter with slow decomposition rate like those of tested *Acacia* spp. (Kimaro *et al.,* 2008) is known to contribute to the build-up of SOC in tropical woodlots and plantations (Jobbady and Jackson, 2000; Kimaro *et al.*, 2007). Native forest soils contained comparatively higher SOC than woodlot soils possibly due to a combination of root activities of grass and tree species. Miombo woodland, which is the dominant forest biome in the study area, is characterized by an open canopy which promotes grass growth on the forest floor. On the other hand, Australian *Acacia* woodlots typically have little or no grass cover due to accumulation of slow decomposing litters on the ground.

Tree-based systems (woodlots and native forests) significantly improved SOC stocks relative to farmland and above the 40 cm depth SOC values on soils from these land uses were consistently higher than SOC in the farmland. This trend suggested that C input from trees was translocated down the soil profile and tree cover contributed to more C in the sub-soils than treeless land-uses. The meta-analysis of global SOC distributions by depth revealed strong effects of vegetation (plant functional) types and clay content on vertical distributions of SOC (Jobbay and Jackson, 2000). Within a given climate, the authors noted that SOC content was much deeper in shrublands followed by grasslands and forest, and it increased with the increase in clay contents in the soil (Jobbay and Jackson, 2000).They found that the vertical distribution of roots affected placement of C within the soil profile while the shoot/root ratio determined the relative amount of C input via litterfall. Thus, the gradual decline in the variations between tree-based systems (woodlots and native forests) and farmland to a 40 cm soil depth suggested that the effects of these vegetation-based processes were confined to the upper soil layers. The effects of root activities on SOC distribution in various land uses were also observed in other tropical tree-based land-use systems (Guo and Gifford, 2002; Saha *et al.*, 2010). For example, Guo and Gifford (2002) associated the substantial increase in SOC to a 100 cm depth in pasture land to root turnover. Other factors influencing vertical distributions of SOC include: climate, SOC leaching from upper to lower soil layers and vertical mixing of SOC by microorganisms (Guo and Gifford, 2002; Jobbay and Jackson, 2000), and species diversity and management (Saha *et al.,* 2010; Tonucci *et al.*, 2011).

Over 50 and 70% of the SOC in the studied land-uses were found within the top 20 cm and 40 cm depths, respectively. These results indicated that a substantial amount of SOC is prone to degradations due to its location in top soils. This is especially true in tree-less systems like agroecosystems where soil may be lost via anthropogenic and natural processes including erosion by water, fire, and tillage (Saha *et al.*, 2010; Solomon *et al.*, 2000) and through microbial transformations (Jobbay and Jackson, 2000). In sandy soils like those in the study site (Kindo *et al.,* 2010) leaching may also account for the loss of SOC in the surface soils (Jobbay and Jackson, 2000) affecting the soil C sequestration potential. Soil C in the sub-soils may be less prone to degradation because of its close association with clay (Tonucci *et al.*, 2011). Higher stability of SOC in silt and clay and micro-sized fractions than in larger soil fractions may be attributed to the formation of organo-mineral complexes from the interactions of clay mineral and humic substances (Brady and Weil, 2004; Solomon *et al.,* 2000). The amount of SOC in macro-sized soil fraction at each depth was about two times greater compared to that found in the micro-sized and in the silt and clay fractions, suggesting that a large proportion of SOC in this site was in the labile fractions which is prone to degradation.

4.5 Conclusions

This study demonstrated a high promise for woodlot AFS to satisfy household fuelwood demands and offset CO_2 emission through alleviating harvesting pressures on native forests. Within eight years, fast growing Australian *Acacia* woodlots produced wood (30 to 62 Mg/ha), which may meet the household fuelwood demand for up to 10 to 20 years and offset up to 31 Mg C/ha of carbon emissions. These results demonstrated the capacity of the woodlot technology in semi-arid conditions to help reduce harvesting pressure on native forests. Besides wood supply, woodlot soils on average accumulated 58 Mg C/ha, which was similar to 64 Mg C/ha found in the native miombo forests, but higher than SOC stocks on farmland (42 Mg C/ha). Trees significantly improved SOC to a 40 cm soil depth. The depth-wise comparisons of SOC stocks by land uses revealed that native forest consistently had the greatest level of C. This improvement was attributed to the combined effects of litter and root turnover from tree and grass species, which are common in the Miombo woodlands. With little grass cover on the ground, *Acacia* woodlots had comparatively lower SOC values than native forest at each sampling depth. Because of the sandy soil texture in the study site, most of SOC stocks (up to 70% in the bulk soil and 50% in fractionated soils) were in the labile fractions and prone to degradation. However, tree-based land use systems may improve organic matter inputs and increase soil aggregation to minimize such losses. Of the tested woodlot species, *A. auriculiformis* showed the highest promise to produce wood and sequestered C in the soil because of higher survival in the study site. Apparently woodlot systems utilizing the appropriate tree species hold high promise to mitigate atmospheric CO_2 increase by reducing harvesting pressure on native forests and offsetting CO_2 emissions through alternative wood production and accumulation of C in the soil. This study therefore recommends establishment of woodlots, especially in the buffer zones separating the farmlands and forest reserves and other niches in agricultural landscapes, to reduce harvesting pressure on native forests and hence offsetting CO_2 emissions.

Acknowledgements

This study was funded by the Natural Science and Engineering Research Council of Canada (NSERC) via the strategic grant from Professor V. Timmer to the woodlot research of the first author, and a grant from the Canadian International Development Agency (CIDA) to the second author under the 2010 Students for Development (SFD) program of the Association of Universities and Colleges of Canada (AUCC). Authors acknowledge support from research collaborators during fieldwork in Tanzania (M.A. Mndolwa and S. Shirima) and laboratory analysis at the University of Saskatchewan, Canada (R. de Freitas). We are also grateful for constructive comments from anonymous reviewers and the book editor.

References

Biran, A., Abbot, J. and Mace, R., 2004. Families and firewood: a comparative analysis of the costs and benefits of children in firewood collection and use in two rural communities in Sub-Saharan Africa. Human Ecology 32: 1-25.

Brady, N.C. and Weil, R.R., 2004. Elements of the nature and properties of soils. Person Education, Inc., Upper Saddle River, NJ, USA.

Carter, M.R. and Gregorich, E.G. (eds.), 2008. Soil sampling and methods of analysis. CRC Press, Boca Raton, FL, USA.

Detwiler, R.P., 1986. Land use change and the global carbon cycle: the role of tropical soil. Biogeochemistry 2: 67-93.

Doran, J.C., Turnbull, J.W., Martensz, P.N., Thomson, L.A.J. and Hall, N., 1997. Introduction to the species'digests. In: Doran, J.C. and Turnbull, J.W. (eds.) Australian trees and shrubs: species for land rehabilitation and farm planting in the tropics. ACIAR Monograph 24. ACIAR, Bruce, Australia, pp. 89-384

Funder, M., 2009. Reducing emissions from deforestation and degradation (REDD): an overview of risks and opportunities for the poor. DIIS Report 2009:21. Danish Institute for International Studies, Copenhagen, Denmark.

Guo, L.B. and Gifford, R.M., 2002. Soil carbon stocks and land use change: a meta analysis. Global Change Biology 8: 345-360.

Hammarstrand, L. and Särnberger, A., 2013. Comparative evaluation of two forest systems under different management regimes in Miombo woodlands: a case study in Kitulangalo area, Tanzania. MSc Thesis. Industrial Ecology Report No. 2013: 4. Department of Energy and Environment, Chalmers University of Technology, Gothenburg, Sweden.

Intergovernmental Panel on Climate Change (IPCC), 2000. Land use, land-use change, and forestry. Cambridge University Press, Cambridge, UK.

Isaac, M.E., Gordon, A.M., Thevathasan, N., Oppong, S.K. and Quashie-Sam, J., 2005. Temporal changes in soil carbon and nitrogen in west African multistrata agroforestry systems: a chronosequence of pools and fluxes. Agroforestry Systems 65: 23-31.

Jobbagy, E.G. and Jackson, R.B., 2000. The vertical distribution of soil organic carbon and its relation to climate and vegetation. Ecological Applications 10: 423-436.

Kaonga, M. and Bayliss-Smith, T.P., 2009. Carbon pools in tree biomass and the soil in improved fallows in eastern Zambia. Agroforestry Systems 76: 37-51.

Karachi, M., Shirima, D. and Lemma, N., 1997. Evaluation of 15 tree leguminous tree and shrubs for forage and wood production in Tanzania. Agroforestry Systems 37: 253-263.

Kimaro, A.A., Timmer, V.R., Chamshama, S.A.O., Mugasha, A.G. and Kimaro, D.A., 2008. Differential response to tree fallows in rotational woodlot systems in semi-arid Tanzania: post-fallow maize yield, nutrient uptake, and soil nutrients. Agriculture, Ecosystems and Environment 125: 73-83.

Kimaro, A.A., Isaac, M.E. and Chamshama, S.O.A., 2011. Carbon pools in tree biomass and soils under rotational woodlot systems in Eastern Tanzania. In: Kumar, B.M. and Nair, P.K.R (eds.) Carbon sequestration potential of agroforestry systems: opportunities and challenges. Advances in Agroforestry 8: 129-143.

Kimaro, A.A., Timmer V.R., Mugasha, A.G., Chamshama, S.A.O. and Kimaro, D.A., 2007. Nutrient use efficiency and biomass production of tree species for rotational woodlot systems in semi-arid Morogoro, Tanzania. Agroforestry Systems 71: 175-184.

Kindo, A.I, Mndolwa, M.A, Edward, E. and Chamshama, S.A.O., 2010. Performance of four-year-old Australian and Papua New Guinean *Acacia* species/provenances at Kongowe, Kibaha, Tanzania. Southern Forests 72: 13-22.

Kwesiga, F., Akinnifesi, F.K., Mafongoya, P.L., Mcdermott, M.H. and Agumya, A., 2003. Agroforestry research and development in southern Africa during the 1990s: review and challenges ahead. Agroforestry Systems 59: 173-186.

Minang, P.A., Bernard, F., Van Noordwijk, M. and Kahurani, E., 2011. Agroforestry in REDD+: opportunities and challenges. ASB Policy Brief 26. ASB Partnership for the Tropical Forest Margins, Nairobi, Kenya.

Mountinho, P., Santillli, M., Schwartzman, S. and Rodriguues, L., 2005. Why ignore tropical deforestation? A proposal for including forests conservation in the Kyoto Protocol. Unasylva 56: 27-30.

Nair, P.K.R., Nair, V.D., Kumar, B. and Haile, S., 2009. Soil carbon sequestration in tropical agroforestry systems: a feasibility appraisal. Environmental Science and Policy 12: 1099-1111.

Negra, C. and Ashton, R., 2009. Roadmap for terrestrial carbon science: research needs for carbon management in agricultural, forestry and other land uses. The Terrestrial Carbon Group Project. Available at: http://www.terrestrialcarbon.org/Terrestrial_Carbon_Group__soil_%26_vegetation_in_climate_solution/Policy_Briefs_files/TCG%20Policy%20Brief%207%20Roadmap%20for%20Terrestrial%20Carbon%20Science.pdf.

Nyadzi, G.I., Otsyina, R.M., Banzi, F.M., Bakengesa, S.S., Gama, B.M., Mbwambo, L. and Asenga, D., 2003. Rotational woodlot technology in northwestern Tanzania: tree species and crop performance. Agroforestry Systems 59: 253-263.

Oelbermann, M., Voroney, R.P. and Gordon, A.M., 2004. Carbon sequestration in tropical and temperate agroforestry systems: a review with examples from Costa Rica and southern Canada. Agriculture, Ecosystems and Environment 104: 359-377.

Ramadhani, T., Otsyina, R. and Franzel, S., 2002. Improving household income and reducing deforestation using rotational woodlots in Tabora district, Tanzania. Agriculture, Ecosystems and Environment 89: 229-239.

Saha, S., Nair, P.K.R., Nair, V.D. and Kumar, B.M., 2010. Carbon storage in relation to soil sized-fractions under tropical tree-based land-use systems. Plant and Soil 328: 433-446.

Scherr, S.J., Shames, S. and Friedman, R., 2012. From climate-smart agriculture to climate-smart landscapes. Agriculture and Food Security 1: 12. Available at http://www.ecoagriculture.org/publication_details.php?publicationID=442.

Sileshi, G., Akinnifesi, F.K., Ajayi, O.C., Chakeredza, S., Kaonga, M. and Matakala, P.W., 2007. Contribution of agroforestry to ecosystem services in the Miombo eco-region of eastern and southern Africa. African Journal of Environment, Science and Technology 4: 68-80.

Solomon, D., Lehmann, J. and Zech, W., 2000. Land use effects on soil organic matter properties of chromic luvisols in semi-arid northern Tanzania: carbon, nitrogen, lignin and carbohydrates. Agriculture, Ecosystems and Environment 78: 203-213.

Sutter, C. and Parreño, J.C., 2007. Does the current Clean Development Mechanism (CDM) deliver its sustainable development claim? An analysis of officially registered CDM projects. Climate Change 84: 75-90.

Takimoto, A., Nair, P.K.R. and Nair, V.D., 2008. Carbon stock and sequestration potential of traditional and improved agroforestry systems in the West African Sahel. Agriculture, Ecosystems and Environment 125: 159-166.

Tonucci, R.G., Nair, P.K.R., Nair. V.D., Garcia, R. and Bernardino, F.S., 2011. Soil carbon storage in silvapasture and related land use systems in the Brazilian Cerrado. Journal of Environmental Quality 40: 833-841.

Section 2.
Greenhouse gas emissions from agroecosystems

Chapter 5.

The impact of different rice cultivars on soil methane emissions

L.S. da Silva[1], D.F. Moterle[2] and J.M.S. de Oliveira[3]

[1]Federal University of Santa Maria, Department of Soil Science, Santa Maria CEP 97105-900, Rio Grande do Sul, Brazil; [2]Federal Institute of Rio Grande do Sul, Campus of Bento Gonçalves, Bento Gonçalves CEP 95700-000, Rio Grande do Sul, Brazil; [3]Federal University of Santa Maria, Department of Biology, Santa Maria CEP 97105-900, Rio Grande do Sul, Brazil; leandrosolos@ufsm.br

Abstract

Anaerobic conditions after the flooding of rice fields leads to the emission of methane (CH_4), however root exudates may reduce the production of this greenhouse gas (GHG). This is because plant aerenchyma may facilitate CH_4 diffusion from the soil into the atmosphere, and the oxygen (O_2) in roots could oxidize CH_4 into carbon dioxide (CO_2). Thus, the potential of soil CH_4 emissions could depend on morphological and physiological characteristics of the various rice cultivars. Determining which rice cultivar minimizes the emission of CH_4, a GHG with a global warming potential 23 times greater than that of CO_2, may play an important role in reducing the overall contribution of rice cultivation to global warming. The objective of this study was to evaluate the CH_4 emission potential from different rice cultivars and their relationship to morphological and physiological plant characteristics. Nineteen rice cultivars were selected, covering a wide variability in plant characteristics such as growth cycle, height, tillering capacity and yield. Plants were grown in plastic pots each containing 7 kg soil (Planosol) derived from a rice paddy under continuously flooded conditions. In predetermined intervals, air was sampled using a static closed chamber and a syringe to analyse CH_4 concentrations. The dynamics of the CH_4 emission rates were similar for all cultivars, with two peaks approximately three weeks after flooding and near flowering. However, the total amount of CH_4 emitted over the rice growing season was different between cultivars (ranging from 8.5 to 26.2 g/m^2). General physiological and morphological plant characteristics did not completely explain the differences in CH_4 emission between rice cultivars, but shoot dry matter production and height were associated with CH_4 emission. Short plant rice cultivars with a short growing season and high yield had an overall lower emission of CH_4.

Keywords: aerenchyma, greenhouse gases, cultivar characteristics, paddy soil, rice yield and biomass

DOI 10.3920/978-90-8686-788-2_5,

5.1 Introduction

Human activity is a major contributor to global warming by enabling higher emission of greenhouse gases (GHG) such as carbon dioxide (CO_2), nitrous oxide (N_2O) and methane (CH_4) to the atmosphere (IPCC, 2007). The negative environmental impact of GHG has motivated scientists to search for mitigation strategies. Apart from the burning of fossil fuels, agricultural activities contribute ~13.5% of the total global GHG emissions as a result of land-use change, use of nitrogen (N) fertilizers, enteric fermentation, and the flooding of rice fields (Johnson *et al.,* 2007). In rice fields, due to anaerobic conditions that are created after flooding, organic carbon (C) is reduced to CH_4 by methanogenic bacteria. However, there are many factors that control the emission of CH_4 from flooded rice fields, which hinder an accurate identification of the main mechanisms. As such, we postulate that rice plants could affect CH_4 emission rates from the soil into the atmosphere.

Rice plants develop intercellular space systems referred to as aerenchyma, which carry oxygen (O_2) from the atmosphere to the roots (Evans, 2003). Aerenchyma provide a channel with low resistance for gas diffusion (Butterbach-Bahl *et al.,* 1997). Although it is not its primary function, aerenchyma also carries CO_2, N_2O and CH_4 from the soil into the atmosphere through a reverse O_2 flow. For this reason, rice plants may contribute up to 90% of the CH_4 transported from the soil into the atmosphere (Butterbach-Bahl *et al.,* 1997). However, due to the large variation in morphological and physiological rice plant characteristics, differences in the potential of CH_4 emission may be expected (Aulakh *et al.,* 2000a-c, 2002; Das and Buruah, 2008; Gogoi *et al.,* 2005; Mitra *et al.,* 1999; Wang *et al.,* 1997a). Root exudates released by plants serve as substrates for methanogenic bacteria (Aulakh *et al.,* 2000a-c), aerenchyma facilitate CH_4 diffusion from the soil into the atmosphere (Nouchi *et al.,* 1990), and oxygen availability in roots may oxidize some of the CH_4 into CO_2 (Aulakh *et al.,* 2000a-c). The effective contribution of each factor determines the potential CH_4 emission rate of each rice cultivar when cultivated under the same growing conditions (Wassmann *et al.,* 1998).

Other studies found that tillering capacity of rice cultivars could influence CH_4 emission rates by increasing aerenchyma channels and accelerating CH_4 transport (Aulakh *et al.,* 2002; Das and Baruah, 2008; Gogoi *et al.,* 2005; Mitra *et al.,* 1999). Leaf area may also influence CH_4 emissions because a larger leaf could develop a comparatively higher number of stomata and micropores which favour CH_4 emission (Nouchi *et al.,* 1990; Wang *et al.,* 1997b). Root volume is also positively correlated with CH_4 emission rates (Mitra *et al.,* 1999). Thus, the identification of different rice cultivars and their relationship to morphological and physiological plant characteristics may be an important tool when choosing rice cultivars with advantageous yield and low CH_4 emission rates. The objective of this study was to evaluate CH_4 emission rates from different rice cultivars grown in a paddy soil under flooded conditions and its relationship to morphological and physiological plant characteristics.

5.2 Materials and methods

The experiment was conducted in a translucent plastic-covered greenhouse at the Department of Soil Science at the Federal University of Santa Maria (UFSM), Brazil. Environmental conditions of the greenhouse included a steady temperature of 25±5 °C by using a fan-driven cooling system. Light and humidity were not controlled in this macrocosm. The experiment consisted of 19 rice cultivars which were selected based on their genetic variability, and morphological and physiological characteristics (Table 5.1).

Approximately 20 seeds of each rice cultivar were germinated. After seven days the germinated seeds were transplanted into plastic pots with a 0.25 m diameter, 0.25 m height, and containing 7 kg of a paddy soil (Planosol). Four days later, only six plants per pot were kept. The study was a complete randomized design (CRD) with three replications per rice cultivar (treatment). Rice plants were under continuously flooded conditions maintaining a water sheet 5 to 10 cm high, which was measured daily using a ruler. The water was used in this study was distilled and manually applied to each pot at the temperature of the greenhouse. Fertilization with N, phosphorus (P) and potassium (K) were applied according to the recommendation of CQFS/NRS (2004) for flooded rice.

To evaluate CH_4 emission rate, initial air samples were taken one week after flooding, and from then on every 15 days up to the flowering phase (six air samples). Three additional air samples were taken after flowering. For each sample, a static closed chamber (Mosier, 1989) was placed on top of the pot (Figure 5.1). After placement of the chamber and prior to sample extraction, a fan was used to homogenize the air inside the chamber. Air samples were extracted from each chamber at time (t) equal to 5, 10, and 15 minutes after covering the plants with the chamber. The air samples were collected in 250 µl borosilicate glass gastight syringes with a luer lock valve. The syringes were previously flushed with N_2 gas to minimize cross-contamination. After sampling, the valve was closed and the syringes were stored in an ice box containing ice gel packs for up to 24 hours. After each gas sampling, air temperature within the chamber was recorded using a pre-installed digital thermometer.

Table 5.1. Characteristics of the different rice cultivars used in this study to quantify differences in CH_4 emissions rates under flooded conditions at the Federal University of Santa Maria, Santa Maria, Brazil (n=3).

Cultivar	Group	Rice growth cycle[1] (days)	Tillering[1]	Type[1]
Arize 1003	Hybrid	125	High	WI[3]
Avaxi CL	Hybrid	115	High	WI
Bluebelle	Japonica	115	Mediun	Intermediate
BRS 7 Taim	Indica	130	High	Modern-Filipino
BRS Atalanta	Indica	100	High	Modern-Filipino
BRS Querência	Japonica	110	High	Modern-american
Carnaroli	Japonica	105 [2]	Mediun	Traditional
BRS Bojuru	Japonica	135	Mediun	Intermediate
EPAGRI 108	Indica	147	High	Modern-Filipino
Farroupilha	Japonica	149 [2]	Mediun	Traditional
IAS Formosa	Japonica	140	Mediun	Traditional
Inov CL	Hybrid	120	High	WI
IRGA 421	Indica	95	High	Modern-Filipino
IRGA 422 CL	Indica	121	High	Modern-Filipino
IRGA 424	Indica	132	High	Modern-Filipino
Koshihikari	Japonica	100 [2]	High	Traditional
SCS 114 Andosan	Indica	140	Mediun	Modern-Filipino
SCS 116 Satoru	Indica	144	High	Modern-Filipino
SCS BRS 113 Tio Taka	Indica	141	Mediun	Modern-Filipino

[1] Data from Marchesan (2006), Terres *et al.* (2004) and cultivar developer information.

[2] Rice growth cycle data from this experiment.

[3] WI = without information.

The CH_4 concentration in air samples was analysed by gas chromatography (Shimadzu GC-2014) with a flame ionization detector operating at 250 °C. The gas chromatograph was equipped with three packed columns operating at 70 °C, a N_2 as a carrier gas with a flow of 26 ml/min, and a direct sampling loop injector of 1 ml. The CH_4 standards used to calibrate the chromatograph were 2.34, 3.35 and 5.16 mg/m^3, respectively (3.31, 4.74 and 7.31 ppm). The CH_4 emission rate was quantified using the following equation:

$$f = \frac{\Delta Q}{\Delta t} \times \frac{PV}{RT} \qquad (1)$$

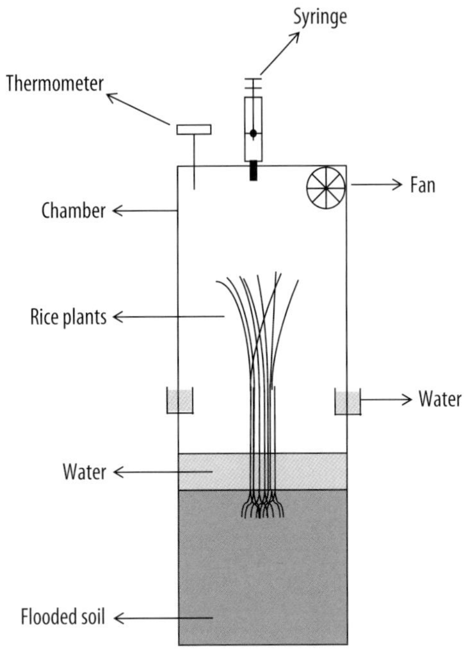

Figure 5.1. Schematic representation of the pot and the chamber growing the different rice cultivars under flooded conditions.

where f is the CH_4 emission rate (µg $CH_4/m^2/h$), Q is the quantity of gas in the chamber at sampling (µg CH_4/chamber), P is the atmospheric pressure (atm) within the chamber, which was assumed to be 1 atm, V is the volume of the chamber (l), Δt is the sampling time, R is the ideal gas constant (0.08205 mol atm/l/K), and T is the temperature (°K) within the chamber at sample time. The rate of gas concentration within the chamber was obtained by a linear regression between gas concentration and sampling time. The total CH_4 emission rate during the rice growth cycle was quantified by integrating the area under the curve obtained by interpolation of the CH_4 emission rate values from each sampling day, expressed as daily averages (Costa *et al.,* 2008), and the experimental period in days (Gomes *et al.,* 2009).

Two plants of each cultivar were selected to be monitored for changes in leaf area, using the following equation:

$$LA\ (cm^2) = (LW \times 0.75) \times LL \qquad (2)$$

where LA is the leaf area in cm^2, LW is the leaf width at half height (cm), and LL is the leaf length (cm). The time required for the plants to come to maturity was considered the rice growing season. After maturity, plant height was manually measured with a ruler, the number of tillers was counted and the plant shoot was cut off, and dried at 60 °C, and weighed. The grain was removed to determine grain yield.

5.3 Results

The general behaviour of CH_4 emission rates during the entire growth cycle was similar between rice cultivars (Figure 5.2). In the initial phase, or 7 days after flooding (DAF), the CH_4 emission rate was almost zero for most cultivars, but peaked at 20 DAF (Figure 5.2). At 34 DAF there was a large drop in the CH_4 emission rate for all cultivars, but after 34 DAF it increased constantly and a new peak was observed near flowering. The peak near the flowering phase varied between the different rice cultivars. After flowering, CH_4 emission rates tended to decrease until the end of the growing season (Figure 5.2).

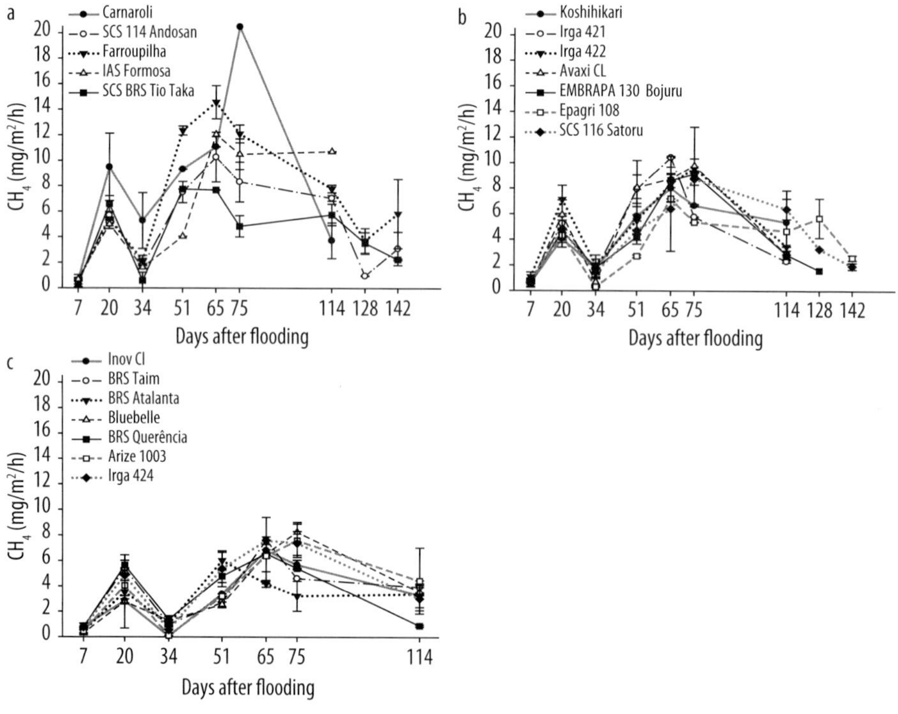

Figure 5.2. Methane emission (mg $CH_4/m^2/h$) rates during the rice cultivar growing cycle after flooding from (a) cultivars with a high CH_4 emission potential, (b) intermediate CH_4 emission potential, and (c) low CH_4 emission potential.

Although behaviour with respect to the pattern of CH_4 emission rates was similar between the various rice cultivars throughout the entire growing season, the actual amount of CH_4 differed (Table 5.2). For example, the highest emission occurred from the Farroupilha cultivar, which had an up to three times greater CH_4 emission rate compared to that of the BRS Atalanta cultivar. Based on total CH_4 emission (mg CH_4/m^2) over the entire rice growing season (Table 5.2 and Figure 5.2), it was possible to separate cultivars in three groups according to high, intermediate, and low

Table 5.2. CH_4 emission during rice growth cycle (total and per kg of rice grain), dry matter production, and grain yield of rice cultivars growing under greenhouse and flooding conditions. Santa Maria, Brazil. Values given in brackets represent the standard error of the mean value.

Cultivar	CH_4 emission		Yield	
	Total (g/m^2)	/kg of rice grain (mg/kg)	Dry matter (g/pot)	Grain (g/pot)
BRS Atalanta	8.5±0.7	240.5±20.2	77.5±1.3	35.4±0.1
Inov CL	9.1±1.5	148.2±19.6	46.3±0.7	61.8±3.2
BRS Querência	9.7±0.2	175.8±1.7	53.3±0.5	55.4±0.4
BRS Taim	10.0±0.1	190.7±7.6	49.8±0.1	52.3±1.3
Bluebelle	10.7±0.1	213.3±9.1	55.1±4.9	50.5±2.0
Arize 1003	10.9±0.9	179.1±17.2	63.1±2.5	61.1±0.6
Irga 424	11.8±0.1	210.6±6.7	49.8±0.8	56.1±1.5
Epagri 108	12.8±0.5	311.2±4.4	90.4±0.2	39.3±2.2
Irga 421	12.9±0.1	546.7±13.4	60.4±1.2	23.5±0.8
Koshihikari	12.2±0.7	240.4±31.8	62.5±0.5	53.9±4.2
BRS Bojuru	13.8±0.6	371.0±5.9	68.3±6.2	37.2±2.2
Avaxi CL	14.8±2.3	235.6±31.4	56.3±4.9	63.0±1.3
Irga 422	14.9±0.2	274.9±14.5	55.4±1.3	54.4±3.4
SCS 116 Satoru	16.4±1.1	407.5±61.0	66.6±6.6	40.3±8.3
SCS BRS Tio Taka	18.7±0.5	423.8±7.6	83.2±2.3	44.0±2.0
SCS 114 Andosan	20.5±1.0	433.2±38.6	79.2±1.6	47.3±1.8
IAS Formosa	24.3±0.1	440.5±33.7	73.5±8.5	55.2±4.3
Carnaroli	25.5±1.6	695.6±44.1	50.8±1.4	36.8±3.1
Farroupilha	26.2±0.2	852.2±34.6	98.1±5.2	30.7±1.4
CV (%)[1]	12.1	19.4	11.1	14.6

[1] CV = coefficient of variation.

emissions. Cultivars Farroupilha, Carnaroli, IAS Formosa, SCS 114 Andosan, and SCS BRS 113 Tio Taka were classified as having high CH_4 emission potential (Figure 5.2a). Cultivars SCS 116 Satoru, Irga 422, Avaxi CL, BRS Bojuru, Koshihikari, Irga 421, and Epagri 108 were classified as having intermediate CH_4 emission potential (Figure 5.2b). Cultivars Irga 424, Arize 1003, Bluebelle, BRS Taim, BRS Querencia, Inov CL, and BRS Atalanta were classified as having low CH_4 emission potential (Figure 5.2c).

The three groups did not present a general pattern in plant characteristics. Both Indica and Japonica, in addition to the traditional and modern-Filipino cultivars (Table 5.1) were classified in the high CH_4 emission potential group (Figure 5.2a). A similar combination of different group and plant type was also observed in the intermediate and low CH_4 emission potential cultivar groups. Hence, general cultivar characteristics should not be used as a determining factor when evaluating CH_4 emission. However, specific characteristic such as shoot dry matter production and plant height had a significant and positive correlation with total CH_4 emission during the entire rice growth cycle (Figure 5.3). The other characteristics evaluated in this experiment were not correlated.

Despite having an apparently high amount of CH_4 emissions, cultivars with high grain yield may be labelled as more efficient with less environmental impact when the quantity of gas per kg of rice grain was considered. A high variation in CH_4 emission per kg of rice grain between cultivars was observed, ranging from 148 mg CH_4/k of rice grain to 852 mg CH_4/kg of rice grain (Table 5.2). Cultivars Arize 1003, Avaxi CL, BRS Atalanta, BRS Querencia, BRS Taim, Inov Cl, Irga 424 and Irga 422 had low CH_4 emission with high grain yield. This positive feature is associated with modern-Filipino type and hybrid cultivars, as well as cultivars with short height and cycle not exceeding 114 days (Tables 5.1 and 5.2).

5.4 Discussion

Low CH_4 emission rates immediately after flooding were expected. This is because a significant production of CH_4 in flooded soil occurs only after the reduction of nitrate, manganese (III and IV), and iron (III) compounds, which occurs ~20 days after flooding Silva *et al.* (2011). Additionally, rice plants initiate tillering in this initial growth phase and their aerenchyma are also not yet fully developed at this time. For this reason, the plant has a low potential to carry the CH_4 produced in the soil to the atmosphere (Aulakh *et al.*, 2000c). The first CH_4 peak at 20 days after flooding (DAF) was could be associated with the decomposition of the labile organic C pool and aerenchyma formation in plants (Aulakh *et al.*, 2002). The drop in CH_4 emission rates

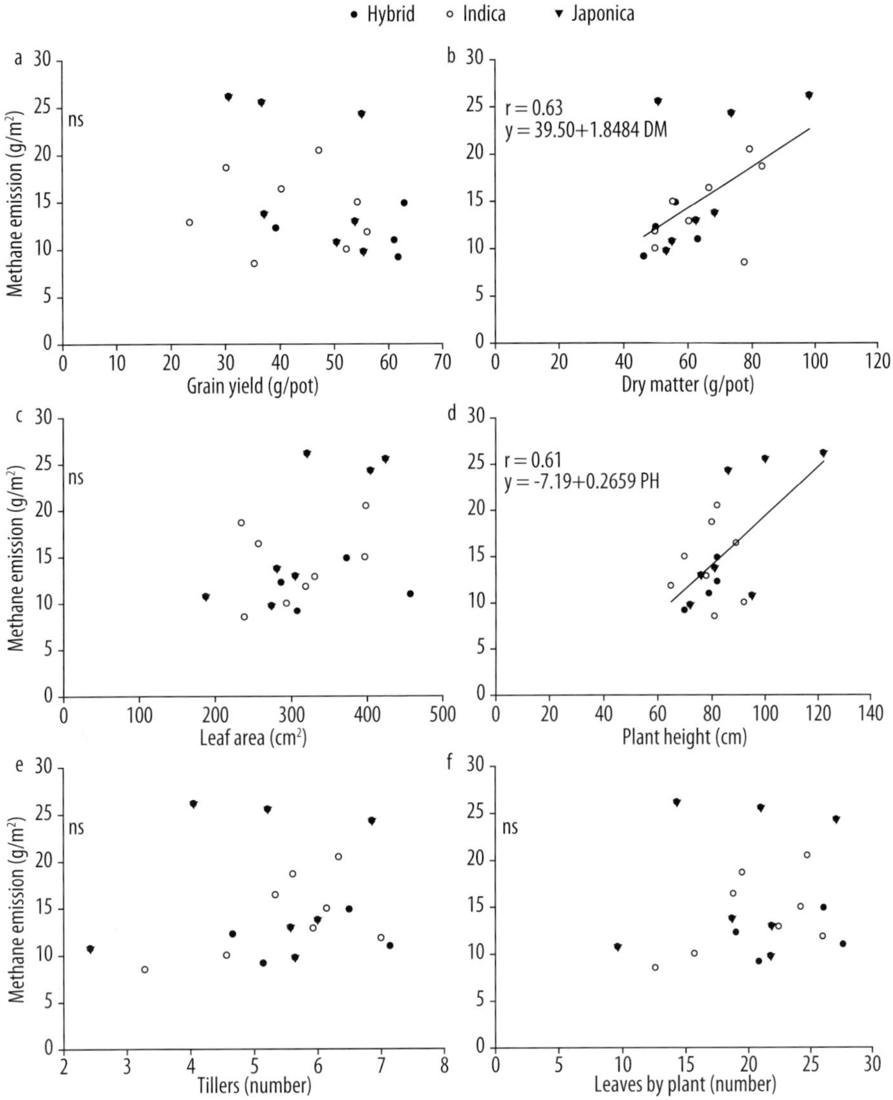

Figure 5.3. Linear regression between total CH_4 emission (g CH_4/m^2) during the rice growth cycle versus (a) grain yield, (b) aboveground biomass, (c) leaf area, (d) plant height, (e) number of tillers, and (f) number of leaves per plant.

after 20 DAF was likely due to the lack of labile C substrates. The peak in CH_4 emissions between 65 and 75 DAF was may be due to root exudation during the flowering phase (Huang *et al.*, 2002). A high concentration of root exudates may have increased the availability of labile organic compounds of low molecular weight, which are substrates for the methanogenic bacteria to produce CH_4 (Le Mer and Roger, 2001). In this phase, aerenchyma were fully mature and functioned as a channel for the transport of CH_4 into the atmosphere (Aulakh *et al.*, 2002). After the flowering phase, there is a significant drop in CH_4 emission rates which was likely caused by the cessation of irrigation and the beginning of plant senescence, resulting in the termination of root exudate production and therefore a source of labile C compounds.

Cultivars were classified into three groups (low, intermediate and high) according to their total CH_4 emission during the entire rice growing cycle. However, neither the origin of cultivars nor plant type could explain this grouping. Mitra *et al.* (1999) found that traditional cultivars had higher rates of CH_4 emission. They associated this with a greater production of biomass and root exudation; processes favouring CH_4 production. In our study, shoot biomass production was significantly correlated with the emission of CH_4. Similarly, others also found a positive correlation between CH_4 emissions and shoot biomass (Aulakh *et al.*, 2002; Das and Baruah, 2008; Gogoi *et al.*, 2005; Lindau and Bollich, 1993; Wang *et al.*, 1997a). According to Wang *et al.*(1997b), higher shoot dry matter production is also associated with greater root development, which increases the release of root exudates. Plant height was also significantly correlated with the CH_4 emission. We speculate that because rice cultivars in this group had a greater height, they also had a greater biomass production which was associated with a greater emission of CH_4.

The number of tillers per plant was not related to total CH_4 emission during the rice growth cycle, which was likely due to the large differences in tillering capacity between the various cultivars. This may also explain the lack of relationship between total CH_4 emission and the number of leaves per plant. Aulakh *et al.* (2002) also found a weak relationship between CH_4 emission and a variety of plant characteristics including the number of leaves per plant using 22 different rice cultivars. However, Mitra *et al.* (1999) and Aulakh *et al.* (2000b) found a significant relationship between tillering capacity and CH_4 emission, however, both of these studies used a small number of cultivars which may have influenced their results.

Results from our study indicated that the modern rice cultivars with a low growth height, a short growing season and a high grain yield had the lowest emission of CH_4. As such, these rice cultivars were capable of addressing two of the most urgent global issues, namely global warming and food security. Modern rice cultivars with these qualities have an overall lower contribution to global warming due to their reduced emission of CH_4 but also due to their ability to produce a high grain yield addressing food insecurity issues.

Acknowledgements

The authors thank the National Council for Scientific and Technological Development (CNPq) and the Foundation of Research Support in Rio Grande do Sul (FAPERGS), Brazil for providing a research grant and a fellowship to L.S. da Silva and D.F. Moterle.

References

Aulakh, M.S., Bodenbender, J., Wassmann, R. and Rennenberg, H., 2000a. Methane transport capacity of rice plants. I. Influence of methane concentration and growth stage analyzed with an automated measuring system. Nutrient Cycling in Agroecosystems 58: 357-366.

Aulakh, M.S., Bodenbender, J., Wassmann, R. and Rennenberg, H., 2000b. Methane transport capacity of rice plants. II. Variations among different rice cultivars and relationship with morphological characteristics. Nutrient Cycling in Agroecosystems 58: 367-375.

Aulakh, M.S., Wassmann, R., Rennenberg, H. and Fink, S., 2000c. Pattern and amount of aerenchyma relate to variable methane transport capacity of different rice cultivars. Plant Biology 2: 182-194.

Aulakh, M.S., Wassmann, R. and Rennenberg, H., 2002. Methane transport capacity of twenty-two rice cultivars from five major Asian rice-growing countries Agriculture. Ecosystems and Environment 91: 59-71.

Butterbach-Bahl, K., Papen, H. and Rennenberg, H., 1997. Impact of gas transport through rice cultivars on methane emission from rice paddy fields. Plant Cell Environmental 20: 1175-1183.

Comissão de Química e Fertilidade do Solo – NRS (CQFS/NRS), 2004. Manual de adubação e calagem para os Estados do Rio Grande do Sul e de Santa Catarina. 10 sd. SBCS – Núcleo Regional Sul/UFRGS, Porto Alegre, Brazil.

Costa, F.S., Bayer, C., Lima, M.A., Frighetto, R.T.S., Macedo, V.R.M. and Marcolin, E., 2008. Variação diária da emissão de metano em solo cultivado com arroz irrigado no Sul do Brasil. Ciência Rural 38: 2049-2053.

Das, K. and Baruah, K.K., 2008. Methane emission associated with anatomical and morphophysiological characteristics of rice (*Oryza sativa*) plant. Physiologia Plantarum 134: 303-312.

Evans, D.E., 2003. Aerenchyma formation. New Phytologist 161: 35-49.

Gogoi, N., Baruah, K.K., Gogoi, B. and Gupta, P.K., 2005. Methane emission characteristics and its relations with plant and soil parameters under irrigated rice ecosystem of northeast India. Chemosphere 59: 1677-1684.

Gomes, J., Bayer, C., Costa, F.S., Piccolo, M.C., Vieira, F.C.B. and Six, J., 2009. Soil nitrous oxide emission as affected by long term tillage, crop rotations and fertilization in a subtropical environment. Soil Tillage Research 101: 36-44.

Huang, Y., Jiao, Y., Zong, L.G., Zheng, X.H., Sass, R.L. and Fisher, F.M., 2002. Quantitative dependence of methane emission on soil properties. Nutrient Cycling in Agroecosystems 64: 157-167.

Intergovernmental Panel on Climate Change (IPCC), 2007. Climate change 2007: the physical science basis: summary for policymakers. Intergovernmental Panel on Climate Change, Geneva, Switzerland.

Johnson, J., Franzluebbers, A.J., Weyers, S.L. and Reicosky, D.C., 2007. Agricultural opportunities to mitigate greenhouse gas emissions. Environmental Pollution 150: 107-124.

Le Mer, J. and Roger, P., 2001. Production, oxidation, emission and consumption of methane by soils: a review. European Journal Soil Biology 37: 25-50.

Lindau, C.W. and Bollich, P.K., 1993. Methane emissions from Louisiana first and ratoon crop rice. Soil Science 156: 42-48.

Marchesan, E., 2006. Características de cultivares de arroz irrigado. UFSM, Santa Maria, Brazil.

Mitra, S., Jain, M.C., Kumar, S., Bandyopadhyay, S.K. and Kalra, N., 1999. Effect of rice cultivars on methane emission. Agriculture, Ecosystems and Environment 73: 177-183.

Mosier, A.R., 1989. Chamber and isotopic techniques. In: Andreae, M.O. and Schimel, D.S. (eds.) Exchange of traces gases between terrestrial ecosystems and the atmosphere. Report of the Dahlem Workshop. John Wiley and Sons, Berlin, Germany, pp.175-187.

Nouchi, I., Mariko, S. and Aoki, K., 1990. Mechanism of methane transport from the rhizosphere to the atmosphere through rice plants. Plant Physiology 94: 59-66.

Silva, L.S., Griebeler, G., Moterle, D.F., Bayer, C., Zschornack, T. and Pocojeski, E., 2011. Dinâmica da emissão de CH_4 em solos representativos do cultivo de arroz irrigado no sul do brasil. Revista Brasileira de Ciência do Solo 35: 473-483.

Terres, A.L.S., Fagundes, P.R.R., Machado, M.O., Magalhães Jr., A.M. and Nunes, C.D.M., 2004. Melhoramento genético e cultivares de arroz irrigado. In: Gomes, A.S. and Magalhães Jr., A.M. (eds.) Arroz irrigado no Sul do Brasil. Embrapa Informação Tecnológica, Brasília, Brazil, pp.161-235.

Wang, B., Neue, H.U. and Samonte, H.P., 1997a. Effect of cultivar difference (IR72, IR65598 and Dular) on methane emission. Agriculture, Ecosystems and Environment 62: 31-40.

Wang, B., Neue, H.U. and Samonte, H.P., 1997b. Role of rice in mediating methane emission. Plant Soil 189: 107-115.

Wassmann, R., Neue, H.U., Bueno, C., Lantin, R.S., Alberto, M.C.R., Buendia., L.V., Bronson, K., Papen, H. and Rennenberg, H., 1998. Methane production capacities of different rice soils derived from inherent and exogenous substrates. Plant and Soil 203: 227-237.

Chapter 6.

Influence of elevated ozone levels on the fate of pyrene in soil grown with wheat (*Triticum aestivum* L.)

F.X. Ai[1], R. Tian[1], Y.Y. Sun[2], Y. Yin[1], R. Ji[1], J.G. Zhu[3] and H.Y. Guo[1]

[1]State Key Laboratory of Pollution Control and Resource Reuse, School of Environment, Institute for Climate and Global Change Research (ICGCR), Nanjing University, Nanjing 210093, China P.R.; [2]State Key Laboratory of Pollution Control and Resource Reuse, Department of Hydrosciences, School of Earth Science and Engineering, Nanjing University, Nanjing 210093, China P.R.; [3]China State Key Laboratory of Soil and Sustainable Agriculture, Institute of Soil Science, Chinese Academy of Science, Nanjing 210008, Jiangsu, China P.R.; hyguo@nju.edu.cn

Abstract

Pollution of the environment with ozone (O_3) and organic pollutants has been steadily increasing over the past decades. Field studies on the synergistic effects of these pollutants are limited. We studied the effects of elevated O_3 levels on pyrene degradation in soil, the bioaccumulation of pyrene, and the oxidative stress on wheat (*Triticum aestivum* L.) using a free-air concentration enrichment system. We grew wheat plants for two growing seasons in pots containing soil spiked with 0, 100 or 200 mg pyrene per kilogram of soil under ambient or elevated (+50%) O_3 levels. Pyrene concentration in soil decreased over time and elevated O_3 further. However, pyrene concentration in wheat shoots and seeds was higher under elevated O_3 levels than under ambient conditions. Pyrene concentration in the soil was correlated to a higher activity of wheat superoxide dismutase, peroxidase, ascorbate peroxidase, and catalase, suggesting that pyrene caused an oxidative stress. Enzymes were higher under elevated O_3 levels, except in soil with 200 mg added pyrene per kilogram of soil. Our results indicated that increasing O_3 levels in the future may on one hand stimulate pyrene biodegradation, but on the other hand lead a higher accumulation in wheat. Thus, synergetic effects of these pollutants may increase their threat to food safety in the future.

Keywords: air pollution, food safety and security, northern hemisphere, polycyclic aromatic hydrocarbons (PAHs), soil pollution

DOI 10.3920/978-90-8686-788-2_6,

6.1 Introduction

Over the past decades, the current average surface ozone (O_3) concentration increased from an estimated pre-industrial value of 10 ppbv to 20-45 ppbv in the mid-latitudes of the northern hemisphere, which is equivalent to a rate of 0.5-2% per year (Morgan *et al.*, 2006; Vingarzan, 2004). The available data from China confirmed this steady increase in O_3 concentration (Wang *et al.*, 2007). For example, monitoring activities from 1999 to 2000 in the Yangtze River delta showed that average daytime O_3 levels in the month of May ranged between 60 and 79 ppbv (Wang *et al.*, 2006). Zhou (2004) reported that 6.1-10.4% of the hourly mean O_3 concentrations exceeded 60 ppbv; and the peak O_3 concentration was as high as 196 ppbv in Yangtze River delta. Global photochemical models projected a further increase in O_3 concentrations by the year 2030 in several parts of Asia (Dentener *et al.*, 2006). Meehl *et al.* (2007) projected an increase of 20-25% in mean global O_3 concentration between 2015 and 2050 and of 40-60% by 2100.

Tropospheric O_3 is a significant air pollutant, and high O_3 concentrations have long been known to affect physiological and biochemical processes as well as biomass growth and the yield of agricultural crops (Feng and Kobayashi, 2009; Fumagalli *et al.*, 2001; Sandermann, 1996; Singh *et al.*, 2010). Wheat (*Triticum aestivum* L.) was recognized as one of the most O_3-sensitive crops (Biswas *et al.*, 2008; Mills *et al.*, 2007; Wang *et al.*, 2007). Recent research suggested that elevated O_3 levels decrease photosynthetic rates and chlorophyll content of wheat, and it also reduces ascorbate in the apoplast and leaf tissue, increases lipid oxidation, and accelerates leaf senescence of this crop (Biswas *et al.*, 2008; Feng *et al.*, 2008, 2010, 2011; Sarkar *et al.*, 2010). Together, this contributed to an overall reduced plant growth (Feng and Kobayashi, 2009; Feng *et al.*, 2008, Sarkar and Agrawal, 2010; Wang *et al.*, 2007). Experiments conducted in China also showed that elevated O_3 concentrations may lead to a significant reduction of wheat biomass (Wang *et al.*, 2007; Zhou, 2004), and thereby negatively affecting global food security as well as food safety and human health.

Polycyclic aromatic hydrocarbons (PAHs) are a class of organic compounds that consist of two or more fused benzene rings and/or pentacyclic molecules arranged in various structural configurations. Because of their hydrophobicity and low water solubility, PAHs are highly recalcitrant and can persist in the environment for a very long time (Bamforth and Singleton, 2005). PAHs are ubiquitous in the natural environment and originate from both natural and anthropogenic sources; the latter source of PAHs is the major cause of environmental pollution (Mueller *et al.*, 1996). Plants accumulate PAHs either by taking up volatilized PAHs from the air through their leaves or by translocation from the root to the shoot (Gao and Collins, 2009; Tao *et al.*, 2009). Due to their carcinogenic, mutagenic, and teratogenic properties (IARC, 1983), and their high persistency in the environment (Johnsen *et al.,* 2005), PAHs are of a particular concern to food safety and human health (Lin *et al.,* 2011).

Wheat is one of the most commonly produced cereals globally, and a reduction in its yield and an overall decrease in its production potential, may negatively influence food security (Lobell *et al.*, 2011). The objective of this study was to evaluate the combined effect of elevated air O_3 levels and pyrene on wheat using free-air concentration enrichment (FACE) to predict its impact on wheat growth.

6.2 Materials and methods

Study site

The O_3-FACE system was located in Xiaoji, Jiangdu County, Jiangsu Province, China P.R. (119°42'E, 32°35'N), where the predominant agricultural production system consists of a rice (*Oriza sativae* L.)-wheat rotation. This region has a subtropical marine climate with a mean annual precipitation of 918 mm, a mean annual temperature of 14-16 °C, a total annual sunshine of more than 2,000 hours, and a frost-free period of approximately 220 days. The soil is Shajiang Aquic Cambisol with a sandy loam texture.

O_3-FACE system

The FACE system consisted of plots with elevated O_3 levels as previously described by Shi *et al.* (2009). Each FACE plot and each ambient plot (ambient O_3 levels) consisted of a 240 m^2 area. The target O_3 concentration for plots with elevated O_3 level was 50% higher than the current ambient O_3 concentration; this level simulated the predicted tropospheric O_3 level in 2100. Each of the plots with elevated O_3 levels was separated from the other plots by at least 70 m to avoid cross-contamination. In our study, one FACE plot and one ambient plot were used. In the plot with elevated O_3 levels, plants were grown within a 14 m in diameter octagonal ring, which consisted of eight 6 m acrylonitrile butadiene styrene (ABS) pipes. Ozone was generated by an O_3 generator (KCF-BT0.2, Jiangsu Koner Ozone Co. Ltd.) from pure O_2 to form about 5% O_3 in 95% O_2, pressurized by an air compressor, and released into the air via tiny holes in the ABS pipes at about 0.5 m above the canopy height. The quantity and direction of the O_3 released was controlled by a proportional integral derivative algorithm with computer feedback that compares the achieved O_3 concentration to the target O_3 concentration of 1.5× ambient O_3 concentration. The system consisted of an O_3 monitor (Thermo Electron 49i, Thermo Scientific Co., USA), a data logger-controller (Campbell CR

10X, Campbell Scientific Co., USA), an anemometer, and a wind vane. O_3 fumigation began at 9:00 a.m. and continued until 4:00 p.m. every day until harvest, except when it rained or when the leaves were wet (Shi *et al.,* 2009). Plants in the plot within the octagonal rings were sampled at least 2 m from the ABS pipes to avoid the much higher O_3 concentrations near the release holes. On the ambient O_3 plot, plants were grown under current O_3 concentrations without the ring structures. Further details of O_3 exposure have been reported by Feng *et al.* (2010, 2011).

Crop cultivation

Yangmai 14 is one of the most common wheat cultivars in China and was chosen as the test cultivar in our study, which consisted of two separate plantings. Dry soil (10 kg) collected from a local farm was placed in a plastic pot (20 cm diameter and 35 cm height). The soil contained less than 1 mg pyrene per kg soil. Pyrene dissolved in acetone was applied to the soil to achieve concentrations of 0, 100, and 200 mg/kg soil (six pots each) and mixed thoroughly. The six control pots contained no added pyrene. Ten wheat seeds were manually sown in each pot. After germination of the seeds, three pots of each pyrene concentration and of the control were randomly placed on the FACE plot, and the three pots each were randomly placed on the ambient plot. Cultivation and crop health measures were similar to those used by local farmers. In the first experiment, the seeds were planted on 15 November, 2008, and O_3 fumigation began on 5 March 2009 at the wheat tillering stage and continued until harvest on 29 May 2009. In the second experiment, the seeds were planted on 17 November 2009, and O_3 fumigation began on 8 March 2010 and continued until harvest on 31 May 2010.

Determination of pyrene

At the grain maturity stage, wheat shoots and soil samples were collected. Shoots were cut immediately above the soil surface, thoroughly washed with tap water and then with deionized water, and stored at -20 °C; three soil cores (2 cm inner diameter) from each pot was collected randomly, and stored at -20 °C. Both plant and soil samples were freeze-dried (Labconco 12L, Labconco Co., USA) for 96 h and then passed through a 2 mm sieve and ground. Samples (10 g) were extracted with 100 ml acetone:hexane (1:1, v:v), concentrated in a rotary evaporator, and dried under a fine stream of nitrogen (N). The residues were dissolved in 0.5 ml methanol. Samples were analyzed by high-performance liquid chromatography on a reverse-phase C_{18} column (4.6×250 mm, 5 μm particle size, Supelco Discovery; kept at 25 °C) with fluorescence detection (HPLC-FLD, Agilent 1200). The mobile phase was methanol:water (75:25, v/v), and the flow rate was 1 ml/min. The excitation wavelength was 237 nm, and the emission wavelength was 385 nm.

Isoenzyme assays

At the wheat jointing stage in 2010, ten leaves were cut randomly in each of the pots, immediately placed into liquid nitrogen, and then stored at -80 °C until further analysis. Enzymes were extracted according to García-Limones *et al.* (2002). Isoenzymes were resolved by native polyacrylamide gel electrophoresis using a high-throughput mini-protein 3 electrophoresis system (Bio-Rad, USA; 8% separating gel, 5% stacking gel). Crude enzyme, corresponding to 23.2 μg of total soluble protein mixed with glycerin and bromophenol blue, was loaded onto each lane. The antioxidant isoenzymes were separated by electrophoresis at a constant voltage of 70 V until the separating gel was reached, and then at 100 V. The running buffer was 25 mM Tris, 192 mM glycine (pH 8.3).

Superoxide dismutase (SOD), peroxidase (POD), and ascorbate peroxidase (APX) isoenzymes in the gels were analyzed according to García-Limones *et al.* (2002). Catalase (CAT) isoenzyme in the gel was analyzed according to Verma and Dubey (2003). Representative photographs were taken with a Canon camera (DS126211). Integrated intensities of activity bands were quantified using Image J software (National Institutes of Health, USA), and then normalized to the control level according to Biemelt *et al.* (2000), denoting the corresponding enzyme activities.

Statistical analysis

Data were expressed as means ± standard error (n=3 pots) and statistically analysed using SPSS (SPSS Inc., Chicago, USA, version 16.0). The data were analysed using a one-way analysis of variance (ANOVA). Tukey's multiple comparison test was performed to distinguish differences in mean values between treatments. Differences between treatments were taken to be statistically significant when $P<0.05$.

6.3 Results

Effects of elevated O_3 levels on pyrene in soil

From 2008 to 2010, the pyrene content in soil spiked with 100 and 200 mg pyrene per kilogram of soil, decreased with time. The decrease occurred most rapidly during the first 70 days. Pyrene concentration in soil without added pyrene remained almost the same. The pyrene concentration of soils spiked with 100 and 200 mg pyrene per kilogram of soil and placed under ambient O_3 levels were significantly higher than the concentration in the same soils placed under elevated O_3 levels at all sampling times (Figure 6.1). For example, at the first sampling, the pyrene concentration in soil spiked with 100 mg pyrene per kilogram of soil and placed under ambient O_3 levels from 2008 to 2010 was 48.68 mg/kg soil, and that placed under elevated O_3 levels was 39.89 mg/kg soil; the pyrene concentration in soil spiked with 200 mg pyrene per kilogram of soil and placed under ambient O_3 levels from 2008 to 2010 was 72.23 mg/kg soil, and that placed under elevated O_3 levels was 65.56 mg/kg soil.

Effects of elevated O_3 on the uptake of pyrene into wheat tissue

Pyrene concentration in wheat shoots spiked with 200 mg pyrene per kilogram of soil and placed under elevated O_3 levels were significantly higher than that in the shoots of wheat grown in the same soil and placed under ambient O_3 levels. However, pyrene concentrations in wheat shoots spiked with 100 mg pyrene per kilogram of soil grown

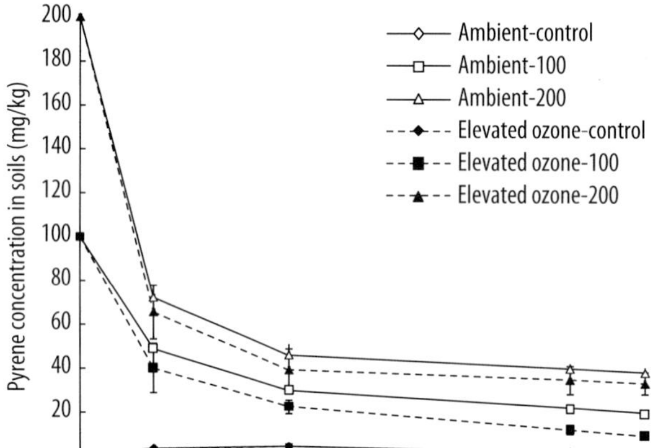

Figure 6.1. Pyrene concentration in soils in plots under ambient and elevated ozone levels from 2008 to 2010 in Xiaoji, Jiangdu County, Jiangsu Province, China P.R. Values represent means ± standard deviation (n=3). Ambient-control means pots with soil without added pyrene placed in ambient ozone plot; Ambient-100 means pots with soil with 100 mg/kg added pyrene placed in ambient ozone plot; Ambient-200 means pots with soil with 200 mg/kg added pyrene placed in ambient ozone plot; Elevated ozone-control means pots with soil without added pyrene placed in elevated ozone plot; Elevated ozone-100 means pots with soil with 100 mg/kg added pyrene placed in elevated ozone plot; Elevated ozone-200 means pots with soil with 200 mg/kg added pyrene placed in elevated ozone plot.

under ambient and elevated O_3 levels were not significantly different (Figure 6.2a, Table 6.1). The concentration of pyrene in wheat shoots (Figure 6.2a) was greatest when grown with 100 mg pyrene per kilogram of soil placed either under ambient or elevated O_3 levels. The pyrene concentration in seeds of wheat grown with 100 mg pyrene per kilogram of soil and under ambient O_3 levels was significantly lower than that in seeds of wheat grown in the same soil and placed under elevated O_3 (Figure 6.2b, Table 6.1).

We calculated the bioaccumulation factors (BAFs) of pyrene in wheat shoots and seeds harvested in 2010 (Table 6.1). The BAF is the ratio of the pyrene concentration of the shoots or seeds to the pyrene concentration in the soil. Without added pyrene in soil, the BAF_{shoots} and the BAF_{seeds} were significantly lower when the plants were grown under ambient O_3 levels than when grown under elevated O_3 levels ($P<0.01$); and the BAF_{shoots} and the BAF_{seeds} spiked with 100 and 200 mg pyrene per kilogram of soil were also lower (but not significantly different) when the plants were grown under ambient O_3 levels when grown under elevated O_3 levels.

Effects of pyrene and elevated O_3 levels on antioxidant isoenzymes in wheat shoots

The values of all four antioxidant isoenzymes (APX, CAT, POD, SOD) evaluated at the jointing stage of wheat shoots in 2010, and grown in soil without added pyrene and

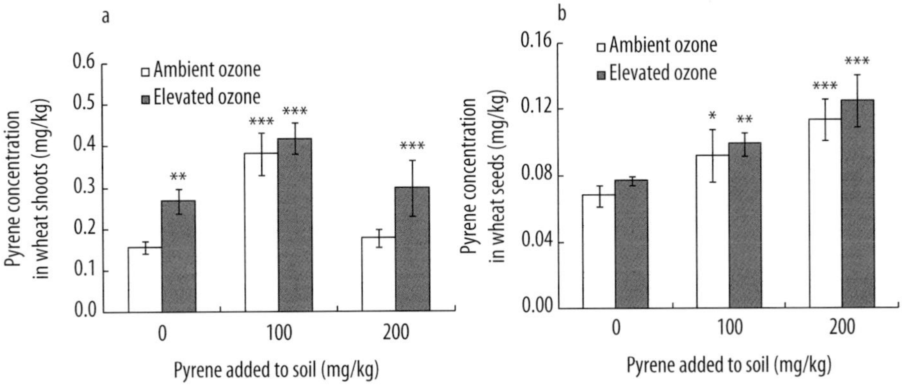

Figure 6.2. Pyrene concentration in (a) wheat shoots and (b) wheat seeds at harvest in 2010 in Xiaoji, Jiangdu County, Jiangsu Province, China P.R. Values represent means ± standard deviation (n=3). Ambient means pots with wheat placed in ambient ozone plot; elevated ozone means pots with wheat placed in elevated ozone plot. Asterisks above the bars indicate a significant difference between various treatments and soil without added pyrene under ambient ozone according to Tukey's tests. * $P<0.05$; ** $P<0.01$; *** $P<0.001$.

Table 6.1. Bioaccumulation factors (BAFs) of pyrene in wheat shoots and seeds harvested in 2010 at Xiaoji, Jiangdu County, Jiangsu Province, China P.R.

Treatment[1]	Pyrene in soil (mg/kg)	Pyrene in shoots (mg/kg)	Pyrene in seeds (mg/kg)	BAF_{shoots}	BAF_{seeds}
Control					
Ambient O_3	0.749±0.021	0.157±0.009	0.068±0.004	0.209±0.0059	0.091±0.0050
Elevated O_3	0.724±0.021	0.269±0.017	0.077±0.001	0.373±0.0296	0.107±0.0033
100 mg pyrene per kg soil					
Ambient O_3	19.765±1.154	0.381±0.029	0.092±0.009	0.019±0.0010	0.005±0.0007
Elevated O_3	9.231±0.515	0.419±0.022	0.099±0.004	0.046±0.0036	0.011±0.0006
200 mg pyrene per kg soil					
Ambient O_3	37.989±1.688	0.177±0.013	0.114±0.007	0.005±0.0001	0.003±0.0001
Elevated O_3	33.446±2.872	0.298±0.039	0.125±0.009	0.009±0.0011	0.004±0.0002

[1] BAF_{shoots} = concentration in shoots/concentration in soils; BAF_{seeds} = concentration in seeds/concentration in soils; Values are means of triplicates ± standard error.

in soil spiked with 100 mg pyrene per kg soil under elevated O_3 levels were significantly higher than the amounts from wheat grown in the same soils under ambient O_3 levels (Figure 6.3). In contrast, the CAT and POD values in the treatment spiked with 200 mg pyrene per kilogram of soil under elevated O_3 levels were significantly lower. In this soil, APX and SOD levels were also lower when the wheat was grown under elevated O_3 levels compared to ambient O_3 levels, but the differences were not significant.

6.4 Discussion

Our results showed that elevated O_3 levels resulted in a faster degradation of pyrene in soil and led to an increased uptake of pyrene in wheat plants grown on pyrene-spiked soils. Additionally, results from our study showed that the antioxidant isoenzymes APX, CAT, POD, and SOD in wheat shoots were strongly affected by combined exposure to pyrene and elevated O_3 levels. O'Mahony *et al.* (2006) studied the effect of O_3 generated from air by electric discharge on farm soils spiked with several types of PAHs, such as phenanthrene. They found that in all soils the initial phenanthrene concentration of 20 ppmv was reduced by more than 50% after the O_3 treatment for just 6 hours. Masten and Davies (1997) evaluated the effect of O_3 as a PAH-oxidizing agent and reported that 95% of the added phenanthrene and 91% of the added pyrene

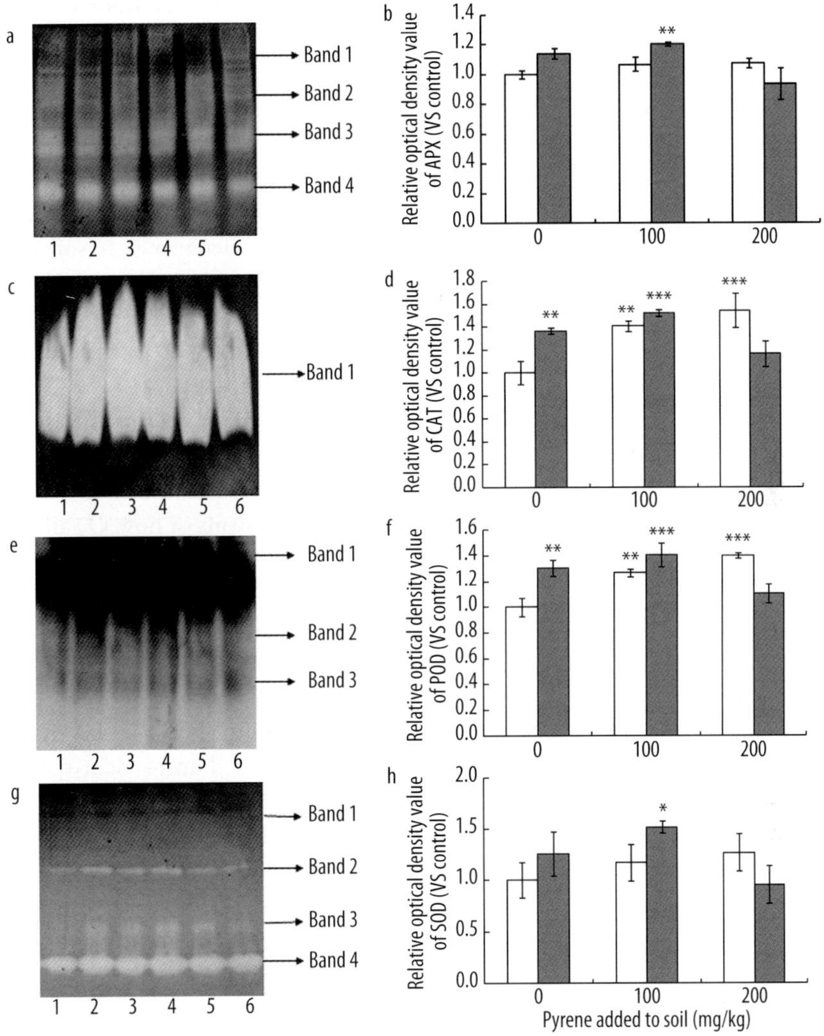

Figure 6.3. Effect of pyrene and elevated O_3 on antioxidative isoenzymes in wheat shoots at the jointing stage in 2010 in Xiaoji, Jiangdu County, Jiangsu Province, China P.R. (a, b) APX; (c, d) CAT; (e, f) POD; (g, h) SOD. (a, c, e, g) Activity staining of isoenzymes in native polyacrylamide gels. Arrows indicate bands of activity. Lanes 1-3: 0 (control), 100, and 200 mg/kg pyrene in soil at ambient O_3 levels, respectively. Lanes 4-6: 0, 100, and 200 mg/kg pyrene in soil at elevated O_3 levels, respectively. (b, d, f, h): Enyzme activities determined by integrating the band intensities. Values represent means ± standard deviation (n=3). White bands means pots with wheat placed in ambient ozone plot, black bands means pots with wheat placed in elevated ozone plot. Asterisks above the bars indicates a significant difference between various treatments and soil without added pyrene under ambient ozone according to Tukey's tests. * $P<0.05$; ** $P<0.01$; *** $P<0.001$.

were removed from soils under higher O_3 levels. In our experiments over two years, we also found that O_3 significantly accelerated the degradation of pyrene in soil.

Compared with ambient O_3 levels, elevated O_3 levels increased the bioaccumulation of pyrene in wheat shoots and seeds (BAF_{shoots} and BAF_{seeds}) significantly when additional pyren was added to the soil. Under both ambient and elevated O_3 levels, the concentration of pyrene in wheat shoots was higher when the plants were grown in soil spiked with 100 mg pyrene per kilogram of soil than when grown in soil with no added pyrene. However, when soil was spiked with 200 mg pyrene per kilogram of soil, the concentration of pyrene in wheat shoots decreased to almost the same level as that in plants grown in soil without added pyrene. This decrease in pyrene concentrations in the shoots at higher pyrene concentrations in the soil might be due to the wheat not being able to grow and develop as it would under current ambient conditions. This may result in an inability to take-up sufficient amount of pyrene from the soil. Little is known about the effect of elevated O_3 levels on the bioaccumulation of PAHs in plants, and further studies are needed to characterize the mechanisms of how O_3 affects the uptake of PAHs by plant tissues.

Many environmental stresses, including O_3 and pyrene, generate oxidative stress in plant tissues by inducing an over-production of reactive oxygen species (ROS), such as the superoxide anion radical ($O_2^{\bullet -}$), singlet oxygen ($^1O_2$), hydrogen peroxide (H_2O_2), and hydroxyl radical ($^{\bullet}OH$) (Liu *et al.*, 2008; Lu *et al.*, 2009; Ryang *et al.*, 2009). To scavenge ROS, plants possess a well-organized anti-oxidative defence system comprising enzymes such as POD, APX, SOD, and CAT (Lu *et al.*, 2009; Verma and Dubey, 2003). In the ascorbate system, ascorbate contributes directly to ROS scavenging via the action of APX (Noctor and Foyer, 1998). POD is widely distributed in all higher plants and protects cells against the destructive influences of H_2O_2 by catalysing its decomposition through oxidation of phenolic and endiolic co-substrates (Lin and Kao, 2002). CAT is present in the peroxisomes of nearly all aerobic cells and is virtually absent from chloroplasts (Dionisio-Sese and Tobita, 1998). It can protect the cell from H_2O_2 by catalyzing its decomposition to O_2 and H_2O (Foyer and Noctor, 2000). Under normal physiological conditions, the ROS generated from metabolism of extraneous chemicals in the plants can be removed well by the antioxidant defense system. However, when ROS generation exceeds the capacity of the cellular antioxidants, it will cause oxidative stress and significant oxidative damage (Cho and Seo, 2005).

Di Cagno *et al.* (2001) reported that O_3 fumigation induces a remarkable increase of POD activity in sunflower (*Helianthus annuus* L.) in a controlled environmental chamber study. Singh *et al.* (2010) reported that the activity of CAT decreased whereas the activity of POD increased in two soybean (*Glycine max* (L.) Merr.) varieties upon exposure to O_3 in an experiment with open-top chambers. Rai and Agrawal (2008) found that the levels of SOD and POD activity, ascorbic acid, and total phenolics in two rice (*Oryza sativa L.*) cultivars were higher when exposed to O_3 in open-top chambers. Our results suggested that the production of ROS increased when wheat was exposed to combined pyrene (100 mg added per kilogram of soil) and elevated O_3 levels. This suggested that wheat plants tried to cope with the oxidative stress induced by these two pollutants by strengthening their antioxidant capabilities, which was supported by an increase in the activities of POD, APX, SOD, and CAT. However, at a higher concentration of added pyrene (200 mg/kg) and with elevated O_3 levels, the APX, POD, CAT, and SOD activities were lower. This may be due to a higher accumulation of pyrene in wheat tissue, which led to a greater accumulation of ROS. As ROS concentrations increased, the ability of these enzymes to eliminate the ROS decreased, leading to the inactivation of the enzymes. Liu *et al.* (2008) reported that when *Arabidopsis thaliana* (L.) Heynh. was exposed to 0.25-1.25 mM phenanthrene for 30 days, POD and APX activities were highest at 0.25 mM phenanthrene and declined at higher concentrations. Such a change in the anti-oxidative defense system in wheat shoots suggested that elevated O_3 levels plus high levels of pyrene had a significantly synergic effect on the oxidative stress in wheat.

In China, most farmlands are polluted with PAHs (Ding *et al.*, 2007; Li *et al.*, 2008; Ye *et al.*, 2006), but these polluted soils produce 14.6×10^8 kg of agricultural products annually (Li *et al.*, 2003). Furthermore, almost the entire population in China currently depends on rice and wheat as staple foods. Our results indicated that elevated O_3 levels plus a high concentration of pyrene in soil leads to high, toxic concentrations of pyrene in wheat tissues and accelerates oxidative stress in wheat. This may negatively affect food quality and safety, which is of particular concern because the O_3 levels are expected to increase in the future.

6.5 Conclusions

Our study used a FACE system which showed that elevated air O_3 levels plus a high concentration of pyrene in soil had a significant synergic effect on oxidative stress in wheat shoots. Our data also showed that although elevated O_3 levels improved the degradation of pyrene in soil, compared with the degradation at ambient O_3 levels, the elevated O_3 levels led to a significantly higher pyrene concentration in shoots and seeds of wheat grown in pyrene-contaminated soils. This suggested that elevated O_3 levels may lead to a reduction in food quality and safety. Future studies should consider the inclusion of a wide range of soil types under field conditions to help predict the combined effect of elevated O_3 levels and PAH-contaminated soils on the uptake of PAHs by crops and their effect on food quality and food safety worldwide.

Acknowledgements

This study was supported by the National Natural Science Foundation of China (Grant No. 21177058), Program for New Century Excellent Talents in University (NCET-12-0266), the International S & T Cooperation Program of China (Grant No. 2010DFA22770), National Natural Science Foundation of China (Grant No. 31261140364) and the Knowledge Innovation Program of Chinese Academy of Sciences (Grant No. KZCX3-SW-440), the China Postdoctoral Science Foundation, Jiangsu Planned Projects for Postdoctoral Research Funds. The main instruments and apparatus of the FACE system were supplied by Japan National Institute for Agro-Environmental Sciences (NIAES) and Japan Agricultural Research Center for Tohoku Region (NARCT).

References

Bamforth, S.M. and Singleton, L., 2005. Bioremediation of polycyclic aromatic hydrocarbons: current knowledge and future directions. Journal of Chemical Technology and Biotechnology 80: 723-736.

Biemelt, S., Keetman, U., Mock, H.P. and Grimm, B., 2000. Expression and activity of isoenzymes of superoxide dismutase in wheat roots in response to hypoxia and anoxia. Plant Cell and Environment 23: 135-144.

Biswas, D.K., Xu, H., Li, Y.G., Liu, M.Z., Chen, Y.H., Sun, J.Z. and Jiang, G.M., 2008. Assessing the genetic relatedness of higher ozone sensitivity of modern wheat to its wild and cultivated progenitors/relatives. Journal of Experimental Botany 59: 951-963.

Cho, U.H. and Seo, N.H., 2005. Oxidative stress in *Arabidopsis thaliana* exposed to cadmium is due to hydrogen peroxide accumulation. Plant Science 168: 113-120.

David, B.L., Wolfram, S. and Justin, C.R., 2011. Climate trends and global crop production since 1980. Science 333: 616-617.

Dentener, F., Stevenson, D., Ellingsen, K., Van Noije, T., Schultz, M., Amann, M., Atherton, C., Bell, N., Bergmann, D., Bey, I., Bouwman, L., Butler, T., Cofala, J., Collins, B., Drevet, J., Doherty, R., Eickhout, B., Eskes, H., Fiore, A., Gauss, M., Hauglustaine, D., Horowitz, L., Isaksen, I.S.A., Josse, B., Lawrence, M., Krol, M., Lamarque, J.F., Montanaro, V., Muller, J.F., Peuch, V.H., Pitari, G., Pyle, J., Rast, S., Rodriguez, J., Sanderson, M., Savage, N.H., Shindell, D., Strahan, S., Szopa, S., Sudo, K., Van Dingenen, R., Wild, O. and Zeng, G., 2006. The global atmospheric environment for the next generation. Environmental Science and Technology 40: 3586-3594.

Di Cagno, R., Guidi, L., De Gara, L. and Soldatini, G.F., 2001. Combined cadmium and ozone treatments affect photosynthesis and ascorbate-dependent defences in sunflower. New Phytologist 151: 627-636.

Ding, A.F., Pan, G.X. and Li, L.Q., 2007. Contents and sources of PAHs in top soils of farmlands in parts of Jiangsu, China. Journal of Ecology and Rural Environment 23: 71-75.

Dionisio-Sese, M.L. and Tobita, S., 1998. Antioxidant responses of rice seedlings to salinity stress. Plant Science 135: 1-9.

Feng, Z.Z. and Kobayashi, K., 2009. Assessing the impacts of current and future concentrations of surface ozone on crop yield with meta-analysis. Atmospheric Environment 43: 1510-1519.

Feng, Z.Z., Kobayashi, K. and Ainsworth, E.A., 2008. Impact of elevated ozone concentration on growth, physiology, and yield of wheat (*Triticum aestivum* L.): a meta-analysis. Global Change Biology 14: 2696-2708.

Feng, Z.Z., Pang, J., Kobayashi, K., Zhu, J.G. and Ort, D.R., 2011. Differential responses in two varieties of winter wheat to elevated ozone concentration under fully open-air field conditions. Global Change Biology 17: 580-591.

Feng, Z.Z., Pang, J., Nouchi, I., Kobayashi, K., Yamakawa, T. and Zhu J.G., 2010. Apoplastic ascorbate contributes to the differential ozone sensitivity in two varieties of winter wheat under fully open-air field conditions. Environmental Pollution 158: 3539-3545.

Foyer, C.H. and Noctor, G., 2000. Oxygen processing in photosynthesis: regulation and signaling. New Phytologist 146: 359-388.

Fumagalli, I., Gimeno, B.S., Velissariou, D., De Temmerman, L. and Mills, G., 2001. Evidence of ozone-induced adverse effects on crops in the Mediterranean region. Atmospheric Environment 35: 2583-2587.

Gao, Y.Z. and Collins C.D., 2009. Uptake pathways of polycyclic aromatic hydrocarbons in white clover. Environmental Science and Technology 43: 6190-6195.

García-Limones, C., Hervás, A., Navas-Cortés, J.A., Jiménez-Diaz, R.M. and Tena, M., 2002. Induction of an antioxidant enzyme system and other oxidative stress markers associated with compatible and incompatible interactions between chickpea (*Cicer arietinum* L.) and *Fusarium oxysporum* f.sp. *ciceris*. Physiological and Molecular Plant Pathology 61: 325-337.

International Agency for Research on Cancer (IARC), 1983. Polynuclear aromatic hydrocarbons. Part 1. Chemical, environmental and experimental data. Monographs on the evaluation of the carcinogenic risk of chemicals to humans Vol. 32. WHO, Lyon, France.

Johnsen, A.R., Wick, L.Y. and Harms, H., 2005. Principles of microbial PAH-degradation in soil. Environmental Pollution 133: 71-84.

Li, X.J., Li, P.J., Lin, X., Gong, Z.Q., Fan, S.X. and Zheng, L., 2008. Spatial distribution and sources of polycyclic aromatic hydrocarbons (PAHs) in soils from typical oil-sewage irrigation area, Northeast China. Environmental Monitoring and Assessment 143: 257-265.

Li, K.Q., Liu, Z.G., Lu, X.L., Yang, J.C., Zhang, Z.J. and Zhu, Q.S., 2003. Uptake and distribution of cadmium in different rice cultivars. Journal of Agro-Environment Science 22: 529-532.

Lin, C.C. and Kao, C.H., 2002. Osmotic stress-induced changes in cell wall peroxidase activity and hydrogen peroxide level in roots of rice seedlings. Plant Growth Regulation 37: 177-183.

Lin, G.F., Weigel, S., Tang, B., Schulz, C. and Shen, J.H., 2011. The occurrence of polycyclic aromatic hydrocarbons in Peking duck: relevance to food safety assessment. Food Chemistry 129: 524-527.

Liu, H., Weisman, D., Ye, Y.B., Cui, B., Huang, Y.H., Colon-Carmona, A. and Wang, Z.H., 2008. An oxidative stress response to polycyclic aromatic hydrocarbons exposure is rapid and complex in *Arabidopsis thaliana*. Plant Science 176: 375-382.

Lobell, D.B., Schlenker, W. and Costa-Roberts, J., 2011. Climate trends and global crop production since 1980. Science 333: 616-620.

Lu, T., He, X.Y., Chen, W., Yan, K. and Zhao, T.H., 2009. Effects of elevated O_3 and/or elevated CO_2 on lipid peroxidation and antioxidant systems in *Ginkgo biloba* leaves. Bulletin of Environmental Contamination and Toxicology 83: 92-96.

Masten, S.J. and Davies, S.H.R., 1997. Efficacy of in-situ ozonation for the remediation of PAH contaminated soils. Journal of Contaminant Hydrology 28: 327-335.

Meehl, G.A., Stocker, T.F., Collins, W.D., Friedlingstein, P., Gaye, A.T., Gregory, J.M., Kitoh, A., Knutti, R., Murphy, J.M., Noda, A., Raper, S.C.B., Watterson, I.G., Weaver, A.J. and Zhao, Z.C., 2007. Global climate projections. In: Solomon, S., Qin, D., Manning, M., Chen, Z., Marquis, M., Averyt, K.B., Tignor, M. and Miller, H.L. (eds.) Climate change: the physical basis. Contribution of working group I to the fourth assessment report of IPCC on climate change. Cambridge University Press, Cambridge, UK.

Mills, G., Buse, A., Gimeno, B., Bermejo, V., Holland, M., Emberson, L. and Pleijel, H., 2007. A synthesis of AOT40-based response functions and critical levels of ozone for agricultural and horticultural crops. Atmospheric Environment 41: 2630-2643.

Morgan, P.B., Mies, T.A., Bollero, G.A., Nelson, R.L. and Long, S.P., 2006. Season-long elevation of ozone concentration to projected 2050 levels under fully open-air conditions substantially decreases the growth and production of soybean. New Phytologist 170: 333-343.

Mueller, J.G., Cerniglia, C.E. and Pritchard, P.H., 1996. Bioremediation of environments contaminated by polycyclic aromatic hydrocarbons. In: Crawford, R.L. and Crawford, P.L. (eds.) Bioremediation: principles and applications. Cambridge University Press, Cambridge, UK, pp. 125-194.

Noctor, G. and Foyer, C.H., 1998. Ascorbate and glutathione: keeping active oxygen under control. Annual Review of Plant Physiology and Plant Molecular Biology 49: 249-279.

O'Mahony, M.M., Dobson, A.D.W., Barnes, J.D. and Singleton, I., 2006. The use of ozone in the remediation of polycyclic aromatic hydrocarbon contaminated soil. Chemosphere 63: 307-314.

Rai, R. and Agrawal, M., 2008. Evaluation of physiological and biochemical responses of two rice (*Oryza sativa* L.) cultivars to ambient air pollution using open top chambers at a rural site in India. Science of the Total Environment 407: 679-691.

Ryang, S.Z., Woo, S.Y., Kwon, S.Y., Kim, S.H., Lee, S.H., Kim, K.N. and Lee, D.K., 2009. Changes of net photosynthesis, antioxidant enzyme activities, and antioxidant contents of *Liriodendron tulipifera* under elevated ozone. Photosynthetica 47: 19-25.

Sandermann, H., 1996. Ozone and plant health. Annual Review of Phytopathology 34: 347-366.

Sarkar, A. and Agrawal, S.B., 2010. Elevated ozone and two modern wheat cultivars: an assessment of dose dependent sensitivity with respect to growth, reproductive and yield parameters. Environmental and Experimental Botany 69: 328-337.

Sarkar, A., Rakwal, R., Agrawal, S.B., Shibato, J., Ogawa, Y., Yoshida, Y., Agrawal, G.K. and Agrawal, M., 2010. Investigating the impact of elevated levels of ozone on tropical wheat using integrated phenotypical, physiological, biochemical, and proteomics approaches. Journal of Proteome Research 9: 4565-4584.

Shi, G.Y., Yang, L.X., Wang, Y.X., Kobayashi, K., Zhu, J.G., Tang, H.Y., Pan, S.T., Chen, T., Liu, G. and Wang, Y.L., 2009. Impact of elevated ozone concentration on yield of four Chinese rice cultivars under fully open-air field conditions. Agriculture Ecosystems and Environment 131: 178-184.

Singh, E., Tiwari, S. and Agrawal, M., 2010. Variability in antioxidant and metabolite levels, growth and yield of two soybean varieties: an assessment of anticipated yield losses under projected elevation of ozone. Agriculture Ecosystems and Environment 135: 168-177.

Tao, Y.Q., Zhang, S.Z., Zhu, Y.G. and Christie, P., 2009. Uptake and acropetal translocation of polycyclic aromatic hydrocarbons by wheat (*Triticum aestivum* L.) grown in field-contaminated soil. Environmental Science and Technology 43: 3556-3560.

Verma, S. and Dubey, R.S., 2003. Lead toxicity induces lipid peroxidation and alters the activities of antioxidant enzymes in growing rice plants. Plant Science 164: 645-655.

Vingarzan, R., 2004. A review of surface ozone background levels and trends. Atmospheric Environment 38: 3431-3442.

Wang, H.X., Zhou, L.J. and Tang, X.Y., 2006. Ozone concentrations in rural regions of the Yangtze Delta in China. Journal of Atmospheric Chemistry 54: 255-265.

Wang, X.K., Manning, W., Feng, Z.W. and Zhu, Y.G., 2007. Ground-level ozone in China: distribution and effects on crop yields. Environmental Pollution 147: 394-400.

Wang, X.K., Zheng, Q.W., Yao, F.F., Chen, Z., Feng, Z.Z. and Manning, W.J., 2007. Assessing the impact of ambient ozone on growth and yield of a rice (*Oryza sativa* L.) and a wheat (*Triticum aestivum* L.) cultivar grown in the Yangtze Delta, China, using three rates of application of ethylenediurea (EDU). Environmental Pollution 148: 390-395.

Ye, B.X., Zhang, Z.H. and Mao, T., 2006. Pollution sources identification of polycyclic aromatic hydrocarbons of soils in Tianjin area, China. Chemosphere 64: 525-534.

Zhou, X.J., 2004. The spatial and temporal variation of trace gases in the lower atmosphere in Yangze Delta as well as the corresponding mechanism. In: Liu, P. and Wang, G.M. (eds.) Study of the interaction between the lower atmosphere and ecosystem in Yangze Delta of China. Monographs of Researches Sponsored by Natural Science Foundation of China-earth Science Series. Meteorology Press, Beijing, China, pp. 48-155.

Chapter 7.

Farmyard manure application mitigates greenhouse gas emissions from managed grasslands in Japan

M. Shimizu[1], R. Hatano[1], T. Arita[2], Y. Kouda[2], A. Mori[3], S. Matsuura[3], M. Niimi[4], M. Mano[5], R. Hirata[5], T. Jin[1], A. Limin[1], T. Saigusa[2], O. Kawamura[4], M. Hojito[3,6] and A. Miyata[5]

[1]Hokkaido University, Kita9 Nishi9, Kita-ku, Sapporo, Hokkaido 060-8589, Japan; [2]Hokkaido Research Organization, 7 Asahigaoka, Nakashibetsu, Shibetsu-gun, Hokkaido 086-1135, Japan; [3]NARO Institute of Livestock and Grassland Science, 768 Senbonmatsu, Nasushiobara, Tochigi 329-2793, Japan; [4]Miyazaki University, Nishi 1-1, Gakuen-kibanadai, Miyazaki, Miyazaki 889-2192 Japan; [5]National Institute for Agro-Environmental Sciences, 3-1-3, Kannondai, Tsukuba, Ibaraki Prefecture 305-8604, Japan; [6]Kitasato University, Higashi 23-35-1, Towada, Aomori 034-8628, Japan; hatano@chem.agr.hokudai.ac.jp

Abstract

Applying manure reduces the requirement for mineral nitrogen (N) fertilizer use and also increases carbon (C) sequestration in soils. This study was conducted to quantify the effect of manure application on the net ecosystem carbon balance (NECB), methane (CH_4) and nitrous oxide (N_2O) emissions at four managed grasslands in different climatic regions in Japan. At each site, two adjoining treatments were established, one for the application of farmyard manure, and the other for the application of mineral fertilizer. Net ecosystem exchange (NEE) of carbon dioxide (CO_2) was quantified using eddy covariance, and CH_4 and N_2O emissions were quantified using static chambers from 2005 to 2007. The manure application rate was based on the potassium demand of grasses. NECB was estimated as NEE minus C applied in manure plus harvested C. NECB in the fertilizer and manure treatments were 1.9±0.9 and -1.8±1.8 Mg C/ha/y, respectively, indicating that the manure plots gained C. Harvested C in the fertilizer and manure treatments were not significantly different (P<0.05). NEE showed a greater CO_2 uptake in the fertilizer treatments (-2.4±1.1 Mg C/ha/y) than in the manure treatments (-1.6±0.7 Mg C/ha/y), but C input through manure application could compensate the shortage in NEE. The emission of CH_4 was close to zero in the fertilizer and manure treatments (0.01±1.25 and -0.22±0.23 kg C/ha/y, respectively). Significant differences (P<0.05) in N_2O emissions were found between the fertilizer and manure treatments (3.6±3.2 and 6.2±3.7 kg N/ha/y, respectively). The global warming potential was significantly higher in the fertilizer treatments than

DOI 10.3920/978-90-8686-788-2_7,

in the manure treatments. Results from this study suggested that the application of manure may mitigate global warming in Japanese grasslands.

Keywords: global warming potential, net ecosystem exchange, carbon dioxide, methane, nitrous oxide

7.1 Introduction

Grasslands are one of the world's most abundant land cover types and comprise ~41% of the earth's terrestrial land area, excluding areas of permanent ice cover (Adams *et al.,* 1990; White *et al.,* 2000). Grassland ecosystems are important ecological and socio-economic resources through their contribution to biodiversity and their use for food, forage, and livestock production (Soussana *et al.,* 2007; White *et al.,* 2000). In order to maintain grassland productivity, particularly in developed countries, mineral fertilizers or livestock manure are applied (Bouwman *et al.,* 2002b). In Japan, livestock husbandry has depended on enormous amounts of imported feed while livestock excreta are in surplus (Mishima, 2001; Mishima *et al.,* 2006). Therefore, manure should be efficiently and appropriately returned to cropped fields as an environmentally sustainable agroecosystem management practice. If managed properly, manure can be an excellent substitute for mineral fertilizer (Bhogal *et al.,* 2011; Bulluck *et al.,* 2002).

Grasslands also contribute to the biosphere-atmosphere exchange of greenhouse gases (GHG), with their fluxes intimately linked to management practices. Grassland soils have a large stock of carbon (C), and thus can act as a net sink for atmospheric CO_2 (Soussana *et al.,* 2004). Primary productivity of grasslands may be improved through the addition of nitrogen (N) fertilizer, which increases belowground C storage (Conant *et al.,* 2001). In addition, manure application increases C input to soil and consequently enhances SOC concentration (Lal, 2004).

Native grasslands can also be a methane (CH_4) sink, where CH_4 is oxidized to CO_2 or assimilated into the microbial biomass in aerobic condition (Mosier *et al.,* 1991). Application of mineral fertilizer may decrease CH_4 uptake (Mosier *et al.,* 1991), and the form of applied N can have a significant effect on CH_4 metabolism. A previous study reported that ammonium-based fertilizer inhibited CH_4 oxidation, whereas the application of farmyard manure that contained more N than the mineral fertilizer had no inhibitory effects (Powlson *et al.,* 1997). However, CH_4 is controlled by soil moisture, soil temperature, soil pH, and the type of plant species, in addition to the application of mineral fertilizer (Kammann *et al.,* 2001; Le Mer and Roger, 2001; Mori *et al.,* 2005; Mosier *et al.,* 1991, 1996; Powlson *et al.,* 1997).

In grassland soil, nitrous oxide (N_2O) is produced as a by-product and intermediate of microbial nitrification or denitrification processes (Bouwman, 1990). These processes require mineral N (ammonium and nitrate) as a substrate, and are controlled by soil moisture content, temperature, pH and organic C (Bouwman, 1990; Maag and Vinther, 1996; Mosier, 1998; Yamulki *et al.,* 1997). The application and type of mineral N fertilizer or livestock manure could augment N_2O emissions (Akiyama and Tsuruta, 2003; Akiyama *et al.,* 2006; Bouwman *et al.,* 2002a; Mori and Hojito, 2012). Manure with a high and easily mineralizable organic C fraction stimulates microbial activity and thus N_2O emissions (Chadwick *et al.,* 2000; Drury *et al.,* 1991).

In order to understand the influence of different grassland management practices, an in-depth knowledge on GHG fluxes is required to foster smart environmental decision-making. In this study, two experimental treatments, one receiving mineral fertilizer and the other receiving manure, were set up in four grasslands straddling different climatic regions in Japan. Net ecosystem C balance (NECB), CH_4 and N_2O emissions were measured for three years. The objective of this study was to quantify the effect of manure application on the NECB, CH_4 and N_2O emission in Japanese managed grasslands.

7.2 Materials and methods

Study site description

This study was conducted in four grasslands located in different climatic regions in Japan, ranging from warm to cool temperate sites. The grasslands were located at Nakashibetsu (NKS; 43°32'N, 144°58'E), Shizunai (SZN; 42°26'N, 142°29'E), Nasushiobara (NSS; 36°55'N, 139°56'E) and Kobayashi (KBY; 31°58'N, 130°56'E). Annual mean air temperature ranged from 5.2 to 15.8 °C, and annual precipitation ranged from 1,160 to 2,595 mm/y at the various sites (Table 7.1).

Soils at all sites were classified as Andosols, and were categorized as well-drained except for the SZN site. At the SZN site, a mottled layer was observed 36 cm below the soil surface, resulting in this soil to be classified as poorly-drained. The pH of the topsoil ranged from 5.3 to 6.1 before the treatments were initiated (Table 7.1). The ranges of soil total C and total N concentration of the topsoil were 36.8-65.2 g/kg and 3.31-5.10 g/kg, respectively (Table 7.1). The dominant grass species were timothy (*Phleum pratense* L.) at the NKS site; reed canary grass (*Phalaris arundinacea* L.) and meadow foxtail (*Alopecurus pratensis* L.) at the SZN site; orchard grass (*Dactylis glomerata* L.) and Italian ryegrass (*Lolium multiflorum* Lam.) at the NSS site; and Italian ryegrass and southern crabgrass (*Digitaria ciliaris* (Retz.) Koeler) at the KBY site (Table 7.1).

Table 7.1. Site locations and their soil chemical and physical characteristics in Nakashibetsu, Shizunai, Nasushiobara and Kobayashi, Japan.

Site	Nakashibetsu	Shizunai	Nasushiobara	Kobayashi
Latitude, longitude	43°32'N, 144°58'E	42°26'N, 142°29'E	36°55'N, 139°56'E	31°58'N, 130°56'E
Mean temperature (°C)	5.2	7.9	12.2	15.8
Mean precipitation (mm)	1,160	1,365	1,561	2,595
Soil type	Andosol	Andosol	Andosol	Andosol
Soil drainage	well-drained	poorly-drained	well-drained	well-drained
Depth of topsoil layer (cm)	25	19	13	15
pH [1]	5.8	5.3	5.3	6.1
Total C (g/kg) [1]	62.5	36.8	49.2	65.2
Total N (g/kg) [1]	4.53	3.31	3.43	5.10
Bulk density (kg/m^3) [1]	658	777	1,026	833
Sand (g/g) [1]	0.58	0.77	0.60	0.67
Silt (g/g) [1]	0.29	0.19	0.22	0.23
Clay (g/g) [1]	0.12	0.037	0.18	0.095
Dominant species	timothy (*Phleum pratense* L.)	reed canary grass (*Phalaris arundinacea* L.) and meadow foxtail (*Aleopecurus pratensis* L.)	orchard grass (*Dactylis glomerata* L.) and Italian ryegrass (*Lolium multiflorum* Lam.)	Italian ryegrass (*Lolium multiflorum* Lam.) and southern crabgrass (*Digitaria ciliaris* (Retz.) Koeler)

[1] Average of the fertilizer and manure plots for topsoil layer (Ap horizon).

All sites were managed as hay meadow, with no animal grazing throughout the study period. Grass was harvested twice per year at the NKS and SZN sites and three or four times per year at the NSS and KBY sites. During the study period, over-seeding was conducted at the NSS site in September 2007 and at KBY in October 2007 to maintain grass productivity.

Management practices

The study was conducted from October 2005 to December 2007 (Table 7.2) and consisted of two treatments at each of the four sites. The first treatment included the application of mineral fertilizer (fertilizer treatment) and the second treatment consisted of cattle manure plus mineral fertilizer (manure treatment). Mineral fertilizer was applied prior to the initiation of grass growth and after each harvest. The mean

Table 7.2. Average annual air temperature (T_a), precipitation (P_{recip}), and mineral fertilizer and manure application rates in Nakashibetsu, Shizunai, Nasushiobara and Kobayashi, Japan.

Year	T_a (°C)	P_{recip} (mm)	Fertilizer N (kg/ha/y)	Manure C (Mg C/ha/y)	Manure N (kg N/ha/y)	Manure C/N	Fertilizer N (kg N/ha/y)
Nakashibetsu			ammonium and urea	beef cattle manure with bark			
2005	5.6	1,008	98	5.3	208	25	74
2006	6.1	1,069	100	4.7	271	17	80
2007	6.2	1,293	100	4.7	242	24	37
Shizunai			ammonium	beef cattle manure with bark			
2005	8.4	1,066	164	5.8	236	25	133
2006	8.1	1,167	183	6.0	310	19	130
2007	9.1	981	74	7.7	331	23	21
Nasushiobara			ammonium	dairy cattle manure with bark			
2005	12.3	1,498	209	1.9	76	24	210
2006	11.7	1,756	198	3.8	195	19	155
			ammonium	dairy cattle manure with sawdust			
2007	12.6	1,747	186	3.6	123	29	157
Kobayashi			ammonium and urea	beef cattle manure with chaff			
2005	16.4	2,130	164	1.9	119	16	141
2006	16.1	3,085	233	3.4	284	12	176
2007	16.8	2,385	218	3.5	242	14	164

application rates of mineral N in the fertilizer treatments varied depending on the study site (Table 7.2) and ranged from 99±1 kg N/ha/y (NKS), 140±58 kg N/ha/y (SZN), 198±12 kg N/ha/y (NSS), to 205±36 kg N/ha/y (KBY). Ammonium-based fertilizer or a combination of ammonium and urea were applied. Beef cattle or dairy cow manure was applied to the manure treatments once per year after the last grass harvest in the NKS, NSS and KBY sites, and in the spring at the SZN site. The application rate of manure ranged from 10 to 46 Mg/ha/y by fresh weight. Differences in manure application rates depended on the site and that site's requirement for potassium (K). The nutrient supply rate from manure was estimated by multiplying the application rate by the mineralization rate, and the difference between the supply rate from manure and the application rate in the fertilizer treatment was supplied with mineral fertilizer. The N mineralization rate was estimated based on Uchida's model (Shiga *et al.,* 1985), which was specifically developed for Japanese conditions. According to this model, the cumulative N mineralization rate $(1 - y_t)$ was estimated to be 0.132 in the applied year (t=1), 0.202 in two years (t=2) and 0.257 in three years (t=3), respectively. The equation is as follows:

$$y_t = a \times 0.01t + c \times 0.63t + f \times 0.955t \quad (1)$$

where y_t is the rate of N remaining in the manure t years after application, t is the years after manure application (inclusive of the applied year), a, c and f are the rates of organic matter fractions with different decomposition rates (a + c + f = 1). In the case of cattle manure, a, c and f were estimated to be 0.04, 0.15 and 0.81, respectively. Therefore, annual N mineralization rates were 13.2, 7.02 and 5.53% of total manure N in the first to third years after manure application, respectively. The mineralization rate of phosphorus (P) and K from the manure was estimated based on a handbook on animal waste management and utilization in Hokkaido published in 2004 (Anonymous, 2004), namely 20% of P and 70% of K are available in the applied year, 10% of P and 10% of K are available the second year, respectively.

Environmental variables

Daily air temperature and precipitation data (Table 7.2) were either obtained from a meteorological station located at the NSS site or from the nearest Automated Meteorological Data Acquisition System (AMeDAS) station managed by the Japanese Meteorological Agency at the NKS, SZN and KBY sites.

Net ecosystem carbon balance measurements

Each treatment at the four different sites was equipped with an eddy covariance system which was installed in the autumn of 2004 for the measurement of net ecosystem exchange (NEE) of CO_2. Hirata *et al.* (2013) reported the annual NEE from 2005 to 2007 at all the sites and treatments, and the data sets were integrated into our study. Briefly, the eddy covariance system consisted of a three-dimensional sonic anemometer-thermometer (CSAT3; Campbell Scientific, Logan, UT, USA) and an open-path infrared CO_2/H_2O analyser (LI7500; LiCor, Lincoln, NE, USA) to measure flux densities of CO_2, latent heat, and sensible heat at a 30 min time step. Flux calculations, corrections and quality checks of the data followed that of Mano *et al.* (2007) and Ueyama *et al.* (2012). An additional filtering criterion was applied based on friction velocity (u*) values, and the common threshold value of 0.05 m/s for u* filtering of all the sites and treatments was determined (Hirata *et al.*, 2008). Data gaps due to inadequate quality or sensor failure were reconstructed for all treatments using a gap-filling strategy. Daytime gaps in NEE were filled using a light-photosynthesis curve expressed by an equation for a nonrectangular hyperbola (Prioul and Chartier, 1977), and night time gaps were filled using the Lloyd and Taylor (1994) equation.

Net ecosystem carbon balance (NECB) in managed grasslands included C input through manure application and C output through crop harvest and grazing as well as NEE (Shimizu *et al.*, 2009). In this study, the sites and treatments were not grazed, and NECB was estimated using the following equation:

$$\text{NECB} = \text{NEE} - \text{C applied in manure} + \text{harvested C} \quad (2)$$

A positive NECB represents a C loss from the ecosystem. The amount of biomass C removed was estimated at every harvest from the dry weight of harvested biomass per unit ground area multiplied by the C content of dry biomass. The biomass harvest data was presented in Hirata *et al.* (2013).

CH_4 and N_2O emissions and global warming potential

CH_4 and N_2O emissions from the soil were measured using static closed chambers with four or six replications per treatment. Shimizu *et al.* (2013) reported the annual CH_4 and N_2O from 2005 to 2007 at all the sites and for each treatment; and these data sets were used in our study. Briefly, CH_4 and N_2O were sampled weekly during the crop growing season, but the frequency was increased to twice per week after fertilization. The measurements were conducted monthly during the non-growing season. CH_4 concentrations were determined using a gas chromatograph equipped with a flame ionization detector (Shimadzu GC-8A, Kyoto, Japan), and N_2O concentrations were determined using a gas chromatograph equipped with an electron capture detector (Shimadzu GC-14B). The CH_4 and N_2O fluxes were quantified by linear regression. Annual CH_4 and N_2O emissions were calculated by means of trapezoidal integration of the areas under the curves.

The global warming potential (GWP) provides a measure of the cumulative radiative forcing of various greenhouse gases relative to some reference gas, usually CO_2, on a specific timescale (IPCC, 2007). We quantified the GWP on a 100-year timescale for all sites and treatments based on the contributions of individual gases, using a conversion factor of 1 for CO_2, 25 for CH_4 and 298 for N_2O (IPCC, 2007).

Statistical analyses

Statistical analyses were performed with R (version 2.15.2, R Development Core Team, Auckland, New Zealand) and Excel Statistics version 5.0 (Esumi, Tokyo, Japan). The differences in NEE, harvested C, NECB, CH_4 emission, N_2O emission and GWP between the sites and treatments were analysed with a two-way factorial analysis of variance (ANOVA), and the Tukey-Kramer multiple comparison test ($P<0.05$). Annual data were used as replicates.

7.3 Results and discussion

Net ecosystem carbon balance

There was no significant difference in harvested biomass C between sites, with values ranging from 2.9 to 5.5 Mg C/ha/y (Table 7.3). Sala *et al.* (1988) reported that there was a strong linear correlation between aboveground net primary production and annual precipitation in a grassland ecosystem. Their conclusion was based on data of 9,498 different natural grasslands in the United States, suggesting that grassland productivity was strongly controlled by precipitation. Gilmanov *et al.* (2007) also reported a strong correlation between the amount of precipitation and gross primary production in European grasslands. However, they noted that this correlation only occurred when precipitation ranged from 300 to 800 mm/y, but found no such correlation when precipitation was greater than 800 mm/y. In our study, annual precipitation ranged from 981 to 3,085 mm/y, suggesting that production may not be limited by precipitation. The harvested biomass C in the mineral fertilizer and manure treatments was not significantly different and ranged from 4.3±0.8 Mg C/ha/y (manure) and 4.1±0.6 Mg C/ha/y (mineral fertilizer) (Table 7.3).

There was no significant difference in NEE between sites, but a significant difference was found between treatments (Table 7.3). The NEE in the mineral fertilizer treatment was -2.4±1.1 Mg C/ha/y compared to -1.6±0.7 Mg C/ha/y in the manure treatment. This indicated more net CO_2 uptake in the fertilizer treatment. However, there was no significant difference in grass production between treatments because the harvested C in both treatments was similar. Therefore, differences in NEE between treatments may be caused by organic matter decomposition. The decomposition of the manure likely led to an increase of organic matter decomposition and therefore decreased net CO_2 uptake. The difference in NEE between treatments was 0.9±0.9 Mg C/ha/y, and the rate of C loss as CO_2 from total amount C applied in manure in three years was estimated as 25±37%. However, the difference in NEE showed no clear correlation with C from the applied manure (Figure 7.1b). This is because the decomposition rate of manure

Table 7.3. Harvested C, net ecosystem exchange (NEE) of CO_2, net ecosystem carbon balance (NECB), CH_4 emission, N_2O emissions and global warming potential (GWP) at the fertilizer (F) and manure (M) treatments in Nakashibetsu (NKS), Shizunai (SZN), Nasushiobara (NSS) and Kobayashi (KBY), Japan.

Site	Year	Harvested C (Mg C/ha/y)		NEE (Mg C/ha/y)		NECB (Mg C/ha/y)		CH_4 emission (kg C/ha/y)		N_2O emission (kg N/ha/y)		GWP (Mg CO_2e/ha/y)	
		F	M	F	M	F	M	F	M	F	M	F	M
NKS	2005	3.6	3.3	-2.3	-0.8	1.3	-2.8	-0.34	-0.25	0.3	1.6	1.3	-2.6
	2006	4.1	4.5	-2.1	-1.6	2.0	-1.8	0.10	-0.10	0.6	9.5	2.1	-0.6
	2007	3.7	3.2	-1.7	-0.9	2.0	-2.4	-0.51	-0.31	1.1	0.7	2.2	-2.3
SZN	2005	5.4	4.0	-3.5	-1.8	1.9	-3.6	-0.11	0.02	2.8	3.6	2.2	-3.2
	2006	5.1	5.0	-3.4	-2.6	1.7	-3.6	-0.02	-0.04	2.9	4.9	2.1	-3.0
	2007	4.0	4.1	-2.8	-1.4	1.2	-5.0	3.89	0.21	1.5	2.2	1.4	-4.7
NSS	2005	4.9	5.2	-3.5	-2.4	1.4	0.9	-0.81	-0.66	4.7	8.3	2.0	2.0
	2006	4.1	4.3	-2.1	-1.6	2.1	-1.1	-0.55	-0.18	9.1	9.0	3.2	0.1
	2007	3.9	3.9	-1.1	-0.1	2.8	0.2	-0.38	-0.41	10.1	11.8	4.1	1.7
KBY	2005	4.5	3.9	-4.4	-2.2	0.1	-0.2	-0.31	-0.27	1.5	10.8	0.3	1.2
	2006	5.5	4.5	-2.1	-1.6	3.5	-0.5	-0.26	-0.24	2.9	5.3	3.8	0.2
	2007	2.9	3.3	-0.5	-1.9	2.4	-2.0	-0.56	-0.46	5.3	6.3	3.1	-1.2
Mean		4.3	4.1	-2.4	-1.6	1.9	-1.8	0.01	-0.22	3.6	6.2	2.3	-1.0
Standard deviation		0.8	0.6	1.1	0.7	0.9	1.8	1.25	0.23	3.2	3.7	1.1	2.1

ANOVA		Harvested C		NEE		NECB		CH_4 emission		N_2O emission		GWP	
	df	*F*-value	*P*-value	*F*-value	*P*-value	*F*-value	*P*-value	*F*-value	*P*-value	*F*-value	*P*-value	*F*-value	*P*-value
Site	3	1.60	0.23	1.33	0.30	7.49	<0.01	2.37	0.11	8.41	<0.01	9.02	<0.01
Treatment	1	0.50	0.49	4.87	<0.05	99.03	<0.01	0.48	0.50	6.51	<0.05	56.35	<0.01
Site × treatment	3	0.23	0.88	0.21	0.89	4.43	<0.05	0.89	0.47	0.48	0.70	3.37	<0.05

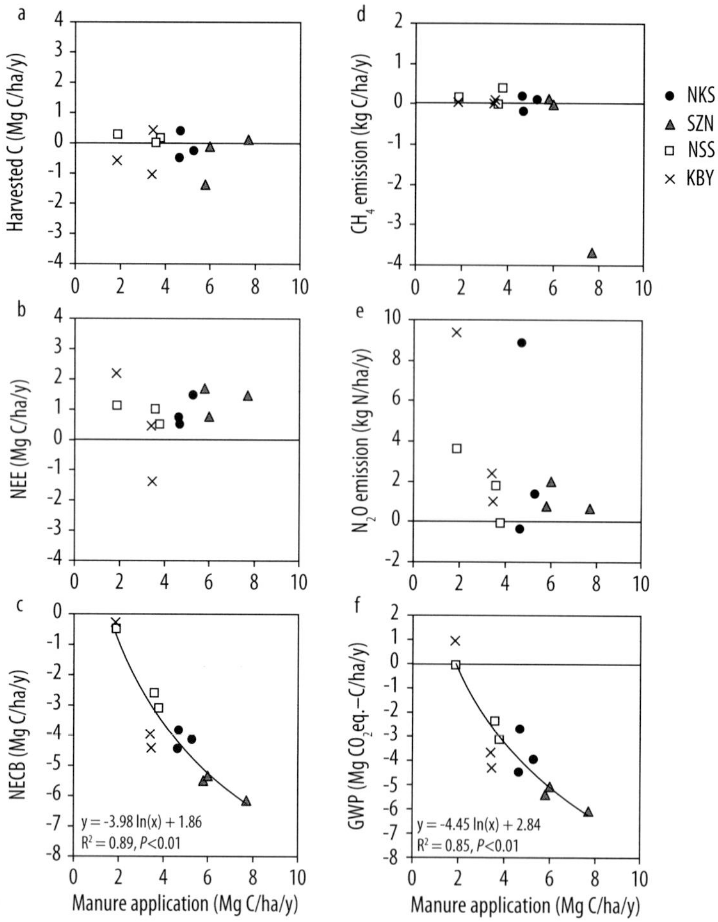

Figure 7.1. (a) The difference in harvested C, (b) net ecosystem exchange (NEE) of CO_2, (c) net ecosystem carbon balance (NECB), (d) CH_4 emissions, (e) N_2O emissions and, (f) global warming potential (GWP) in plots amended with mineral fertilizer or manure in Nakashibetsu (NKS), Shizunai (SZN), Nasushiobara (NSS) and Kobayashi (KBY), Japan.

may be intricately affected by climate and quality of the manure. Higher temperatures promote organic matter decomposition (Kirschbaum, 1995). Murwira *et al.* (1990) reported that manure decomposition increased with increasing moisture content up to an optimum level and then decreased. In addition, the chemical composition of the manure can also influence the rate of manure decomposition. For example, Delve *et al.* (2001) found that the rate of manure mineralization was negatively correlated to the manure's C/N ratio. Additionally, Toma and Hatano (2007) found that CO_2 emission from residues with a low C/N ratio was high.

There were significant differences in NECB between sites (P<0.01), and between treatments (P<0.01) (Table 7.3). There was a significant interaction between sites and treatments (P<0.05). For all the sites, the NECB in the mineral fertilizer treatments was higher than that in the manure treatments. The NECB in the fertilizer and manure treatments was 1.9±0.9 and -1.8±1.8 Mg C/ha/y, respectively. These results indicated that the grassland at all sites may be a C source to the atmosphere without manure application. The NECB in the manure treatments were highest at the NSS site, followed by the KBY, NKS and SZN sites. The NECB between the manure and fertilizer treatments was negatively correlated with the application rate of C from manure (y = -3.98 ln(x) + 1.86; R^2=0.89; P<0.01) (Figure 7.1c). This indicated that an increasing application rate of C from manure may decrease NECB. However, increasing application rate of manure over the long-term may result in the excessive accumulation of some nutrients, a greater runoff of P, and leaching of N (Edmeades, 2003). Therefore, it is important to consider the quantity of nutrients removed by crops during growth and harvest compared to that added by manure.

CH_4 emissions

There was no significant difference in CH_4 emissions between sites, with values ranging from -0.8 to 0.1 kg C/ha/y (Table 7.3). CH_4 production and consumption may be strongly controlled by soil moisture (Van den Pol-Van Dasselaar *et al.,* 1998). Kammann *et al.* (2001) observed that CH_4 uptake rates decreased with the increasing mean soil moisture at a semi-wet grassland site near Giessen, Germany. In our study, soil at the SZN site was poorly drained (Table 7.1), and the soil moisture was greater compared to the other sites (Shimizu *et al.,* 2013). The large CH_4 emissions at the SZN site, which ranged from -0.1 to 3.9 kg C/ha/y, could have been caused by the high soil moisture content.

The CH_4 emission in the fertilizer and manure treatments were 0.01±1.25 and -0.22±0.23 kg C/ha/y, respectively, and there was no significant difference between treatments (Table 7.3). The application of fertilizer N can reduce CH_4 uptake (Mosier *et al.*, 1991; Mosier *et al.,* 1996). However, in our study, there was no correlation between CH_4 emission and the application rate of N from mineral fertilizer, including control treatments that received neither mineral fertilizer nor manure (Shimizu *et al.,* 2013). Kammann *et al.* (2001) also reported that no inhibition of CH_4 oxidation occurred regardless of the amount of N fertilizer applied, while the average CH_4 uptake rates were decreased with the increasing mean soil moisture. However, at SZN site, the CH_4

emission in the fertilizer treatment was ~20 times higher than that in the manure treatment in 2007. At the SZN site, the annual net CH_4 emission was related to high CH_4 emissions after heavy rainfall (Shimizu *et al.*, 2013). Increased organic matter inputs and stimulation of soil microbial and faunal activity may favour improved soil aggregation and increased soil porosity, infiltration capacity and hydraulic conductivity (Haynes and Naidu, 1998). We speculated that increased hydraulic conductivity might block the formation of hotspots of CH_4 production in the manure treatment at the SZN site.

N_2O emissions

There was a significant difference in N_2O emission between sites ($P<0.01$) (Table 7.3). The N_2O emission in the NSS site ranged from 4.7 to 11.8 kg N/ha/y, and was followed by KBY (from 1.5 to 10.8 kg N/ha/y), SZN (from 1.5 to 4.9 kg N/ha/y) and NKS (from 0.3 to 9.5 kg N/ha/y) sites. Most likely the higher N_2O emissions in the NSS and KBY sites were due to higher precipitation resulting in an enhanced denitrification rate. Shimizu *et al.* (2013) reported that background N_2O emission and fertilizer induced N_2O emission were positively correlated to precipitation. In soils, N_2O is produced as a by-product and intermediate of microbial nitrification and denitrification processes, respectively (Bouwman, 1990). Shimizu *et al.* (2010) reported that N_2O at the SZN site was produced mainly by denitrification. Clayton *et al.* (1997) and Dobbie *et al.* (1999) also reported that high N_2O emissions occurred at a high moisture content, and suggested that higher N_2O fluxes were most likely due to processes of denitrification. In addition, Mori and Hojito (2011) reported that the cumulative precipitation during the 18-day period before and the 12-day period after N application was an important control factor for N_2O emissions. This indicated that the amount of precipitation and the timing of N application, with respect to precipitation, may control N_2O emission, and may have contributed to the inter-annual and inter-site differences observed in our study.

N_2O emission from the manure treatment was 6.2±3.7 kg N/ha/y, and 3.6± 3.2 kg N/ha/y from the mineral fertilizer treatment. Compared to N_2O emission induced by mineral fertilizer, relatively low N_2O emissions (<1% of the N applied) were detected from for animal manures applied to grassland (Akiyama and Tsuruta, 2003; Chadwick *et al.*, 2000; Mori and Hojito, 2011; Velthof *et al.*, 1997). Akiyama and Tsuruta (2003) reported that N_2O emissions were 0.08 and 0.05% of the applied N for dried cattle excreta (C/N ratio of 24.3) and for cattle manure (C/N ratio of 15.9), respectively. However, in our study, mineral fertilizer in the manure treatments was used to prevent excessive application of K from manure, and this was also reflective of actual farming practices used in these regions of Japan. Available C from manure and N from mineral fertilizer may have enhanced the denitrification rate and N_2O production. This may have increased N_2O emissions in the manure treatments. Jin *et al.* (2010) reported that the activity of denitrification enzymes at the SZN site was higher in the manure treatment than in the mineral fertilizer treatment due to a higher soil pH.

In this study, the NEE was lower in the manure treatment than the mineral fertilizer treatment due to the slow rate of decomposition of manure (Ginting *et al.,* 2003). This suggested that manure application may increase organic C and N in soil, but increasing organic matter stocks in soils may promote CO_2 emissions from soil with high turnover of C and N. It will also promote N_2O emissions by providing substrates for nitrification and denitrification. Jager *et al.* (2011) reported that CO_2 and N_2O emission from arable soils increased with increasing soil C and N stocks when manure was applied annually over the long-term (Baer and Blair, 2008).

Global warming potential

There were significant differences in the GWP between the sites (P<0.01), and between the treatments (P<0.01) (Table 7.3). In addition, there was a significant interaction between sites and treatments (P<0.05). The GWP was higher in the mineral fertilizer treatments than in the manure treatments. The NECB was a major contributor to the GWP (Figure 7.2). In the mineral fertilizer treatment, the harvested biomass C was greater than the NEE, leading to a positive GWP (2.3±1.1 Mg CO_2eq C/ha/y). However in the manure treatments, the sum of the C applied in manure and NEE was greater than harvested biomass C, leading to a negative GWP (-1.0±2.1 Mg CO_2eq C/ha/y). N_2O emission contributed significantly to GWP, while the contribution of CH_4 emission was negligible.

In the fertilizer treatment, there was no significant difference in GWP between sites. On the other hand, the GWP in the manure treatments was highest in the NSS site followed by the KBY, NKS and SZN sites. The difference of GWP between the manure and fertilizer treatments had a negative relationship with the application rate of manure C (y = -4.45 ln(x) + 2.84; R^2=0.85; P<0.01) (Figure 7.1f). This indicated that an increasing application rate of manure C may mitigate global warming at least at the site-scale. However, manure C application cannot be effective in GWP mitigation at low application. This suggested that the optimal manure C application rate for this region of Japan is 1.9 Mg C/ha/y.

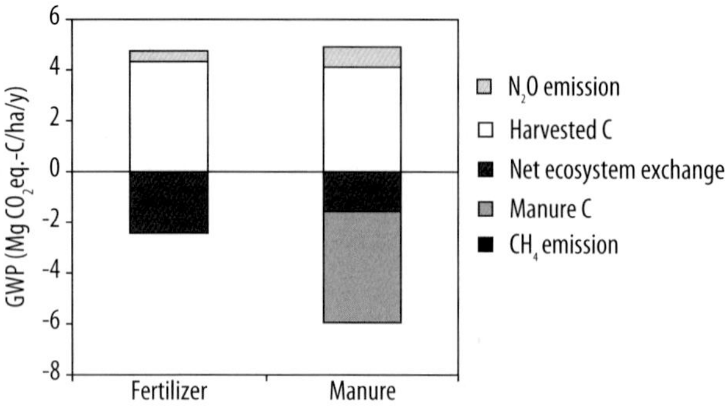

Figure 7.2. Contribution of the each global warming potential (GWP) component in plots that received mineral fertilizer and manure. The GWP of CH_4 was close to zero in the fertilizer and manure plots (0.0001±0.011 and -0.002±0.002 Mg CO_2eq.-C/ha/y, respectively).

Acknowledgements

This study was partly supported by a research grant provided by the Project entitled 'Establishment of good practices to mitigate Greenhouse Gas emissions from Japanese grasslands' (FY 2004-2009), which was implemented by the Japan Grassland Agriculture and Forage Seed Association (GAFSA) and funded by the Japan Racing Association. M. Oikawa and T. Mitamura (GAFSA) provided support throughout the study. This study was partly funded by the Ministry of Agriculture, Forestry and Fisheries, Japan through a research project entitled 'Development of technologies for mitigation and adaptation to climate change in Agriculture, Forestry and Fisheries' (FY 2010-2014).

References

Adams, J.M., Faure, H., Fauredenard, L., Mcglade, J.M. and Woodward, F.I., 1990. Increases in terrestrial carbon storage from the last glacial maximum to the present. Nature 348: 711-714.

Akiyama, H. and Tsuruta, H., 2003. Effect of organic matter application on N_2O, NO, and NO_2 fluxes from an Andisol field. Global Biogeochemical Cycles 17: 1100-1111.

Akiyama, H., Yan, X.Y. and Yagi, K., 2006. Estimations of emission factors for fertilizer-induced direct N_2O emissions from agricultural soils in Japan: Summary of available data. Soil Science and Plant Nutrition 52: 774-787.

Anonymous, 2004. Handbook of animal waste management and utilization in Hokkaido. Hokkaido Prefectural Experiment Stations and Hokkaido Animal Research Center, Sapporo, Japan, pp. 64-67.

Baer, S.G. and Blair, J.M., 2008. Grassland establishment under varying resource availability: a test of positive and negative feedback. Ecology 89: 1859-1871.

Bhogal, A., Nicholson, F.A., Young, I., Sturrock, C., Whitmore, A.P. and Chambers, B.J., 2011. Effects of recent and accumulated livestock manure carbon additions on soil fertility and quality. European Journal of Soil Biology 62: 174-181.

Bouwman, A.F., 1990. Exchange of greenhouse gases between terrestrial ecosystems and the atmosphere. In: Bouwman A.F (ed.) Soils and the greenhouse effect. John Wiley and Sons, Chichester, UK, pp. 61-127.

Bouwman, A.F., Boumans, L.J.M. and Batjes, N.H., 2002a. Emissions of N_2O and NO from fertilized fields: summary of available measurement data. Global Biogeochemical Cycles 16: 1058-1067.

Bouwman, A.F., Boumans, L.J.M. and Batjes, N.H., 2002b. Modeling global annual N_2O and NO emissions from fertilized fields. Global Biogeochemical Cycles 16: 1080-1091.

Bulluck, L.R., Brosius, M., Evanylo, G.K. and Ristaino, J.B., 2002. Organic and synthetic fertility amendments influence soil microbial, physical and chemical properties on organic and conventional farms. Applied Soil Ecology 19: 147-160.

Chadwick, D.R., Pain, B.F. and Brookman, S.K.E., 2000. Nitrous oxide and methane emissions following application of animal manure to grassland. Journal of Environmental Quality 29: 277-287.

Clayton, H., McTaggart, I.P., Parker, J., Swan, L. and Smith, K.A., 1997. Nitrous oxide emissions from fertilised grassland: a 2-year study of the effects of N fertiliser form and environmental conditions. Biology and Fertility of Soils 25: 252-260.

Conant, R.T., Paustian, K. and Elliott, E.T., 2001. Grassland management and conversion into grassland: effects on soil carbon. Ecological Applications 11: 343-355.

Delve, R.J., Cadisch, G., Tanner, J.C., Thorpe, W., Thorne, P.J. and Giller, K.E., 2001. Implications of livestock feeding management on soil fertility in the smallholder farming systems of sub-Saharan Africa. Agriculture, Ecosystems and Environment 84: 227-243.

Dobbie, K.E., McTaggart, I.P. and Smith, K.A., 1999. Nitrous oxide emissions from intensive agricultural systems: variations between crops and seasons, key driving variables, and mean emission factors. Journal of Geophysical Research 104 (D21): 26891-26899.

Drury, C.F., McKenney, D.J. and Findlay, W.I., 1991. Relationships between denitrification, microbial biomass and indigenous soil properties. Soil Biology and Biochemistry 23: 751-755.

Edmeades, D.C., 2003. The long-term effects of manures and fertilisers on soil productivity and quality: a review. Nutrient Cycling in Agroecosystems 66: 165-180.

Gilmanov, T.G., Soussana, J.F., Aires, L., Allard, V., Ammann, C., Balzarolo, M., Barcza, Z., Bernhofer, C., Campbell, C.L., Cernusca, A., Cescatti, A., Clifton-Brown, J., Dirks, B.O.M., Dore, S., Eugster, W, Fuhrer, J., Gimeno, C., Gruenwald, T., Haszpra, L., Hensen, A., Ibrom, A., Jacobs, A.F.G., Jones, M.B., Lanigan, G., Laurila, T., Lohila, A., Manca, G., Marcolla, B., Nagy, Z., Pilegaard, K., Pinter, K., Pio, C., Raschi, A., Rogiers, N., Sanz, M.J., Stefani, P., Sutton, M., Tuba, Z., Valentini, R., Williams, M.L. and Wohlfahrt, G., 2007. Partitioning European grassland net ecosystem CO_2 exchange into gross primary productivity and ecosystem respiration using light response function analysis. Agriculture, Ecosystems and Environment 121: 93-120.

Ginting, D., Kessavalou, A., Eghball, B. and Doran, J.W., 2003. Greenhouse gas emissions and soil indicators four years after manure and compost applications. Journal of Environmental Quality 32: 23-32.

Haynes, R.J. and Naidu, R., 1998. Influence of lime, fertilizer and manure applications on soil organic matter content and soil physical conditions: a review. Nutrient Cycling in Agroecosystems 51: 123-137.

Hirata, R., Miyata, A., Mano, M., Shimizu, M., Arita, T., Kouda, Y., Matsuura, S., Niimi, M., Mori, A., Saigusa, T., Hojito, M., Kawamura, O. and Hatano, R., 2013. Carbon dioxide exchange at four intensively managed grassland sites across different climate zones of Japan and the influence of manure application on ecosystem carbon and greenhouse gas budgets. Agricultural and Forest Meteorology 177: 57-68.

Hirata, R., Saigusa, N., Yamamoto, S., Ohtani, Y., Ide, R., Asanuma, J., Gamo, M., Hirano, T., Kondo, H., Kosugi, Y., Li, S.G., Nakai, Y., Takagi, K., Tani, M. and Wang, H.M., 2008. Spatial distribution of carbon balance in forest ecosystems across East Asia. Agricultural and Forest Meteorology 148: 761-775.

Intergovernmental Panel on Climate Change (IPCC), 2007. Climate changes 2007: the physical science basis. Cambridge University Press, Cambridge, UK.

Jager, N., Stange, C.F., Ludwig, B. and Flessa, H., 2011. Emission rates of N_2O and CO_2 from soils with different organic matter content from three long-term fertilization experiments-a laboratory study. Biology and Fertility of Soils 47: 483-494.

Jin, T., Shimizu, M., Marutani, S., Desyatkin, A.R., Iizuka, N., Hata, H. and Hatano, R., 2010. Effect of chemical fertilizer and manure application on N_2O emission from reed canary grassland in Hokkaido, Japan. Soil Science and Plant Nutrition 56: 53-65.

Kammann, C., Grunhage, L., Jager, H.J. and Wachinger, G., 2001. Methane fluxes from differentially managed grassland study plots: the important role of CH_4 oxidation in grassland with a high potential for CH_4 production. Environmental Pollution 115: 261-273.

Kirschbaum, M.U.F., 1995. The temperature-dependence of soil organic-matter decomposition, and the effect of global warming on soil organic-c storage. Soil Biology and Biochemistry 27: 753-760.

Lal, R., 2004. Soil carbon sequestration to mitigate climate change. Geoderma 123: 1-22.

Le Mer, J. and Roger, P., 2001. Production, oxidation, emission and consumption of methane by soils: a review. European Journal of Soil Biology 37: 25-50.

Lloyd, J. and Taylor, J.A., 1994. On the temperature-dependence of soil respiration. Functional Ecology 8: 315-323.

Maag, M. and Vinther, F.P., 1996. Nitrous oxide emission by nitrification and denitrification in different soil types and at different soil moisture contents and temperatures. Applied Soil Ecology 4: 5-14.

Mano, M., Miyata, A., Yasuda, Y., Nagai, H., Yamada, T., Ono, K., Saito, M. and Kobayashi, Y., 2007. Quality control for the open-path eddy covariance data. Journal of Agricultural Meteorology 63: 125-138.

Mishima, S., 2001. Recent trend of nitrogen flow associated with agricultural production in Japan. Soil Science and Plant Nutrition 47: 157-166.

Mishima, S., Taniguchi, S. and Komada, M., 2006. Recent trends in nitrogen and phosphate use and balance on Japanese farmland. Soil Science and Plant Nutrition 52: 556-563.

Mori, A. and Hojito, M., 2011. Nitrous oxide and methane emissions from grassland treated with bark- or sawdust-containing manure at different rates. Soil Science and Plant Nutrition 57: 138-149.

Mori, A. and Hojito, M., 2012. Effect of combined application of manure and fertilizer on N_2O fluxes from a grassland soil in Nasu, Japan. Agriculture, Ecosystems and Environment 160: 40-50.

Mori, A., Hojito, M., Kondo, H., Matsunami, H. and Scholefield, D., 2005. Effects of plant species on CH_4 and N_2O fluxes from a volcanic grassland soil in Nasu, Japan. Soil Science and Plant Nutrition 51: 19-27.

Mosier, A.R., 1998. Soil processes and global change. Biology and Fertility of Soils 27: 221-229.

Mosier, A.R., Parton, W.J., Valentine, D.W., Ojima, D.S., Schimel, D.S. and Delgado, J.A., 1996. CH_4 and N_2O fluxes in the Colorado shortgrass steppe: 1. Impact of landscape and nitrogen addition. Global Biogeochemical Cycles 10: 387-399.

Mosier, A.R., Schimal, D., Valentine, D., Bronson, K. and Parton, W., 1991. Methane and nitrous oxide fluxes in native, fertilized and cultivated grasslands. Nature 350: 330-332.

Murwira, H.K., Kirchmann, H. and Swift, M.J., 1990. The effect of moisture on the decomposition rate of cattle manure. Plant and Soil 122, 197-199.

Powlson, D.S., Goulding, K.W.T., Willison, T.W., Webster, C.P. and Hutsch, B.W., 1997. The effect of agriculture on methane oxidation in soil. Nutrient Cycling in Agroecosystems 49: 59-70.

Prioul, J.L. and Chartier, P., 1977. Partitioning of transfer and carboxylation components of intracellular resistance to photosynthetic CO_2 fixation: a critical analysis of the methods used. Annals of Botany 41: 789-800.

R Development Core Team, 2012. R: a language and environment for statistical computing. R Foundation for Statistical Computing, Vienna, Austria. Available at: http://www.R-project.org.

Sala, O.E., Parton, W.J., Joyce, L.A. and Lauenroth, W.K., 1988. Primary production of the central grassland region of the United-States. Ecology 69: 40-45.

Shiga, H., Ohyama, N., Maeda, K. and Suzuki, M., 1985. An evaluation of different organic materials based on their decomposition pattern in paddy soils. Bulletin of National Agricultural Research Center 5: 1-19.

Shimizu, M., Hatano, R., Arita, T., Kouda, Y., Mori, A., Matsuura, S., Niimi, M., Jin, T., Desyatkin, A.R., Kawamura, O., Hojito, M. and Miyata, A., 2013. The effect of fertilizer and manure application on CH_4 and N_2O emissions from managed grasslands in Japan. Soil Science and Plant Nutrition 59: 69-86.

Shimizu, M., Marutani, S., Desyatkin, A.R., Jin, T., Hata, H. and Hatano, R., 2009. The effect of manure application on carbon dynamics and budgets in a managed grassland of Southern Hokkaido, Japan. Agriculture, Ecosystems and Environment 130: 31-40.

Shimizu, M., Marutani, S., Desyatkin, A.R., Jin, T., Nakano, K., Hata, H. and Hatano, R., 2010. Nitrous oxide emissions and nitrogen cycling in managed grassland in Southern Hokkaido, Japan. Soil Science and Plant Nutrition 56: 676-688.

Soussana, J.F., Fuhrer, J., Jones, M. and van Amstel, A., 2007. The greenhouse gas balance of grasslands in Europe. Agriculture, Ecosystems and Environment 121: 1-4.

Soussana, J.F., Loiseau, P., Vuichard, N., Ceschia, E., Balesdent, J., Chevallier, T. and Arrouays, D., 2004. Carbon cycling and sequestration opportunities in temperate grasslands. Soil Use and Management 20: 219-230.

Toma, Y. and Hatano, R., 2007. Effect of crop residue C:N rate on N_2O emissions from Gray Lowland soil in Mikasa, Hokkaido, Japan. Soil Science and Plant Nutrition 53: 198-205.

Ueyama, M., Hirata, R., Mano, M., Hamotani, K., Harazono, Y., Hirano, T., Miyata, A., Takagi, K. and Takahashi, Y., 2012. Influences of various calculation options on heat, water and carbon fluxes determined by open- and closed-path eddy covariance methods. Tellus B 64: 19048-19056.

Van den Pol-Van Dasselaar, A., Van Beusichem, M.L. and Oenema, O., 1998. Effects of soil moisture content and temperature on methane uptake by grasslands on sandy soils. Plant and Soil 204: 213-222.

Velthof, G.L., Oenema, O., Postma, R. and Van Beusichem, M.L., 1997. Effects of type and amount of applied nitrogen fertilizer on nitrous oxide fluxes from intensively managed grassland. Nutrient Cycling in Agroecosystems 46: 257-267.

White, R., Murray, S. and Rohweder, M., 2000. Pilot analysis of global ecosystems: grassland ecosystems. World Resources Institute, Washington, DC, USA.

Yamulki, S., Harrison, R.M., Goulding, K.W.T. and Webster, C.P., 1997. N_2O, NO and NO_2 fluxes from a grassland: effect of soil pH. Soil Biology and Biochemistry 29: 1199-1208.

Section 3.
Nutrient dynamics and microbe interactions

Chapter 8.

Dynamics of gross nitrogen transformations related to particle-size soil organic fractions in the southeastern Pampa of Argentina

C. Videla[1], P.C. Trivelin[2], G.A. Studdert[1] and J.A. Bendasolli[2]
[1]*Facultad de Ciencias Agrarias, UNMDP Ruta 226 Km 74.5 CC 276, 7620 Balcarce, Argentina;* [2]*CENA, University of São Paulo, Lab. de Isótopos Estáveis, Av. Centenário, n° 303, C.P. 96, 13400-970 Piracicaba, SP, Brazil; cvidela@balcarce.inta.gov.ar*

Abstract

We analysed the effect of soil management and depth on potential gross nitrogen mineralization (GMR), nitrification (GNR) and consumption rates (GCR), and their relationships with organic matter particle-size fractions (>200 μm, light and heavy, 50-200 μm, light and heavy, 20-50 μm and <20 μm). The experimental site was under conventional tillage for 18 years and subsequently converted to no till (NT), pasture (PP) or continued under conventional tillage (CT). Two sampling depths (0-10 and 10-20 cm) were analysed. Gross processes were estimated through ^{15}N dilution technique. GMR values ranged between 0.532 to 2.598 mg N/kg/d. The PP-GMR for the 0-10 cm layer was higher than CT at same depth and NT for 10-20 cm depth. No tillage showed similar results as PP, but at lower rates. Ammonium consumption was an important process for the three management systems and analysed depths. For all management systems, it was found that 82-95% of organic-carbon (C) and 87-97% of total nitrogen (N) was located in the <50 μm fraction, indicating a high physical protection of organic matter (OM) and suggesting low turnover rates. Pasture accumulated 112% more C and 116% more N than CT in the >50 μm light fraction. No-tillage was not significantly different from CT for all the analysed variables. Management systems did not modify the <50 μm fraction. Simple relationships between C, N and the C/N ratio of the OM fractions explained less than 60% of the variability on GMR and GCR. The C content in the 50 to 200 μm and >200 μm light fraction were frequently associated with GMR and GCR. When management systems were analysed separately, multiple regression models explained more than 90% of the variability. For more conservative management practices (PP and NT), the C and N contents in labile fractions (50-200 μm and >200 μm) accounted for most of the variability of GMR and GCR, while for CT, the finest and slow recycling fractions explained GMR variability.

Keywords: soil management, ^{15}N dilution technique, pasture, no tillage, conventional tillage

8.1 Introduction

Soil organic matter (SOM) decomposition and concomitant mineralization and immobilization processes are key components of the nitrogen (N) cycle in the soil-plant system. These processes are highly complex because SOM is a heterogeneous mixture of fractions ranging from partially decayed plant and animal residues recently incorporated into the soil, to very recalcitrant humic substances. Soil microbial biomass is an important component of the soil organic fraction that regulates the transformation of nutrients; and its activity and size are affected by fluctuations in temperature, moisture, pH, energy status, and substrate availability.

Net mineralization is used as an *in situ* indicator of N availability in the soil for crops (Hatch *et al.*, 1990; Stenger *et al.*, 1995), which results from a higher gross rate of mineralization (ammonification and nitrification) compared to the rate of immobilization. Through the use of the ^{15}N isotope dilution technique, it was demonstrated that gross mineralization and immobilization, which determine N availability, are processes partly independent of each other (Davidson *et al.*, 1991). Factors controlling N consumption and production do not affect net and gross N transformation rates to the same extent (Davidson *et al.*, 1992). As such, the estimation of gross N fluxes in soil allows for a better understanding of the processes controlling the N cycle and, also, N availability for plants.

When soil management practices are modified, it takes a relatively long time to detect changes in SOM content. However, some soil properties may show high sensitivity to changes in soil quality as a result of changes in land management practices. For example, several SOM fractions were proposed as potential indicators to quantify N availability to plants. This is because their measurement has proven to be sensitive to changes in soil management practices. Nitrogen content in the 150-3,000 μm light fraction (<1.13 g/cm^3) was reported as highly correlated to net N mineralization in soils under seven different agroecosystem management practices (Barrios *et al.*, 1996). Carbon concentration in the particulate organic fraction (POC) in the 212-2,000 μm (Fabrizzi *et al.*, 2003) and of 53-2,000 μm (Eiza *et al.*, 2005) size fractions were proposed as sensitive indicators on the effects of soil management in the southeastern Pampa of Argentina. Accoe *et al.* (2004) related SOM fractions with gross N processes under pastures of different ages, and found that the heavy fraction of macro-organic matter (>1.37 g/cm^3) could be used as an early indicator of SOM accumulation after the conversion of agricultural lands to permanent grasslands. They also showed that total soil organic N content explained 93% of the variability in the gross mineralization rate (GMR) (Accoe *et al.*, 2004).

The Mollisols (Phaeozems) of the southeastern Pampa of Argentina originally had high levels of soil organic C content (SOC), ranging between 25 to 45 g/kg in the top 20 cm upper soil layer (Durán *et al.,* 2011). Continuous cropping under conventional tillage (CT) decreased both labile and humified SOC fractions and has changed their relative proportion in the soil (Eiza *et al.*, 2005; Fabrizzi *et al.*, 2003; Sainz-Rozas *et al.,* 2011). Depending on the crop sequence, equilibrium SOC levels without erosion and under CT ranged between 23 and 35 g/kg in the eastern part of this region (Caride *et al.*, 2012; Studdert and Echeverría, 2000; Studdert *et al.*, 1997). As a consequence, continuous cropping under CT led to physical, chemical and biochemical soil degradation (Darwich, 1991; Duran *et al.*, 2011), along with nutrient deficiencies, especially N, for wheat (*Triticum aestivum* L.) (Berardo, 1994; Echeverría and Videla, 1998) and maize (*Zea mays* L.) crops (Andrade *et al.*, 2002).

An alternative practice proposed for this region, in order to improve SOM levels, is the inclusion of 3 to 4 years of pasture, after 6 to 7 years of cropping (Studdert *et al.*, 1997). Due to the input of plant debris from senescing grasses and slower decomposition rates, pastures may increase levels of SOM similar to those obtained at the beginning of the cropping period (Studdert *et al.*, 1997). An increase in crop N availability for crops was also observed, especially in the first years after the conversion of pastures into cropping, as a result of N mineralization in the labile fractions that originated from pasture residues (roots and aboveground biomass) (Ernst and Siri-Prieto, 2009). Another alternative to reduce soil disturbance and degradation includes no-tillage (NT), which was successfully adopted in Argentina (AAPRESID, 2013) and is widely used in the southeastern Pampa. A reduction or the elimination of soil tillage promotes aggregate formation and stabilization (Six *et al.,* 2000), and contributes to the modification of soil properties including bulk density, pore size distribution, temperature, moisture and soil air (Álvarez and Steinbach, 2009; Aparicio and Costa, 2007; Ernst and Siri-Prieto, 2009). The adoption of NT has also shown to increase SOC and soil N content, as well as microbial biomass C and N, leading to stratification of these fractions (Alvarez *et al.*, 1995; Franzluebbers, 2010; Jagadamma and Lal, 2010; Nelson *et al.,* 2006). However, a meta-analysis of 56 studies comparing NT versus CT concluded that NT did not result in a greater potential in SOC and N sequestration in the upper 30 cm in Mollisols (Puget and Lal, 2005). Similar results were reported by Eiza *et al.* (2005) and Diovisalvi *et al.* (2008) in the top 20 cm of Mollisols from the southeastern Pampa. Puget and Lal (2005) and Steinbach and Álvarez (2006) pointed out that a lower SOM content under CT, may result in a higher the capacity to sequester C and N after NT implementation. However, all studies did report an increase in SOM stock in the top 5 cm when converting from CT to NT.

Notwithstanding the increment of soil total N content due to either pasture establishment or continuous use of NT, ability of Mollisols in the southeastern Pampa to supply N to crops is more obvious when the soils are tilled (Domínguez *et al.*, 2001). Mollisols of the southeastern Pampa protect SOM strongly and it is necessary to break-up soil aggregates in order to expose these fractions. This was demonstrated by Mandiola *et al.* (2011), who showed that management systems with less tillage-induced disruption (pasture and NT) led to higher SOC and particulate organic C (POC) contents in macroaggregates and to an increase in macroaggregate stability. On the other hand, despite the better N supply, continuous disruption of macroaggregates by tillage altered aggregate formation dynamics and reduced SOC content (Six *et al.*, 2004); although this may be counterbalanced with adequate C inputs (Domínguez *et al.*, 2009). As such, sustainable soil management should take this delicate equilibrium into account. This would promote the accumulation of N and its supply to crops, leading to a reduction in the required amount of mineral fertilizer addition, and an accumulation of SOC and aggregate stability. An evaluation of gross N cycling process rates will further contribute to our knowledge of SOC and soil organic N dynamics and provide additional insight into sustainable management strategies.

The objectives of this study were to investigate the effect of management systems and soil depth on rates of gross N mineralization, nitrification and consumption using the ^{15}N-isotope dilution technique. We also analysed the effect of management systems on the distribution of organic matter in different particle-size fractions, and evaluated the relationship of C and N content between the whole soil and of the particle-size fractions, and gross N mineralization and consumption rates.

8.2 Materials and methods

Site description and soil sampling

Soil samples were taken from a long-term field experiment, established in 1976 at the Experimental Station of the National Institute of Agricultural Technology (INTA) located in Balcarce (37°45' S, 58°18' W), Buenos Aires Province, Argentina. The site was located 130 m over sea level, with an average annual precipitation of 950 mm, and an average annual temperature of 13.7 °C. The experimental area was located on a complex soil mixture classified as a Luvic Phaeozem which was part of the Mar del Plata series, and a Petrocalcic Phaeozem, which was part of the Balcarce series (INTA, 1979). The petrocalcic horizon was below 0.7 m (INTA 1979). The entire study site had a 2% slope. From 1976 to 1993 the plots were under CT using a variety of crop sequences (Table 8.1). Since 1994, three different management systems were implemented, including CT, NT and pasture (Pp), using a randomized complete block

Table 8.1. Crop sequences prior to the initiation of the current study at the Experimental Station of the National Institute of Agricultural Technology (INTA) located at Balcarce, Buenos Aires Province, Argentina.

Year[1]	Treatment[2]					
	1		2		3	
1976	P		P		Pp	
1977	W		W		Pp	
1978	0	P	Ov	P	P	
1979	W		W		W	
1980	0	P	Ov	P	0	P
1981	W		W		W	
1982	0	P	Ov	P	Pp	
1983	W		W		Pp	
1984	0	P	Ov	P	P	
1985	W		W		W	
1986	0	P	Ov	P	0	P
1987	W		W		W	
1988	0	P	Ov	P	Pp	
1989	W		W		Pp	
1990	0	P	Ov	P	P	
1991	W		W		W	
1992	0	P	Orc		0	P
1993	W		Orc		W	

[1] In 1994 sequences 1, 2 and 3 were switched to maize-soybean-wheat under NT, maize-soybean-wheat under CT and PP, respectively.

[2] P = potato (*Solanum tuberosum* L.); W = wheat; 0=oat (*Avena sativa* L.); Ov = oat + vetch (*Vicia sativa* L.); Orc = oat + red clover (*Trifolium pretense* L.); Pp = pasture.

design with three replications, and a plot size of 10×17.5 m. The crop sequence was maize – soybean (*Glycine max* L. Merr.) – spring wheat. The permanent pasture was composed of a mixture of tall fescue (*Festuca arundinacea* Schreb.) and white clover (*Trifolium repens* L.). The soil was analysed in 1994 (0-20 cm) showing a pH of 6.3 (soil (1): water (1.25)); a cation exchange capacity of 33.1 cmol (+)/kg; a clay content of 265 g/kg; a silt content of 337 g/kg an available P of 22 mg/kg (Bray and Kurtz I); a SOC content of 27.1 g/kg; and a soil total N content of 2.87 g/kg. In December 2000, soil samples were collected at depths of 0-10 and 10-20 cm using a stainless steel soil sampler, mixed, air dried, sieved through 2 mm sieve and stored until further analysis.

Isotopic dilution experiment

This experiment was carried out under laboratory conditions, at 25±2 °C temperature. In order to restore the soil biological activity and to minimize the creation of artifacts due to soil handling, soil samples were re-moistened to 60% of their maximum holding capacity and pre-incubated for 2 weeks before the initiation of the ^{15}N labelling experiment.

Potential gross mineralization and consumption were measured using the pool dilution method (Kirkham and Bartholomew, 1954). Six disposable pots (5 cm diameter, 7 cm height) were filled with 80 g of soil from each management system, soil depth and replication, and covered with perforated Parafilm® to enable gas exchange. At day 0 of the incubation, pots containing the soil were injected with a 3.5 ml $(^{15}NH_4)_2SO_4$ solution (10% enrichment), resulting in 10 μg ^{15}N/g soil. The $(^{15}NH_4)_2SO_4$ was added to the soil using a multiple point injection device containing seven hypodermic syringes with needles (18G, 1.5 inch) (Monaghan, 1995). The needles were inserted into the soil, up to 0.5 cm from the bottom of the pot. The solution was slowly injected into the soil while at the same time the needles were slowly pulled out of the soil in order to warrant a homogeneous distribution of the labelled solution within the soil mass. At 24 h (t_0) and 72 h (t_f) after ^{15}N injection, 60 g of fresh soil from three pots corresponding to each treatment replicate was removed and extracted with 200 ml of 2M KCl (60 min shaking) and filtered through Whatman No. 1 filter paper for NH^+_4, NO^-_3 and ^{15}N-analyses. Moisture content was determined gravimetrically by drying the soil at 105 °C for 48 h. Sampling times (t_0 and t_f) were selected according to a previous experiment, that determined that GMR remained constant for a 3-4 days after ^{15}N injection. However, t_0 was an adequate time for the labelled solution to be distributed homogenously in the soil and to prevent GMR underestimation associated with heterogeneous distribution (Luxhøi *et al.*, 2003).

Potassium chloride extracts were analysed for inorganic N using the steam distillation method in the presence of MgO and Devarda's alloy (Bremner, 1965). The samples were prepared for ^{15}N analysis using the polytetrafluoroethylene (PTFE) diffusion technique (Sørensen and Jensen, 1991). 60 ml of KCl extract was weighed in a plastic vessel, a PTFE envelope containing a quartz filter paper disc of 5 mm diameter acidified with 10 μl of 2.5 M $KHSO_4$ was added to the solution, approximately 0.2 g of MgO was then added to the vessel, which was closed immediately, shaken at 50 rpm over three days in the dark at 30 °C. The envelope was removed and the disc dried in a desiccator containing silica gel and an acid trap. The vessels were left open overnight. On the following day, an additional acidified disc was added to the vessel with 0.2 g of Devarda's alloy. The vessels were again closed and shaken for additional three days after which the disc was removed and dried as described before. Discs were capped into tin capsules for ^{15}N analysis. The $^{14}N/^{15}N$ ratios of NH_4^+ and NO_3^- derived N captured on the acidified discs were measured with a mass spectrometer (Europa Scientific Ltd., Krewe, UK) connected to a C and N elemental analyser (Barrie and Prosser, 1996). Gross transformation rates were calculated using the Kirkham and Bartholomew (1954) equations as presented by Barraclough *et al.* (1985):

$$GMR = \theta \cdot \frac{\ln(A^*_0 - A^*_f)}{\ln(A_0 / A_f)} \quad (1)$$

$$CR = GMR - \left(\frac{A_f - A_0}{t}\right) \quad (2)$$

where GMR is gross mineralization rate (mg N/kg/d); A^*_f is atom % ^{15}N excess of NH_4^+ pool at time f; A^*_0 is atom % ^{15}N excess of NH_4^+ pool at time 0; θ is rate of change of ammonium pool calculated as $(A_f - A_0)/t$, being A_0 and A_f the ammonium pool size (mg/kg) at t_0 and t_f, respectively; t is time between the first and the second extraction ($t_f - t_0$, d), and CR is NH_4^+ consumption rate (mg N/kg/d).

For potential gross nitrification rate (GNR) calculation, the average abundance of ammonium pool [A*] was calculated with the following equation (Barraclough, 1991):

$$[A^*] = \frac{1}{t} \cdot A^*_0 \cdot A_0 \frac{GMR}{\theta} \quad (3)$$

where [A*] is the average abundance of ammonium pool in the time t.

Particle-size fractionation of soil organic matter

A total of 30 g of air dried and sieved (2 mm) soil from each treatment and replication was dispersed in 100 ml of 5 g/l sodium hexametaphosphate [$(NaPO_3)_6$], placed in the refrigerator for 16 hours and subsequently shaken for 2 hours at 50 rpm. The dispersed soil was passed through 200 and 50 μm sieves and, and rinsed until the deionised water that passes through the sieve is free from particles (less than 1 litre). The material retained on the two sieves was allowed to settle in deionised water to separate the light fraction (LF) from the heavy fraction (HF) by flotation. The soil that passed through the 50 μm sieve was ultrasonicated at 240 W for 10 min and then sieved through a 20 μm sieve. The following six fractions were obtained: >200 μm light fraction (>200LF) composed mainly of roots fragments and aboveground residues; >200 μm heavy fraction (>200HF) composed mostly of sand particles; 50-200 μm light fraction (50-200LF) dominated by organic material not recognizable with optical microscope; 50-200 μm high fraction (50-200HF) composed of organo-mineral complexes; 20-50 μm fraction (20-50F) composed of organic-mineral complexes retained on the 20 μm sieve after ultrasonication; and <20 μm fraction (<20F) that passed through the 20 μm sieve and included mineral particles of clay and silt sizes, very stable microaggregates, and amorphous organic matter, humified and strongly bound to inorganic particles. The mass of all fractions was calculated using the bulk density values of 1.10, 1.19 and 1.27 g/cm^3 for PP, CT and NT, respectively at a soil depth of 0-10 cm, and 1.12, 1.16 and 1.25 g/cm^3 for PP, CT and NT, respectively at a soil depth of 10-20 cm. All fractions were oven dried, weighed and ground to a fine powder, and analysed for total C and N and $\delta^{13}C$ with a mass spectrometer (Europa Scientific Ltd., Krewe, UK) connected to a C and N elemental analyser (Barrie and Prosser, 1996).

Statistical analysis

Nitrogen transformation rates were analysed using SAS/ANOVA procedure (SAS Institute, 1985), and means were compared using Tukey's multiple comparison test with a significance level of 0.05. The model used was a split plot, with the management system as principal plot and the soil depth as subplot. Pearson's correlation coefficient (r) was used to assess the relationship between gross N transformation rates, C and N contents, and C/N ratios for the whole soil and for each fraction. Stepwise multiple linear regression analysis (backward procedure, significance level for the *F* value was 0.1 for removal from the model) was used to study the relationship between gross N transformation rates and C and N contents of the whole soil and the fractions.

8.3 Results and discussion

Mineral nitrogen and isotopic enrichment

The size and ^{15}N-enrichment of NH^+_4-N and NO^-_3-N pools at 0 and 3 days after the ^{15}N injections are shown in Figure 8.1. For both soil layers, NH^+_4-N contents were always below 10 mg N/kg for all management systems and for both soil depths (Figure 8.1a,c), and decreased between the first and the second extraction after ^{15}N injection; indicating ammonium consumption between extractions. The low levels of NH_4^+-N indicated that consumption processes occurred actively. Nitrification would have been one of the main processes of NH_4^+-N consumption because if ammonium was immobilized by microorganisms, the concentration of NO^-_3-N should be very low, but this was not observed (Figure 8.1a,c). On the other hand, ammonia volatilization was assumed to be negligible due to the slightly acid soil pH and the high soil buffer capacity (Videla, 1994). Moreover, we assumed that ammonium fixation was negligible, because only 0.01% of the NH_4^+-N added to Phaeozems of Balcarce was fixed to clays (L. Picone, personal communication).

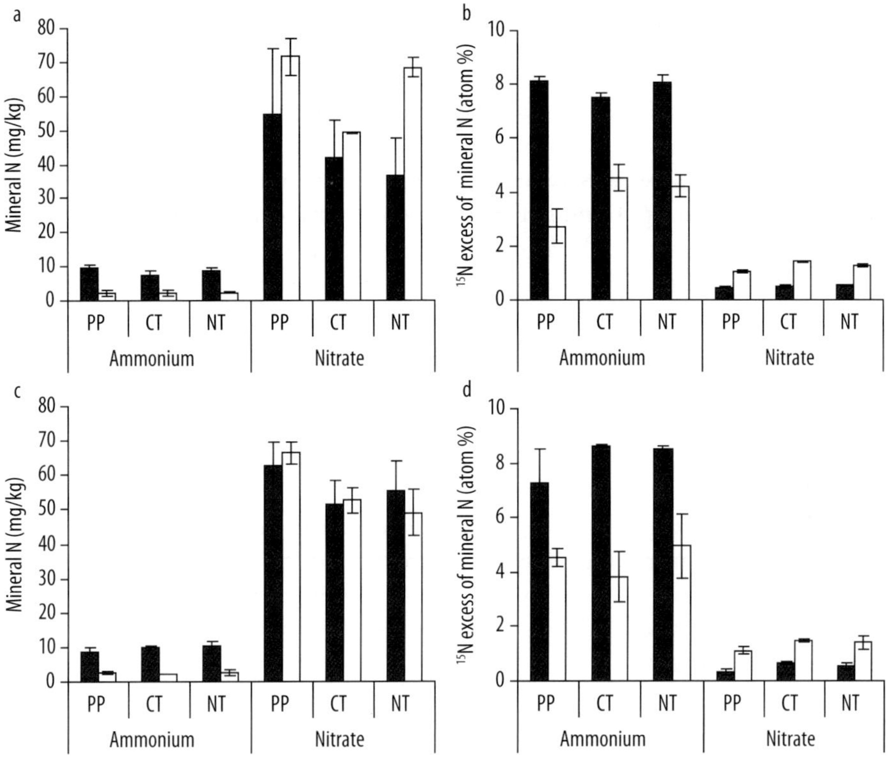

Figure 8.1. (a) Mineral nitrogen content (0-10 cm); (b) ^{15}N excess showing t_0 = dark columns (first extraction) and t_f = white columns (second extraction) at a soil depth of 0-10 cm; (c) mineral nitrogen content (10-20 cm); (d) ^{15}N excess showing t_0 = dark columns (first extraction) and t_f = white columns (second extraction) at a soil depth of 10-20 cm under permanent pasture (PP), conventional tillage (CT) and no-tillage (NT). Error bars show ± standard error; n=27.

The added ^{15}N recovered in the inorganic N pool at t_0, which represented the abiotic consumptive processes, was 99%, and confirmed abiotic processes of ammonium fixation and ammonia volatilization, and showed that microbial immobilization did not play a significant role in these soils. On the other hand, the selected t_0 was sufficient to promote uniform distribution, by diffusion, of the ^{15}N label within the sample and to prevented possible underestimation of GMR associated with heterogeneity (Luxhoi *et al.*, 2003).

Nitrate-N contents were initially high (accumulated during the pre-incubation period) and increased after ^{15}N injection (Figure 8.1a and 8.1c), possibly due to substrate stimulation. In experiments to measure net mineralization through aerobic incubation,

Navarro *et al.* (1980) concluded that the nitrification process in the Phaeozems of the Pampa was limited by the ammonium production rate. In this experiment, the addition of ^{15}N-labelled ammonium would have provided a substrate to the nitrifiers and therefore would have stimulated nitrification. In spite of this, it is necessary to keep in mind that the difference between production (gross mineralization) and removal (consumption) led to rapid ammonium depletion (Figure 8.1a). This suggested that the actual nitrification rate may have been limited by the ammonium supply, which also coincided with results from Navarro *et al.* (1980). To confirm this, an independent measurement of GNR, through labelling the nitrate pool with ^{15}N, would be required.

At t_0, the ^{15}N ammonium pool enrichment was 7-8%, and did not show any significant differences among management practices and soil depths (Figure 8.1b,d). The ^{15}N enrichments decreased by 2 to 4% at t_f depending on the rate of $^{14}NH_4^+$-N production from SOM mineralization. For the nitrate pool, ^{15}N isotopic enrichment at t_0 was 0.4-0.5%, whereas at t_f it was 1.0-1.2%. This indicated that some of the ^{15}N label was assimilated by nitrifying bacteria, which allowed for the quantification of the nitrification rate.

Gross transformation rates

The potential gross transformation rates (GMR, NR and GCR) are shown in Figure 8.2. Notwithstanding the low NH_4^+-N level measured, it was possible to quantify the GMR through the change in the ammonium pool enrichment. This ratio of change was greater than 0.2 in 96% of the evaluated samples. Takahashi (2001) suggested that this is the threshold at which the uncertainties in GMR estimations may increase.

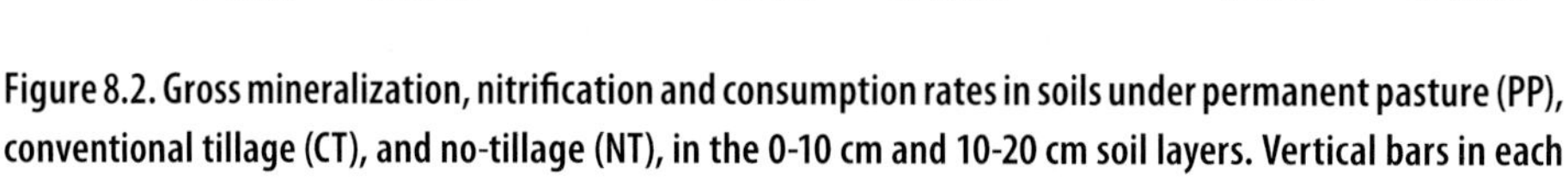

Figure 8.2. Gross mineralization, nitrification and consumption rates in soils under permanent pasture (PP), conventional tillage (CT), and no-tillage (NT), in the 0-10 cm and 10-20 cm soil layers. Vertical bars in each column indicate standard error of the mean (Tukey, *P*<0.05; n=27).

Mean GMR ranged from 0.532 to 2.598 mg N/kg/d where 0.276 and 3.771 mg N/kg/d were extreme values. The type of management system and its interaction with soil depth affected the GMR ($P<0.05$), where PP (1.465 mg N/kg/d) was significantly higher than NT (0.907 mg N/kg/d) and CT (1.093 mg N/kg/d). The PP-GMR (0-10 cm) was significantly higher ($P<0.05$) than CT at same depth and for NT at 10-20 cm (Figure 8.2). Higher GMR values observed under PP than under cropping agreed with the generally higher GMR reported under the original vegetation, either grass or forest, than for ecosystems disturbed by agriculture (Corre *et al.*, 2006; Zaman and Chang, 2004). This suggested a decline in the SOM active fraction as a result of cultivation (Buyanosky *et al.*, 1994; Studdert *et al.*, 2010), and plays a significant role in the topsoil, where the effects of management practices are greatest (Domínguez *et al.*, 2009; Eiza *et al.*, 2005; Puget and Lal, 2005).

Gross nitrification rates had values greater than those of GMR, ranging from 1.05 to 3.173 mg N/kg/d and with extreme values of 0.324 and 4.417 mg N/kg/d. However, no significant effect in gross nitrification rates was detected between management practices and soil depth. Consumption rates (Figure 8.2) were not affected either by the management system nor sampling depth. However, the mean consumption rates were greater than their respective GMR, being 3.395 mg N/kg/d the average value and 0.688 and 8.08 4 mg N/kg/d the extreme values. It is expected that GMR would be greater than GNR (Chen and Stark, 2000; Compton and Boone, 2002; Davidson *et al.*, 1991), but the opposite relationship was reported by Luxhøi *et al.* (2003), Recous *et al.* (1999) and Watson and Mills (1998). Results from our study coincided with those reported by Booth *et al.* (2005) who found that GCR was approximately 113% of mineralization across a wide range of ecosystems, indicating the affinity of nitrifying bacteria for the ammonium produced by SOM mineralization. Alternatively, the mineralization process may occur through a direct route. For example, it was proposed that when the direct route is operative (Barak *et al.*, 1990; Barraclough, 1997), only the excess N with respect to microorganism metabolic requirements is released into the ammonium pool. If this was the case, the dilution of the added ^{15}N isotope would be lower and the GMR underestimated. Finally, it should be kept in mind that GNR was calculated indirectly by ammonium pool enrichment. Luxhøi *et al.* (2003) concluded that ^{15}N-NH^{+}_{4} may have stimulated nitrification under low initial ammonium content. In our experiment, the ammonium pool was always less than 10 mg/kg, which may have led to an overestimation by substrate stimulation.

The GNR represented between 50 and 75% of GCR, for PP and NT management practices, indicating that nitrification was the dominating sink for NH^{+}_{4}-N. Cookson *et al.* (2007) reported that this may be due to the large accumulation of dissolved organic N (DON) in grassland soils which provided conditions encouraging nitrification and sink for NH_4^+. This was supported by the availability of dissolved organic carbon (DOC), which was lowest in arable and grassland soils. However DON accumulated and mineralization was higher than immobilization, which provided an opportunity for nitrification to occur (Cookson *et al.,* 2007).

Gross mineralization rates and GCR were significantly correlated (r=0.749, $P<0.0001$), as well as GNR and GMR (r=0.233, $P=0.0028$), and GNR and GCR (r=0.225, $P=0.004$). The close relationship between GMR and GCR may be due to a high microbial assimilation, dependent on ammonium production through gross N mineralization. This suggested competition between microorganisms and plants for available N was high (Corre *et al.*, 2006). Gross mineralization rates and GNR were positively correlated, which confirmed that the nitrification process was encouraged by an increase in mineralization.

Mass, C and N contents and $\delta^{13}C$ in the particle-size fractions of soil organic matter

The different soil management systems influenced soil mass distribution among the different particle-size fractions at 0-10 cm (Figure 8.3a) and the 10-20 cm (Figure 8.3b) depths. The LF was the most affected (>200LF and 50-200LF). Permanent pasture showed higher contents of these two fractions than in CT and NT management practices at a depth of 0-10 cm (Figure 8.3a), whereas differences were not as pronounced at the 10-20 cm depth (Figure 8.3b). As expected, the mineral-dominated heavy fractions (>200HF and 50-200HF (Figure 8.3a,b)) did not respond to the different soil management practices, because management does not change the distribution of soil mineral particles, unless it is related to erosion processes, which was not the case in our study. Likewise, there were no significant differences between management practices in the mass of the 20-50F (Figure 8.3a,b). However, both soil depths showed a significantly higher content in the <20F under NT than under CT and PP. It is difficult to understand this difference, because it was not expected that very fine and highly humified organo-mineral fractions, resistant to degradation, would respond to changes in soil management practices (Six *et al.*, 2004; Tisdall and Oades, 1982). Results from several studies suggested that the chemical composition of clay- and silt-sized organo-mineral complexes showed little influence by changes in land management practices (Christensen, 2001).

There was no effect of the different management practices on either C or N contents or in the $\delta^{13}C$ of all particle-size fractions at the 10-20 cm depth. Therefore, these results are not presented. At a depth of 0-10 cm, PP had a significantly greater ($P<0.05$) C and N content than NT and CT in the >200LF and 50-200LF fractions, and a greater N content in the <20 μm fraction (Table 8.2). For similar soil conditions,

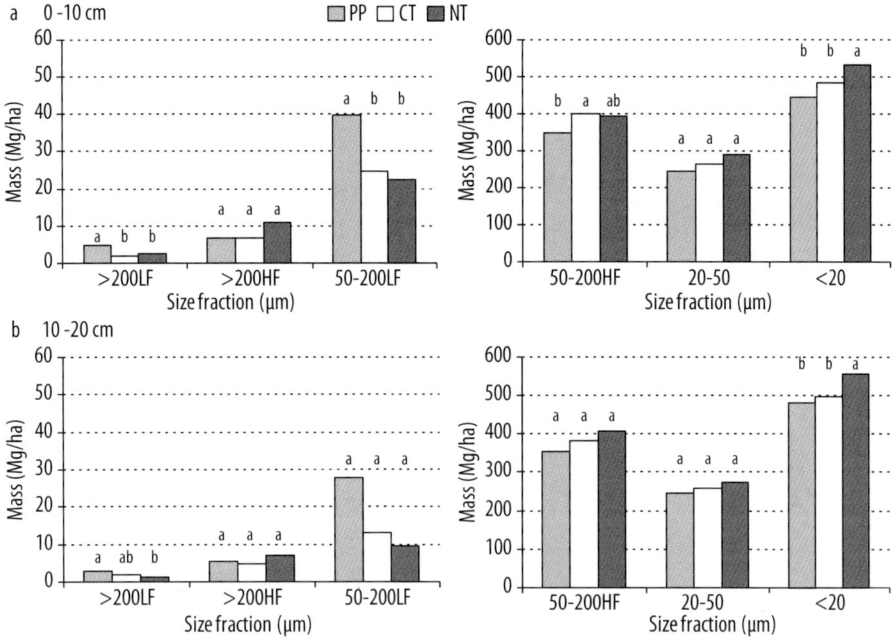

Figure 8.3. Soil mass distribution after fractionation of soil under permanent pasture (PP), conventional tillage (CT) and no-tillage (NT) management systems at (a) 0-10 cm depth and (b) 10-20 cm depth. Different letters above columns indicate significant differences (*P*<0.05) among management system means for each soil fraction.

increases in SOC in the light, medium and heavy densimetric fractions under pasture, in comparison with CT and NT, were observed but the greatest differences that explained greater C mineralization, occurred in medium and light fractions (Alvarez *et al.*, 1998). The input of dead plant material in the short-term, after conversion of arable lands to permanent pasture tended to accumulate in the particulate organic matter (POM) fraction (>200LF) indicating that these active fractions, with faster recycling rates, could serve as a reliable and relatively easily detectable indicator of changes in SOM induced by changes in management practices. Additionally, increases in N content in the <20F is not commonly observed when land management practices change from crop production to pasture. This is because these fractions have very slow recycling rates and contain small microaggregates stabilized by microbial products, root exudates, polyvalent cations and other persistent binding agents (Christensen 2001). Despite this, $\delta^{13}C$ results confirmed the incorporation in this fraction of soil organic N derived from pasture residues, which is exclusively of C3 origin.

Table 8.2. Total carbon (C) and nitrogen (N) contents, δ^{13}C and C/N ratio for particle size fractions in the 0-10 cm soil layer under different management systems in Balcarce, Argentina.

Fraction[1] (μm)	Management system[2]	Total C[3] (kg/ha)	Total N[3] (kg/ha)	δ^{13}C (‰)	C:N
>200LF	PP	1,308.5[a]	70.5[a]	-26.71[a]	18.55[a]
	CT	444.5[b]	28.4[b]	-27.40[a]	15.64[a]
	NT	590.9[b]	35.1[b]	-23.77[a]	16.90[a]
50-200LF	PP	2,612.9[a]	225.4[a]	-25.51[a]	11.58[a]
	CT	1,399.3[b]	108.9[b]	-22.56[b]	12.86[a]
	NT	1,735.0[b]	128.4[b]	-21.98[b]	13.53[a]
20-50F	PP	3,204.9[a]	248.3[a]	-23.95[a]	12.91[a]
	CT	3,172.8[a]	270.1[a]	-22.41[b]	11.75[a]
	NT	2,222.6[a]	192.6[a]	-22.39[b]	11.55[a]
<20F	PP	19,963.0[a]	2,273.8[a]	-22.04[a]	8.79[a]
	CT	22,091.0[a]	1,974.2[b]	-21.33[b]	11.19[a]
	NT	23,584.2[a]	2,172.2[ab]	-21.15[b]	10.86[a]

[1] LF = light fraction; HF = heavy fraction; F = fraction.

[2] PP = pasture; CT = conventional tillage; NT = no-tillage.

[3] Values of each fraction followed by different lower case letters are significantly different ($P<0.05$, Tukey n=3).

After six years of implementing the different agroecosystem management practices, there were no statistical differences in the C and N contents between CT and NT in any of the separated fractions (Table 8.2). Though, a tendency towards a higher C and N accumulation in the light fractions (>200LF and 50-200LF) under NT than under CT were observed (Table 8.2). This implies a greater proportion of easily mineralizable C and N under NT, which was evidenced too by higher GMR (Figure 8.2).

It is currently accepted that under NT, lower soil temperatures (Fabrizzi *et al.*, 2005) and a reduction in soil disturbance encourage C sequestration (Lal *et al.*, 2003), although these changes in SOM can not be easily detected over the short-term when adopting NT. Puget and Lal (2005) concluded that continuous NT on Chernozems did not produce a strong positive effect on whole soil C and N sequestration potential, but it did produce an increase in soil C content at a soil depth of 0-5 cm when converting from CT to NT. On the other hand, for the Phaeozems in the southeastern Pampa, Eiza *et al.* (2005) reported higher POC content at a 0-5 cm depth after 6 years of NT or after 9 years of switching from CT to NT (maize-soybean-wheat sequence). Likewise, for Phaeozems of the same region, Fabrizzi *et al.* (2003) concluded that NT produced an increase in SOC and soil organic N in the surface layer of a degraded soil, mainly in the POM (53-212 μm and 212-2,000 μm). They also found that greater C and N contents under NT were associated with higher potentially mineralizable N, however there was no effect of NT on SOC and organic N contents of a non-degraded soil (Fabrizzi *et al.*, 2003). Steinbach and Álvarez (2006) reported that for the Phaeozems in the Argentine Pampa, the higher SOC content under CT, the lower the increase in SOC after NT implementation.

In our study, most C and N (85-93% and 89-93% of total SOC and soil organic N, respectively) was found in the <50 μm fractions for all the management practices. These results agreed with those reported by Christensen (2001), who observed that for temperate arable soils, 50 to 75% of the SOM is present in the clay sized fraction while the silt and sand sized fractions accounted for 20 to 40% and <10%, respectively. The particulate SOM fraction was composed of amorphous and highly polymerized organic compounds which were strongly bound to clay and silt particles, and had very slow turnover rates (20-40% every 20 years) (Feller and Beare, 1997) and a high resistance to mineralization. Therefore, despite the relatively high SOC and soil organic N contents of the Phaeozems from the southeastern Pampa, the contribution of available N through mineralization is generally not sufficient to produce high crop yields. This is particularly notorious if the soil is not disturbed, suggestion that strategic tillage operations may be an alternative tool to promote N mineralization in addition to the safe use of fertilizers (Domínguez *et al.*, 2001).

The soil used in this study was developed under a mixture of C_3 and C_4 grass species, and therefore its original $\delta^{13}C$ characteristic value is approximately -20.1‰ (Andriulo *et al.*, 1999). The $\delta^{13}C$ of the <20 μm fraction was similar to the characteristic value of the whole soil in its original condition (Table 8.2). The decrease in $\delta^{13}C$ associated with the increase in the fraction size (Table 8.2), was due to the recent cropping history at this study site. The inclusion of mostly C_3 species in the crop sequence led to a discrimination against $^{13}CO_2$ sequestration and the subsequent decrease in $\delta^{13}C$ of fractions with high turnover rates (Boutton, 1996).

The $\delta^{13}C$ under PP was significantly more depleted than that under NT and CT for the 50-200LF, 20-50F and <20F fractions (Table 8.2). This was likely due to the presence of C_3 species (tall fescue and white clover), whereas CT and NT treatments incorporated a maize crop (C_4) every three years. The C/N ratios were higher in the fractions of greater size, decreasing with the decrease in fraction particle size. There were no significant differences between soil management practices in the C/N ratio for each fraction (Table 8.2).

Carbon contents in the >200LF, 50-200LF, 50-200HF and 20-50F fractions were highly correlated with the respect to N content (r ranging between 0.88 and 0.97, $P<0.0001$). However, the correlation between whole soil SOC and organic N contents, was not significant (r=0.0939, $P=0.7109$). Whole soil SOC content correlated with the C content in all the fractions, except >200HF, because this fraction is mainly composed of sand. Total organic N had a high correlation (r=0.885, $P<0.0001$) with N content, but only in the <20F fraction, indicating that in this fraction contained most of the organic N. On the other hand, correlations between total organic N and N content in >200LF (r=0.428, $P=0.076$) and 50-200LF (r=0.464, $P=0.053$) fractions were weak, but these fractions also had relatively high N content (Table 8.2).

Relationships between gross transformation rates and soil organic matter content

Pearson's correlation coefficients between C and N contents and C/N ratio for each fraction, GMR and GCR are shown in Table 8.3. Correlations with >200HF and 50-200HF fraction were not significant and therefore, they were not included in Table 8.3. These two fractions are mainly minerals, and as such it was expected they would participate, on a small-scale, in short-term mineralization processes, which was also supported by a lack in correlation with the calculated gross rates (Feller and Beare, 1997).

Gross mineralization rate was strongly correlated with the total organic N content of the whole soil (r=0.601) than GCR (r=0.505), whereas these processes did not correlate with SOC content (Table 8.3). These results do not agree with those reported by Accoe *et al.* (2004), who found that total C and N contents explained more than 80% of the GMR, the GNR and the immobilization rate in soils converted from continuous arable cropping to permanent grassland. Accoe *et al.* (2004) studied soils 6 to 50 years after a change in land management practice, showing a much greater range in SOC and organic N content compared to that of our study (data not shown).

Carbon content in the >200LF fraction was significantly correlated with GMR and GCR (Table 8.3). This fraction was composed of roots and aboveground residues with very high mineralization rates (Feller and Beare, 1997). This fraction is known as particulate organic matter (POM) or as the labile fraction, and responds rapidly to changes in soil management (Cambardella and Elliott, 1992). Fabrizzi *et al.* (2003) reported a high correlation between potentially mineralizable-N and POM-C and POM-N fractions, indicating that most of the mineralizable N comes mainly from the POM fraction.

Table 8.3. Pearson's correlation coefficients between gross mineralization and consumption rates, and the C and N contents and C:N ratios of the whole soil and its particle size fractions.

Variable	Gross mineralization rate[1]	Gross consumption rate[1]
C_{total}	0.339	-0.030
N_{total}	0.601 ***	0.505 *
C_{total}:N_{total}	-0.142	-0.399
$C_{>200LF}$	0.702 ***	0.457 *
$C_{200-50LF}$	0.581 **	0.412 *
C_{50-20F}	0.339	-0.067
$C_{<20F}$	0.082	-0.319
$N_{>200LF}$	0.636 ***	0.395
$N_{200-50LF}$	0.613 ***	0.485 **
N_{50-20F}	0.192	-0.207
$N_{<20F}$	0.373	0.434
C:$N_{>200LF}$	0.449	0.234
C:$N_{200-50LF}$	-0.115	-0.297
C:N_{50-20F}	0.322	0.449
C:$N_{<20F}$	-0.251	-0.484 *

[1] Pearson's correlation coefficients significant at * 0.05, ** 0.01 and *** 0.001 probability levels.

Carbon and N contents in the >200LF and 50-200LF fractions were strongly correlated with GMR and weakly correlated with GCR. These fractions are considered as POM and are often used to express early changes in SOM content. The C and N contents of minor sized fractions were neither correlated with GMR nor with GCR. The C/N ratio of the <20F fraction was related negatively to GCR.

Results of multiple linear regression analysis to relate GMR and GCR with a combination of C and N contents in the individual SOM fractions or in the whole soil and their C/N ratios are shown in Table 8.4. When all the soil management and soil depths were included in the analysis, C content in the >200LF fraction explained almost 50% of GMR variability, and N content in the <20F and C content in the >200HF fraction had minor contributions (Table 8.4). The particulate fraction is known to have high recycling rates, and it is frequently mentioned in the literature that it plays a role in net mineralization rates or potentially mineralizable N (PMN). Fabrizzi *et al.* (2003) reported a high correlation between PMN and POM-C and POM-N fractions, indicating that most of the mineralizable N comes mainly from

Table 8.4. Multiple linear regression models for the variation of the gross mineralization (GMR) and consumption (GCR) rates.[1]

	Parameter	Coefficient	R^2 partial	R^2 model	*P*-value
GMR	Constant	-0.91386		0.7247	0.0003
	$C_{>200LF}$	0.03105	0.4924		
	$N_{<20F}$	0.02952	0.1819		
	$C_{>200HF}$	-0.0781	0.0504		
GCR	Constant	-0.46393		0.2551	0.0325
	N_{total}	0.05634	0.2551		

[1] As influenced by C and N content and C:N ratios in the whole soil (subscript 'total') and in LF>200 µm (subscript '>200LF'), HF>200 µm (not shown), HF 50-200 µm (not shown), LF 50-200 µm (not shown), 20-50 µm (not shown) and <20 µm (subscript '<20F') fractions.

the POM fraction. The strong relation between gross and net mineralization (Accoe *et al.*, 2004; Zaman and Chang, 2004), indicated that these may be the same factors that regulate ammonium production and consumption processes.

The N content in <20F fraction was included in GMR equation (Table 8.4), although with low partial regression coefficients. This fraction contains silt and clay sized particles, and organic matter with an amorphous structure, which is also highly humified and protected in the organo-mineral complex, resulting in very low mineralization rates. Additionally, the chemical composition of the OM associated with minerals is not strongly influenced by changes in land management practices, but it may be affected when the soil is very degraded (Christensen, 2001; Dominguez *et al.*, 2009; Eiza *et al.*, 2005; Cambardella and Elliott, 1992). For these reasons, the results obtained from GMR were unexpected and it is recommended that this be investigated further.

The models (Table 8.4) explained only 72.5% of GMR and 25.5% of GCR variability. These values were low when compared to those found by Accoe *et al.* (2004), which explained more than 93% of the variability in the GMR and 88% of the of gross immobilization rates of soils under permanent pastures. In order to explain the larger proportion of the variability in GMR and GCR, we performed stepwise multiple linear regression analyses for each soil management system (PP, CT and NT). The new linear equations for GMR and GCR showed high regression coefficients for PP and CT (Table 8.5), explaining more than 90% of the variability of both independent variables.

However, under NT, the R^2 for GMR was 0.567, whereas for GCR, no variable reached the significance level of 10%, and therefore it was not possible to fit a model (Table 8.5).

For GMR under CT, the greatest variability was explained by N content of <20F fraction (79.7%), although this fraction did not present a significant Pearson's coefficient with GMR (Table 8.3). Carbon content in the <20F fraction was not modified by the different soil management practices evaluated in this study, but the N content was. Nitrogen content in the <20F fraction showed a positive coefficient in the model (Table 8.5), and suggested that it may be mineralized by microorganisms. The SOM protected by its association with clay and silt sized mineral particles (<20 μm), is partially hydrolysable (Six *et al.*, 2002). However, the exposure of this fraction to tillage encouraged microbial activity and mineralization. This particle size fraction also contains another biochemically protected fraction with a very slow recycling rate, which is considered non-hydrolysable (Leavitt *et al.,* 1996; Paul *et al.,* 1995). This fraction is akin to that referred to as the 'passive' SOM pool (Parton *et al.,* 1987). Its chemical composition is so complex that it remains in the soil for centuries (Paul *et al.,* 2001). The presence of this fraction may explain why soils southeast of Buenos Aires, in spite that they have been managed with conventional techniques of cultivation for many years still have relatively high levels of SOM, but provide a low amount of N to crops (Studdert *et al.,* 1997) due to a low GMR.

Under NT, the N content in the >200LF fraction explained only 56% of the GMR variability, despite it was only observed a trend in respect to the increase of this fraction (Table 8.2). Despite this, the >200LF fraction is an important and readily available source of N, and its C and N contents were strongly affected by soil management, and presented significant Pearson's coefficients with GMR and GCR (Table 8.3). In the soil under PP, C content in 50-200HF and >200HF fractions, C/N ratio in the >200LF, and C/N ratio of the whole soil were included in the model explaining 99.9% of the GMR variability. These results were similar to those reported by Accoe *et al.* (2004) and Fabrizzi *et al.* (2003) who found a close relationship between particulate fractions and N availability.

We did not find a simple relationship between GMR and GCR with the analysed organic matter fractions. The PP included C and N contents of the >200 LF and 50-200LF fractions, also called macroorganic matter. This fraction has high recycling rates and contributes from 3 to 21.9% of the gross mineralization (Monaghan and Barraclough, 1995). The presence of these fractions in the PP soil, 6 years after establishment of pasture, indicated an improvement in OM quantity and quality. On the contrary, the GMR and GCR in CT were frequently related to the C and N contents in the <50 μm fractions with slower recycling rates, which may be due to a low

Table 8.5. Models of multiple linear regression for the gross mineralization (GMR) and gross consumption rates (GCR) variation in conventional tillage, no tillage and permanent pasture managements systems.[1]

Conventional tillage					Permanent pasture					No tillage				
Parameter	Coefficient	R^2 partial	R^2 model	*P*-value	Parameter	Coefficient	R^2 partial	R^2 model	*P*-value	Parameter	Coefficient	R^2 partial	R^2 model	*P*-value
Gross mineralization rate														
Constant	-0.5032		0.9896	0.0011	Constant	-1.142		0.999	0.0024	Constant	0.563		0.567	0.084
$N_{<20F}$	0.0855	0.797			$C_{50\text{-}200HF}$	0.177	0.949			$N_{>200LF}$	0.589	0.567		
$C{:}N_{>200HF}$	0.0292	0.193			$C{:}N_{>200LF}$	0.046	0.033							
					$C_{>200HF}$	-3.264	0.015							
					C:N	0.065	0.003							
Consumption rate														
Constant	4.433		0.8999	0.0039	Constant	1.425		0.986	0.0017	No variable reached 0.15 significance level				
$C_{>200HF}$	-2.256	0.899			$C_{50\text{-}200HF}$	0.439	0.915							
$C{:}N_{total}$	-0.403	0.071												

[1] As influenced by C and N contents and C:N ratios in the whole soil (subscript 'total') and in LF >200 μm (subscript '>200LF'), HF >200 μm (subscript '>200HF'), HF 50-200 μm (subscript '50-200HF'), LF 50-200 μm (subscript '50-200LF'), 20-50 μm (subscript '20-50F') and <20μm (subscript '<20F') fractions.

availability of labile fractions from continuous tillage. No-tillage is intermediate to PP and CT with respect to soil disturbance, indicating that this management practice may improve soil condition. However, the time frame under which this site was under NT may have been insufficient in order to establish any significant differences.

Nonetheless, it is necessary to point out that in spite of the differences in the fitted models, NT did not present significant changes in N and C accumulated stocks. The impact of NT on C and N in Phaeozems should also be viewed with caution because NT does not always lead to significant increases in C and N in the short- to medium-term. Puget and Lal (2005) did not find significant increases in N stocks for this type of soil, after 8 years of NT implementation, although an increase in SOC was observed in the top 5 cm of soil. Probably for this soil order, the mechanisms and ways for SOM recovery are more complex and dependent on local conditions, including inherent soil conditions and climate. Thus, the generalizations about the effect of the NT on C accumulation must be very prudently communicated.

8.4 Conclusions

The application of the ^{15}N dilution technique helped in the quantification of N processes in soils under different management practices, appearing as a promising method for an in-depth evaluation of N dynamics. The quantification of GNR by labelling the ammonium pool did not provide reliable results on the possibility of substrate stimulation. As such, further studies are recommended which should include the labelling of the nitrate pool. Permanent pasture accumulated a greater amount of C and N at the 0-10 cm depth in the fractions above 50 μm (particulate organic matter), while NT only showed a trend, while CT remained unchanged. Gross mineralization rates showed the same behaviour and had a high correlation with C and N content in the labile fraction. No-till did not influence soil C and N content, nor its distribution in the OM particle sized fractions. The different management practices did not modify the composition of C and N in fractions <50 μm, confirming its slow mineralization rate. Multiple linear regressions between C, N and C/N ratio of the OM fractions explained only 72% of the GMR and CGR variability when PP, CT and NT were analyzed together. The C content in the >200 μm LF and 50-200 μm fractions were associated with GMR and GCR. The variability of the GMR and CR was better explained when multiple regression equations were separated by management systems. For CT, C and N contents of the smallest fractions were frequently included in the equations, while for PP the labile fraction >50μm those were included more frequently in the models. Overall, NT presented an intermediate between CT and PP management practices, revealing only a trend in an improvement in soil conditions.

Acknowledgements

Funding was provided by the Universidad Nacional de Mar del Plata and Antorchas-Vitãe Foundation. The first author received a fellowship from CAPES (Brazil) – SPU (Argentina) agreement. Special thanks for assistance in laboratory work are due to Sandra M. Nicoleti and Juliana G. Giovannini. We thank Miguel Luiz Baldessin for his help with the isotope analyses.

References

Accoe, F., Boeckx, P., Busschaert, J., Hofman, G. and Van Cleemput, O., 2004. Gross transformations rates and net mineralisation rates related to the C and N contents of soil organic matter fractions in grasslands soils of different age. Soil Biology and Biochemistry 36: 2075-2087.

Alvarez, C.R., Alvarez, R., Grigera, M.S. and Lavado, R.S., 1998. Associations between organic matter fractions and the active soil microbial biomass. Soil Biology and Biochemistry 30: 767-773.

Álvarez, R. and Steinbach, H., 2009. A review of the effects of tillage systems on some soil physical properties, water content, nitrate availability and crops yield in the Argentine Pampas. Soil and Tillage Research 104: 1-15.

Alvarez, R., Díaz, R., Barbero, N., Santanatoglia, O. and Blotta, L., 1995. Soil organic carbon, microbial biomass and CO_2-C production from three tillage systems. Soil and Tillage Research 33: 17-28.

Andrade, F.H., Echeverría, H.E., González, N.S. and Uhart, S.A., 2002. Requerimientos de nutrientes minerales. In: Andrade, F.H. and Sadras, V.O. (eds.) Bases para el Manejo del Maíz, el Girasol y la soja. Editorial Medica Panamericana Sa, Buenos Aires, Argentina, pp. 211-237.

Andriulo, A., Guerif, J. and Mary, B., 1999. Evolution of soil carbon with various cropping sequences on the rolling pampas. Determination of carbon origin using variations in natural ^{13}C abundance. Agronomie 19: 349-364.

Aparicio, V. and Costa, J.L., 2007. Soil quality indicators under continuous cropping systems in the Argentinean Pampas. Soil and Tillage Research 96: 155-165.

Argentinean Association of Producers under No-tillage (AAPRESID), 2013. Statistics of surface under no-tillage in Argentina. Available at: http://www.aapresid.org.ar.

Barak, P., Molina, J.A.E., Hadas, A. and Clapp, C.E., 1990. Mineralization of amino acids and evidence of direct assimilation of organic nitrogen. Soil Science Society American Journal 54: 769-774.

Barraclough, D., 1991. The use of mean pool abundances to interpret ^{15}N tracer experiments. Plant and Soil 131: 89-96.

Barraclough, D., 1997. The direct or MIT route for nitrogen immobilization: a ^{15}N mirror image study with leucine and glycine. Soil Biology and Biochemistry 29: 101-108.

Barraclough, D., Geens, E.L., Davies, G.P. and Maggs, J.M., 1985. Fate of fertilizer nitrogen. 3. The use of single and double labeled N-15 ammonium-nitrate to study nitrogen uptake by ryegrass. Journal of Soil Science 36: 593-603.

Barrie, A. and Prosser, S.J., 1996. Automated analysis of light-element stable isotopes by isotope ratio mass spectrometry. In: Boutton, T.W. and Yamasaki, S. (eds.) Mass spectrometry of soils. Marcel Dekker Inc., New York, NY, USA, pp.1-46.

Barrios, E., Buresh, R.J. and Sprent, J.I., 1996. Nitrogen mineralization in density fractions of soil organic matter from maize and legume cropping systems. Soil Biology and Biochemistry 28: 1459-1465.

Berardo, A., 1994. Aspectos generales de fertilización y manejo del trigo en el área de influencia de la Estación Experimental INTA-Balcarce. Boletín Técnico No. 28. INTA, San José de Balcarce, Argentina.

Booth, M.S., Stark, J.M. and Rastetter, E., 2005. Controls on nitrogen cycling in terrestrial ecosystems: a synthetic analysis of literature data. Ecological Monographs 75: 139-157.

Boutton, T.W., 1996. Stable carbon isotope ratios of soil organic matter and their use as indicators of vegetation and climate change. In: Boutton T.W. and Yamasaki, S. (eds.) Mass Spectrometry of Soils Marcel Dekker Inc., New York, NY, USA, pp. 47-82.

Bremner, J.M., 1965. Inorganic forms of nitrogen. In: Black, C.A. and Evans, D.D. (eds.) Methods of soil analysis. American Society of Agronomy, Madison, WI, USA, pp. 1149-1178.

Buyanosky, G.A., Aslam, M. and Wagner, G.H., 1994. Carbon turnover in soil physical fractions. Soil Science Society of America Journal 58: 1167-1173.

Cambardella, C.A. and Elliot, E.T., 1992. Particulate soil organic matter changes across a grassland cultivation sequence. Soil Science Society of American Journal 56: 777-783.

Caride, C., Piñeiro, G. and Paruelo, J.M., 2012. How does agricultural management modify ecosystem services in the argentine Pampas? The effects on soil C dynamics. Agriculture, Ecosystems and Environment 154: 23-33.

Chen, J. and Stark, J.M., 2000. Plant species effects and carbon and nitrogen cycling in a sagebrush-crested wheatgrass soil. Soil Biology and Biochemistry 32: 47-57.

Christensen, B.T., 2001. Physical fractionation of soil and structural and functional complexity in organic matter turnover. European Journal of Soil Science 52: 345-353.

Compton, J.E. and Boone, R.D., 2002. Soil nitrogen transformations and the role of light fraction organic matter in forest soils. Soil Biology and Biochemistry 34: 933-943.

Cookson, W.R., Osman, M., Marschner, P., Abaye, D.A., Clark, I., Murphy, D.V., Stockdale, E.A. and Watson, C.A., 2007. Controls on soil nitrogen cycling and microbial community composition across land use and incubation temperature. Soil Biology and Biochemistry 39: 744-756.

Corre, M.D., Dechert, G. and Veldkamp, E., 2006. Soil nitrogen cycling following Montane forest conversion in Central Sulawesi, Indonesia. Soil Science Society of American Journal 70: 359-366.

Darwich, N.A., 1991. Estado actual y manejo de los recursos naturales en la región pampeana húmeda sur. In: Seminario juicio a nuestra agricultura. Hacia el desarrollo de una agricultura sostenible. Hemisferio Sur, Buenos Aires, Argentina, pp.51-62.

Davidson, E.A., Hart, S.C. and Firestone, M.K., 1992. Internal cycling of nitrate in soils of a mature coniferous forest. Ecology 73: 1148-1156.

Davidson, E.A., Hart, S.C., Shanks, C.A. and Firestone, M.K., 1991. Measuring gross mineralization, immobilization, and nitrification by 15N isotopic pool dilution in intact soil cores. Journal of Soil Science 42: 335-349.

Diovisalvi, N.V., Studdert, G.A., Domínguez, G.F. and Eiza, M.J., 2008. Fracciones de carbono y nitrógeno orgánicos y nitrógeno anaeróbico bajo agricultura continua. Ciencia del Suelo 26: 1-11.

Domínguez, G.F., Diovisalvi, N.V., Studdert, G.A. and Monterubbianesi, M.G., 2009. Soil organic C and N fractions under continuous cropping with contrasting tillage systems on mollisols of the southeastern pampas. Soil and Tillage Research 102: 93-100.

Domínguez, G., Studdert, G., Echeverría, H. and Andrade, F., 2001. Sistemas de cultivo y nutrición nitrogenada en maíz. Ciencia del Suelo 19: 47-56.

Durán, A., Morrás, H., Studdert, G. and Liu, L., 2011. Distribution, properties, land use and management of Mollisols in South America. Chinese Geographical Science, 21: 511-530.

Echeverría, H.E. and Videla, C.C., 1998. Eficiencia fisiológica y de utilización de nitrógeno en trigo en la Región Pampeana argentina. Ciencia del Suelo 16: 83-87.

Eiza, M.J., Fioriti, N., Studdert, G.A. and Echeverría, H.E., 2005. Fracciones de carbono orgánico en la capa arable: efecto de los sistemas de cultivo y de la fertilización nitrogenada. Ciencia del Suelo 23: 59-68.

Ernst, O. and Siri-Prieto, G., 2009. Impact of perennial pasture and tillage systems on carbon input and soil quality indicators. Soil and Tillage Research 105: 260-268.

Fabrizzi, K.P., García, F.O., Costa.J.L. and Picone, L.I., 2005. Soil water dynamics, physical properties and corn and wheat responses to minimum and no-tillage systems in the southern Pampas of Argentina. Soil and Tillage Research 81: 57-69.

Fabrizzi, K.P., Morón, A. and García, F.O., 2003. Soil carbon and nitrogen organic fractions in degraded vs non-degraded mollisols in Argentina. Soil Science Society of American Journal 67: 1831-1841.

Feller, C. and Beare, M.H., 1997. Physical control of soil organic matter dynamics in the tropics. Geoderma 79: 69-116.

Franzluebbers, A.J., 2010. Achieving soil organic carbon sequestration with conservation agricultural systems in the Southeastern United States. Soil Science Society of America Journal 74: 347-357.

Hatch, D.J., Jarvis, S.C. and Phillips, L., 1990. Field measurement of nitrogen mineralization using soil core incubation and acetylene inhibition of nitrification. Plant and Soil 124: 97-107.

Instituto Nacional de Tecnología Agropecuaria (INTA), 1979. Carta de Suelos de la República Argentina. Hoja: 3757-31, Balcarce. Secretaría de Agricultura, Ganadería y Pesca, Buenos Aires, Argentina.

Jagadamma, S. and Lal, R., 2010. Distribution of organic carbon in physical fractions of soils as affected by agricultural management. Biology and Fertility of Soils 46: 543-554.

Kirkham, D. and Bartholomew, W.V., 1954. Equations for following nutrient transformations in soil, utilizing tracer data. Soil Science Society of American Proceedings 18: 33-34.

Lal, R., Follett, R.F. and Kimble, J.M., 2003. Achieving soil carbon sequestration in the United States: a challenge to policy makers. Soil Science 168: 827-845.

Leavitt, S.W., Follett, R.F. and Paul, E.A., 1996. Estimation of slow- and fast-cycling soil organic carbon pools from 6N HCl hydrolysis. Radiocarbon 38: 231-239.

Luxhøi, J., Nielsen, N.E. and Jensen, L.S., 2003. Influence of $^{15}NH^+_4$-application on gross N turnover rates in soil. Soil Biology & Biochemistry 35: 603-606.

Mandiola, M., Studdert, G.A., Dominguez, G.F. and Videla, C., 2011. Organic matter fraction distribution in aggregate sizes of a mollisol under contrasting managements. Journal of Soil Science and Plant Nutrition 11: 47-57.

Monaghan, R. and Barraclough, D., 1995. Contributions to gross N mineralization from ^{15}N-labelled soil macroorganic matter fractions during laboratory incubation. Soil Biology and Biochemistry 27: 1623-1628.

Monhagan, R., 1995. Errors in estimates of gross rates of nitrogen mineralization due to non-uniform distributions of ^{15}N label. Soil Biology and Biochemistry 27: 855-859.

Navarro, C.A., Echeverría, H.E., González N.S. and Iglesias, M.A., 1980. Cinética de las reacciones de amonificación y nitrificación en los suelos del sudeste bonaerense. In: IX Reunión Argentina de la Ciencia del Suelo, Paraná, Entre Ríos, 1980. Actas. Tomo II, pp. 431-437.

Nelson, M.A., Griffith S.M. and Steiner, J.J., 2006. Tillage effects on nitrogen dynamics and grass seed crop production in western Oregon, USA. Soil Science Society American Journal 70: 825-831.

Parton, W.J., Schimel, D.S., Cole, C.V. and Ojima, D.S., 1987. Analysis of factors controlling soil organic matter levels in great plains grasslands. Soil Science Society of America Journal. 51: 1173-1179.

Paul, E.A., Collins, H.P. and Leavitt, S.W., 2001. Dynamics of resistant soil carbon of Midwestern agricultural soils measured by naturally occurring ^{14}C abundance. Geoderma 104: 239-256.

Paul, E.A., Horwath, W.R., Harris, D., Follett, R., Leavitt, S.W., Kimball, B.A. and Pregitzer, K., 1995. Establishing the pool sizes and fluxes in CO_2 emissions from soil organic matter turnover. In: Lal, R., Kimble, J., Levine, E. and Stewart, B.A. (eds.) Soils and global change. CRC Press, Boca Raton, FL, USA, pp. 297-304.

Puget, P. and Lal, R., 2005. Soil organic carbon and nitrogen in a Mollisol in central Ohio as affected by tillage and land use. Soil and Tillage Research 80: 201-213.

Recous, S., Aita, C. and Mary, B., 1999. *In situ* changes in gross N transformations in bare soil after addition of straw. Soil Biology and Biochemistry 31: 119-133.

Sainz-Rozas, H.R., Echeverría, H.E. and Angelini, H., 2011. Niveles de materia orgánica y pH en suelos agrícolas de la Región Pampeana y extrapampeana Argentina. Ciencia del Suelo 29: 29-37.

SAS Institute, 1985. User's Guide. Statistics, V. 5. SAS Inst., Cary, NC, USA.

Six, J., Bossuyt, H., Degryze, S. and Denef, K., 2004. A history of research on the link between (micro) aggregates, soil biota, and soil organic matter dynamics. Soil and Tillage Research 79: 7-31.

Six, J., Feller C., Denef, K., Ogle, S.M., De Moraes Sa, J.C. and Albrecht, A., 2002. Soil organic matter, biota and aggregation in temperate and tropical soils – effects of no-tillage. Agronomie 22: 755-775.

Six, J., Paustian, K., Elliot. E.T. and Combrink, C., 2000. Aggregate and soil organic matter dynamics under conventional and no-tillage systems. Soil Science Society of American Journal 64: 681-689.

Sørensen, P. and Jensen, E.S., 1991. Sequential diffusion of ammonium and nitrate from soil extracts to a polytetraethylene trap for 15N determination. Analytica Chimica Acta 252: 201-203.

Steinbach, H. and Álvarez, R., 2006. Changes in soil organic carbon contents and nitrous oxide emissions after introduction of no-till in Pampean agroecosystems. Journal of Environmental Quality 35: 3-13.

Stenger, R., Priesack, E. and Beese, F., 1995. Rates of net nitrogen mineralization in disturbed and undisturbed soils. Plant and Soil 171: 323-332.

Studdert, G.A. and Echeverría, H.E., 2000. Crop rotations and nitrogen fertilization to manage soil organic carbon dynamics. Soil Science Society of American Journal 64: 1496-1503.

Studdert, G.A., Domínguez, G.F., Agostini, M.A. and Monterubbianesi M.G., 2010. Cropping systems to manage Southeastern Pampas' Mollisol health. I. Organic C and mineralizable N. In: Liu, X., Song, C., Cruse, R.M. and Huffman, T. (eds.) Proceedings of the international symposium on soil quality and management of world mollisols – new advances in research and management of world mollisols. July 13-16, 2010. Harbin, Heilongjiang, China P.R., pp. 199-200.

Studdert, G.A., Echeverría, H.E. and Casanovas, E.M., 1997. Crop-pasture rotation for sustaining the quality and productivity on a typic argiudoll. Soil Science Society of American Journal 61: 1466-1472.

Takahashi, S., 2001. Comparison of gross nitrogen mineralization rates by zero-order models. Soil Science Society of American Journal 65: 244-246.

Tisdall, J.M. and Oades, J.M., 1982. Organic matter and water stable aggregates in soils. Journal of Soil Science 33: 141-163.

Videla, C.C., 1994. La volatilización de amoníaco: una vía de pérdida de nitrógeno en sistemas agropecuarios. Boletín Técnico INTA 131: 16 pp.

Watson, C.J. and Mills, C.L., 1998. Gross nitrogen transformations in grassland soils as affected by previous management intensity. Soil Biology and Biochemistry 30: 743-753.

Zaman, M. and Chang, S.X., 2004. Substrate type, temperature, and moisture content affect gross and net mineralization and nitrification rates in agroforestry systems. Biology and Fertility of Soils 39: 269-279.

Chapter 9.

Development of a soil health index based on the ecological soil functions for organic carbon stabilization with application to alluvial soils of northeastern Italy

A. Ferrarini[1], F. Fornasier[2] and C. Bini[3]

[1]Institute of Agronomy, Genetics and Field Crops, Università Cattolica del Sacro Cuore, via Emilia Parmense 84, 20122 Piacenza, Italy; [2]Research Centre for the Soil-Plant System, Council for Research and Experimentation in Agriculture, Via Trieste 23, 34170 Gorizia, Italy; [3]Department of Environmental Sciences, Informatics and Statistics, Ca' Foscari University, Santa Marta, Dorsoduro 2137, 30121 Venice, Italy; andrea.ferrarini@unicatt.it

Abstract

There is a need to assess soil health quantitatively to assist with sustainable soil management in agroecosystems. The objective of this research was to develop a farm-level soil health index (SHI) to help identify the most sustainable management practices including C sequestration and agroecosystem resilience to climate change. For all indexing phases we used the Soil Management Assessment Framework. The SHI was created using a minimum data set (MDS) of physical, chemical and biochemical soil indicators chosen via expert opinion (EO-MDS) (24 indicators) and via the results of PCA (PCA-MDS) (16 indicators). Each observed value was converted into a score (0 to 1) by using site-specific non-linear scoring algorithms. Indicator scores were incorporated into a composite SHI which quantified the ecological performance of 5 soil functions: habitat and biodiversity, water movement and availability, filtering and buffering, nutrient cycling, physical stability and support, and long-term C stabilization. For both MDS evaluations, we assessed agroecosystems managed organically (OS) or conventionally (CS). Results from our study showed that OS had significantly higher (P<0.05) SHI scores compared to the CS. SHI, with EO-MDS, was more efficient (P=0.028) than the PCA-MDS (P=0.039) when determining the effect of soil management practices on soil health. Non-linear transformation was a useful technique and represented soil ecological functionality effectively. Five out of six soil functions had significantly greater (P<0.05) ecosystem performance in OS compared to CS, except for water movement and availability. The benefit for OS lead to a greater soil health status, allowing the system to be more resilient to climate change, to efficiently provide nutrients to plants, and to sustain an active soil food web for

M. Oelbermann (ed.) **Sustainable agroecosystems in climate change mitigation**
DOI 10.3920/978-90-8686-788-2_9, © Wageningen Academic Publishers 2014

the long-term stabilization of C. The SHI may help optimize and provide extensive insight into all aspects of soil health which are intricately linked to soil-based ecological functions at the farm level.

Keywords: soil health index, soil ecological functions, soil carbon stabilization, climate change mitigation, vegetable production.

9.1 Introduction

Soil is a non-renewable resource, which provides a number of environmental, social and economic services (Doran and Zeiss, 2000; Karlen *et al.,* 2003). Major short-term impacts on soil functions are derived from land-use change and agroecosystem management practices (Powlson *et al.*, 2011a,b). However, climate change may be expected to have an impact on soil processes over the long-term (Schils *et al.,* 2008). As such, evaluating the impact of different management practices on ecological functions may be necessary to evaluate the sustainability of agricultural systems.

To mitigate climate change, one option may be to increase the level of soil organic matter (SOM) which leads to the long-term sequestration of soil carbon (C) (Lal, 2011a). Soil organic matter plays a key role in the terrestrial C budget. Increasing SOM persistence (Schmidt *et al.*, 2011) and understanding its accessibility (Dungait *et al.*, 2012) is crucial for agricultural sustainability. Soil organic matter and therefore soil organic C (SOC) increase under conservative tillage systems in comparison to systems under conventional tillage. This increase is due to a greater amount of C-rich macroaggregates (>250 mm) (Six *et al.*, 2000), and a reduced macroaggregate turnover rate, which promotes microaggregate stabilization (Six *et al.,* 2002). As such, stabilized C pools influence key soil properties and processes, including soil health, resistance and resilience to disturbance, and mitigation to climate change, positively (Lal, 2011b).

Over the past decade, the concept of soil health was discussed frequently in the scientific literature. The notion of soil health includes the capability of supporting the production of quality food and fibre to meet human and livestock needs (Kibblewhite *et al.,* 2008). The knowledge of soil functioning in relation to management practices and climate change issue has led soil scientists to evaluate a set of soil health indicators representative of fundamental soil ecological functions (Bastida *et al.,* 2008; Dominati *et al.,* 2010; Faber and Wensen, 2012). Such tools need to be flexible with respect to the selection of soil functions and should be suitable for the management goals of interest to the land owner (Wienhold *et al.*, 2009). A soil health index should be based on soil physical, chemical and biological properties (Carter, 2002; Schindelbeck *et al.,* 2008, Lal, 2011a).

The soil health concept includes the soil's impact on the environment, which is the basis for the Soil Management Assessment Framework (SMAF) (Andrews *et al.*, 2004), and provides site-specific interpretations of soil quality indicators. Other assessment tools currently used include the agroecosystem performance assessment tool (Liebig *et al.*, 2004), the Cornell Soil Health Assessment (Gugino *et al.*, 2009), and the ecosystem performance index (Rutgers *et al.*, 2012). The SMAF, in comparison to other soil quality assessment tools, was found to be more sensitive in discriminating the effects of land management practices on soil and water resources (Zobeck *et al.*, 2008). The SMAF employs soil health indicator data to assess management effects on soil functions using a three-step process that includes indicator selection, indicator interpretation, and integration into a final index (Andrews *et al.*, 2004). The fundamental phase is the creation of a scoring curve and its validation step. The scoring approach is used to standardize the relationship between a soil function and a soil indicator (Andrews and Carrol, 2001). This relationship dictates the shape of an indicator's scoring curve and normalizes indicator measurements to a value between 0 and 1 (Hussain *et al.,* 1999). Karlen and Stott (1994) proposed three general shapes for soil quality scoring curves: more-is-better, less-is-better and optimum range. The more-is-better curve shows an upper asymptotic sigmoidal curve where soil properties are associated with improved soil quality at higher levels such as that for SOC and aggregate stability. Less-is-better curve (lower asymptote) indicates poor soil quality at high levels such as that for bulk density. The mid-point optimum curve represents a positive influence on soil quality up to an optimal level beyond which their influence is negative such as pH and porosity. Different site-specific scoring algorithms for physical, biological, and chemical soil indicators were previously developed and used in soil quality studies (Wienhold *et al.,* 2009).

The objective of this research was to further develop and apply the SMAF tool to evaluate the effects of conventional agricultural systems (CS) and organic agricultural systems (OS) on soil health in order to develop a farm-level SHI based on ecological soil functions. Such an evaluation will advance our understanding of complex soil ecosystem relationships between soil health indicators and aid in the interpretation of soil quality under vegetable production systems. The overall goal was to obtain more information about the potential to use this farm-level index to determine environmental services such as C sequestration and resilience to climate change.

9.2 Materials and methods

Study site

The study area (Figure 9.1) was located in northeastern Italy, in the Brenta river floodplain (45°30'N and 12°5'E). Within this region two different farms one representing a CS and the other an OS were used in this study. The soils at both sites were classified as Cambisols with a fine-silty texture and mesic moisture regime (ARPAV, 2004). The parent material was a calcareous silt. Each farm had 2 plots, and each plot was 2 ha in size. The plots were arranged in a randomized split plot design with four subplots (20×20 m). For the CS, an annual crop rotation of maize (*Zea mays* L.) and red Italian chicory (*Cichorium intybus* L.) began 15 years ago. This agroecosystem was tilled to a 55 cm depth using a mouldboard plough and harrow. The CS also used mineral fertilizers applied at a rate of 130-170 kg N/ha/y, 55-155 kg P/ha/y and 185 kg K/ha/y in addition to a high rate of pesticide application consisting of fungicides, herbicides and insecticides. The same crops and crop rotation were used in the OS, which was initiated 14 years ago, but with an alfalfa (*Medicago sativa* L.)/grass (graminoids) mixture cover crop every third year. The plots were chisel ploughed to a 15 cm depth in addition to the use of cow manure which was composted and applied at 40 Mg/ha/y (wet weight). No chemical inputs in terms of mineral fertilizers or pesticides occurred in the OS.

Soil sampling and analysis

Five samples were collected randomly from each subplot using a soil corer with a 5 cm inner diameter. The five samples were bulked into a composite sample and stored in

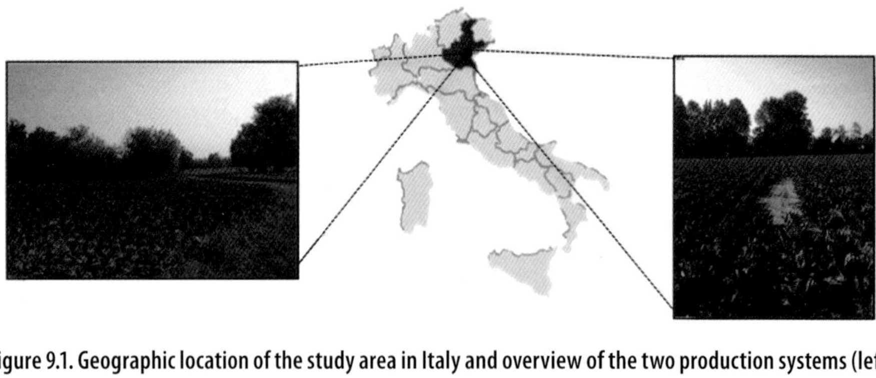

Figure 9.1. Geographic location of the study area in Italy and overview of the two production systems (left: organic system; right: conventional system).

sterile polypropylene bags (3.5 l). All soil was sampled to a 5 cm depth in the middle July 2011, approximately one week after an irrigation event. For all soil samples, 500g were presieved using an 8 mm mesh size to remove stones and coarse organic matter and to perform the soil aggregates analysis. A 250 g subsample was stored at field capacity at 4 °C for biochemical analysis the remaining 250 g was air-dried and passed through a 2 mm sieve. All soil analyses were carried out at the Pedology and Biochemistry Lab at Ca' Foscari University, Venice, Italy, except for soil enzymatic activities and DNA analyses which was conducted at the Soil-Plant System Laboratories in Gorizia, Italy.

The laboratory analyses were carried out, according to the Italian soil analyses manual (MIPAAF, 2000), for bulk density (BD), water-filled pore space (WFPS), porosity, total Kjedahl nitrogen (N), pH, cation exchange capacity (CEC), extractable Olsen phosphorus (P), electrical conductivity (EC), SOC, basal and quantitative soil respiration (28 days incubation). Potentially mineralizable N (µg N/g_{dwsoil}/week) was determined through a short-term incubation (7 days) using a KCl extraction and spectrophotometric determination of the initial and final content of NH_4^+ (Bloem *et al.*, 2006). Active C was determined by oxidation of the soil sample with 0.02 M $KMnO_4$ solution, shaking, centrifugation and spectrophotometric analysis at 550 nm (Culman *et al.*, 2012).

Six soil enzymatic activities involved in C, N, P, and sulphur (S), were analysed using acetate esterase, ß-glucosidase, leucine-amminopeptidase, chitinase, alkaline phosphatase and arylsulfatase. These activities were determined on soil extracts using microplates with fluorogenic substrates following enzyme desorption (Fornasier and Margon, 2007). Double strand DNA (dsDNA) was extracted using the same procedure as for enzymes but using 0.12 M pH 7.8 Na_3PO_4 buffer as extractant (Ascher *et al.*, 2009). The dsDNA in the supernatant was quantified fluorimetrically using PicoGreen reagent (Life Technologies).

As for microarthropods, a 10 cm diameter soil core was taken from each subplot. Microarthropods were extracted after 14 days using a Berlese-Tullgren funnel and stored in 70% ethyl alcohol. We quantified the soil biological index (QBS-ar) based on the presence of microarthropod groups at the order level according to its adaptation to the soil environment (eco-morphological index proposed by Parisi *et al.*, 2005). Earthworm density was determined in the field with commercial mustard extraction (3 g/l) and hand-sorting as described in Pelosi *et al.* (2009).

Soil aggregate and carbon fractionation

Mean weighted diameter (MWD), stable macroaggregate index (SmAI) and stable aggregate index (SAI) were determined through wet sieving using two 80 g subsamples with two pre-treatments: (1) slaking and (2) capillary re-wetting with subsequent slaking (Marquez *et al.,* 2004). After wet sieving, three aggregate size fractions were obtained: small macroaggregate (Ma) (0.25-2.0 mm), microaggregate (mA) (53-250 μm) and silt+clay (s+c) (<53 μm). A second fractionation to obtain different C fractions was carried out on the air-dried small macroaggregate fraction using a microaggregate isolator (Six *et al.,* 2000). Briefly, the first physical fractionation was carried out using 50 glass beads and flushing microaggregates and finer particles through the 250 μm mesh screen into a 53 μm sieve in order to obtain microaggregates occluded within macroaggregates (mM). The material collected in the 250 μm mesh was the non-protected coarse particulate organic matter (POM) within macroaggregates (inter-mM-cPOM). The material in the 53 μm sieve was subsequently wet sieved to obtain water-stable microaggregates (mM). On a 5 g subsample of mM a 1.85 g/cm^3 sodium polytungstate solution was used for further densitometric fractionation (Six *et al.*, 2000). Density flotation was done to collect the fine unprotected POM within the macroaggregate fraction (inter-mM-fPOM). The heavy fraction was dispersed overnight by shaking and subsequently was passed through a 53 μm sieve, in order to collect microaggregate-protected POM (intra-mM-POM) and the microaggregate-derived mineral fractions (mineral mM). All C fractions were oven-dried at 60 °C and the C content was determined using a Carlo Erba CHN 1000 (Cernusco, Italy) elemental analyser and the C stock was quantified according to Denef *et al.* (2007).

Soil health assessment framework

The development of the soil health index (SHI) followed three steps of SMAF (Andrews *et al.*, 2004): (1) indicator selection; (2) interpreting indicators in terms of soil function with a non-linear scoring algorithm; and (3) combining indicator scores into an composite SHI based on ecological soil functions. In the first step, we selected a minimum data set (MDS) of soil health indicators either by expert opinion (EO-MDS) or by Principal Component Analysis (PCA-MDS) to better represent the ecological soil functions proposed in Andrews *et al.* (2004). These ecological soil functions included physical stability and support (PSS), water movement and availability (WMA), nutrient cycling (NC), filtering and buffering (FB), habitat and biodiversity (HB). We also included a fifth soil function named long-term C stabilization (C stab). For the EO-MDS we selected indicators that were previously identified in the literature (Karlen and Stott, 1994; Lima *et al.* 2013; Romaniuk *et al.*, 2011; Schindelbeck *et al.*, 2008; Wienhold *et al.*, 2009). The PCA-MDS was selected by performing a standardized PCA (R statistical package). As described in Andrews *et al.* (2002), we assumed that PCAs with high Eigenvalues best represented variation in soil management effects (Table 9.1). Only PCAs with Eigenvalues ≥1 were examined. For the final PCA-MDS, only factors having absolute values within 10% of the highest factor loading were retained. To eliminate the redundant variables, we applied a multivariate correlation coefficients analysis as described in Shukla *et al.* (2006).

The second step involved the transformation of the observed values into unitless values (0 to 1). This transformation was carried out using three main scoring curves including more-is-better, less-is-better, and mid-point optimum (Karlen and Stott, 1994). The experimental data were scored using the SMAF Excel spreadsheet as previously outlined in SMAF applications (Andrews *et al.*, 2004; Karlen *et al.*, 2008). Algorithms were mathematically developed and validated taking into consideration the impact of inherent soil and site, crop, and climate conditions. We used the SMAF scoring algorithms for bulk density, water-filled pore space, ß-glucosidase, potentially mineralizable N, SOC, stable aggregate index, pH and electrical conductivity. The thresholds obtained from the scoring step, the site-specific factors and the type of curve shapes are reported in Table 9.2. For the indicators where a SMAF scoring curve was not developed, a general non-linear scoring curve (NLSC), as described in Karlen and Stott (1994) was used:

$$\text{NLSC: score (y)} = \frac{1}{[1 + ((\text{B} - \text{LT})/(x - \text{LT}))^{2\text{S} \times (\text{B} + x - 2\text{LT})}]} \qquad (1)$$

where B is the baseline value of the soil indicator, and the score equals 0.5, L is the lower threshold, S is the slope of the tangent to the curve at the baseline, x is the soil indicator analytical value. These thresholds were taken from the literature (Table 9.2). We modelled our scoring curves, applying a stepwise procedure to develop the site-specific factors, introducing only one new factor class at a given time. The following indicators were thus transformed: stable macroaggregate index, porosity, active C, respiration, earthworms, QBS-ar, total N and extractable P. Although no well-suited set of thresholds for the remaining indicators was available from the literature, we developed a new scoring curve. We selected a normal distribution curve for each indicator data set and subsequently generated a cumulative normal distribution curve and normalized the Y axis on a scale of 0 to 1. This yielded the scoring curves for stable macroaggregate index, mean weight diameter, porosity, dsDNA soil content, CEC, and four soil enzymatic activities (Table 9.3). With respect to the long-term stabilization of C, due to the absence of any inclusion of C fractions in soil quality indexing studies, we decided to leave the selection and transformation phases *a posteriori*.

Table 9.1. PCA results of data with significant differences between management systems (*P*<0.05).

Principal components	PC1	PC2	PC3	PC4	PC5
Eigenvalues[a]	**5.92**	**2.12**	**1.43**	**1.13**	0.75
variance %	58.42	13.66	7.31	5.57	4.08
Cumulative var %	58.42	72.08	79.39	84.96	89.04
Factors loadings[b,c]					
Soil organic carbon	***-0.266***	**0.076**	0.042	-0.026	-0.097
Active carbon	**-0.254**	***0.143***	**0.088**	0.013	-0.016
Poten miner N	**-0.260**	-0.005	**-0.100**	0.044	0.121
dsDNA soil content	***-0.167***	**-0.333**	**-0.165**	**0.095**	-0.248
QBS-ar	**-0.246**	-0.024	0.035	-0.036	-0.071
Earthworms density	**-0.263**	**-0.072**	0.050	***-0.130***	-0.082
Soil respiration	**-0.195**	**-0.227**	**-0.236**	***0.189***	0.146
Alkaline phospatase	**-0.256**	**0.052**	**0.088**	***-0.156***	0.073
Arylsulfatase	**-0.044**	**0.264**	***-0.560***	**0.141**	0.129
Chitinase	**-0.044**	**0.126**	***-0.579***	**0.230**	-0.280
β-glucosidase	**0.062**	**0.304**	-0.113	***-0.403***	-0.358
Leucine aminopeptidase	**-0.259**	**-0.153**	**-0.087**	-0.004	-0.039
Total N	**-0.265**	0.037	0.014	-0.021	-0.070
Extractable-P	**0.151**	***0.427***	**0.174**	0.009	-0.066
CEC	0.016	**0.216**	**-0.355**	***-0.600***	0.321
pH	**0.044**	**0.414**	**0.096**	**0.316**	-0.452
Electrical conductivity	**-0.247**	***0.103***	**0.125**	-0.013	-0.048
Bulk density	**0.229**	***-0.208***	**-0.092**	**-0.109**	-0.225
Stable aggregate index	**-0.251**	0.022	**0.059**	***-0.151***	-0.215
Porosity	**0.229**	***-0.208***	**-0.092**	**-0.109**	-0.226
Water filled pore space	**-0.148**	***0.318***	**0.080**	**0.286**	0.372
Stable macro-aggr index	**-0.252**	-0.009	-0.036	0.023	-0.165
Mean weighted diameter	**-0.234**	**0.016**	0.081	***-0.300***	-0.136

[a] Boldface Eigenvalues correspond to the PCs examined for the final index

[b] Boldface factor loadings are considered variables highly weighted.

[c] Bold-italic factor loadings correspond to the indicators included in the PCA-MDS.

Table 9.2. Soil health indicators, scoring curve shape, thresholds, site-specific factors and references for the non-linear scoring curves used in this research. All of the indicators were evaluated before for normality *t*-test ($P<0.05$), otherwise they were log transformed.

Indicator	Scoring curve	Lower threshold	Baseline	Upper threshold	Lower baseline	Optimum	Upper baseline	Slope	Site specific factor	References[1]
Bulk density (g/cm^3)	less is better	1.2	1.8	2.1	-	-	-	-1.6041	*c* value [2]	SMAF algorithm; Andrews *et al.* (2004)
Stable aggregate index (%)	more is better	30	45	60	-	-	-	0.0879	SOC_{obs} [3]	SMAF algorithm; Andrews *et al.* (2004)
Stable macroaggregate index (%)	more is better	0	8	18	-	-	-	0.1097	SOC_{obs}	Marquez *et al.* (2004)
Mean weight diameter (mm)	more is better	0.1	0.8	1.2	-	-	-	2.180		Best managed soil
Porosity (%)	optimum	20	-	80	40	50	60	0.128	*c* value	Karlen and Stott (1994)
Water-filled pore space (%)	optimum	15	-	105	30	60	90	0.0398	*c* value	SMAF algorithm; Wienhold *et al.* (2009)
Active carbon (mg/kg)	more is better	0	420	857	-	-	-	0.00299	SOC_{obs}	Culman *et al.* (2012)
Pot. mineralizable N (mg N/g_{ds}·week)	more is better	0	3.7	8	-	-	-	0.326	*c* value	SMAF algorithm; Andrews *et al.* (2004)
Soil organic carbon (g/kg)	more is better	0	4.2	8.7	-	-	-	0.3041	*c* value	SMAF algorithm; Andrews *et al.* (2004)
Respiration (g $C\text{-}CO_2/g$)	more is better	2	6.6	10.1	-	-	-	0.311	humidity [4]	Luo and Zhou (2006)
Earthworms ($\#/m^2$)	more is better	0	70	150	-	-	-	0.0076	-	Peigné *et al.* (2009)
QBS-ar	more is better	0	50	100	-	-	-	0.0169	-	Parisi *et al.* (2005)
dsDNA soil content (ng/g_{ds})	more is better	0	1,704	3,547	-	-	-	0.00049	-	Best managed soil
β-glucosidase (nM 4-Methylumbelliferone/g_{ds})	more is better	0	14.7	35.7	-	-	-	0.3632	SOC_{obs}, pH, N_t	SMAF algorithm; Stott *et al.* (2010)
Soil enzymatic activities [5]	more is better	-	-	-	-	-	-	-	-	Best managed soil
Total N (mg/kg)	more is better	1	3	5	-	-	-	0.1619	-	SMAF algorithm; Karlen and Stott (1994)
Electrical conductivity (mS/cm)	optimum	0	-	2.5	0.7	1	1.3	0.8456	crop[6]	SMAF algorithm; Andrews *et al.* (2004)
CEC ($cmol^{(+)}/kg$)	more is better	10	32.2	48.5	-	-	-	0.0507	crop[6]	Best managed soil
pH	optimum	5	-	8	6	6.5	7	1.9648	crop[6]	SMAF algorithm; Andrews *et al.* (2004)
Extractable-P (mg/kg)	trapezoidal	7.5	-	300	-	30-75	-	0.5032	-	Morari *et al.* (2008)

[1] If available in literature a SMAF algorithm or a well-suited threshold its reference was reported, if not we generated a cumulative distribution function, applying the principle of *best managed soil.*

[2] Texture, climate and soil organic carbon (SOC) classes used to determine SMAF indicator value (their combination gives *c* value), as presented in Potter *et al.* (2006) and Stott *et al.* (2010).

[3] Measured value of Soil Organic Carbon (g/kg).

[4] Field water gravimetric content (gH_2O/g_{ds}).

[5] See Table 9.3.

[6] Crop requirements chosen among those involved in organic (OS) and conventional (CS) system crop rotations.

Table 9.3. Scoring algorithms for soil enzymatic activities, generated from cumulative distribution function and iterative stepwise procedure, for the inclusion of site-specific factors. The starting data set was the normal distribution of enzymatic activity values.

Enzymatic activity	Algorithm[1]	Fixed parameters			Site-specific factors[2]	R^2	*P*-value
		a	b	c	d		
Alkaline phospatase	y=a+b·cos(cx+d)	0.505	0.445	0.026	*f*(pH·SOC·2.481)	0.934	0.014
Arylsulfatase	y=a+b·cos(cx–d)	0.498	0.436	0.144	*f*(pH·SOC)	0.895	0.032
Chitinase	y=a+b·cos(cx+d)	0.504	0.446	0.112	*f*(pH·SOC·3.213)	0.866	0.043
Leucin-amminopeptidase	y=a+b·cos(cx+d)	0.496	0.435	0.036	*f*(pH·N_t · 2.967)	0.833	0.049
Acetate esterase	y=a+b·cos(cx+d)	0.498	0.44	0.002	*f*(pH·SOC · 2.256)	0.875	0.025

[1] y: enzymatic activity score; x: analytical value.

[2] Site-specific factors are the soil sample observed values; SOC = soil organic carbon.

The third phase was the integration of indicator scores into a composite farm-level SHI. We calculated the scores for six ecological soil functions. An example of how SHI is structured is presented in Table 9.4. Briefly, each soil function was weighted and integrated according to Equation 2.

$$SHI = 0.20 \times s(NC) + 0.15 \times s(WMA) + 0.20 \times s(PSS) + 0.10 \times s(BH) + 0.10 \times s(FB) + 0.25 \times s(Cstab) \qquad (2)$$

where numbers are the numerical weights assigned empirically to each soil function, s(NC) is the score for the soil's ability to sustain plant growth, s(WMA) is the score for the soil's ability to facilitate water entry, transfer and absorption, s(PSS) is the rating for the soil's ability to resist degradation and extreme events of climate change, s(FB) is the score for the soil's ability to filter and buffer high concentration of nutrients and pollutants, s(BH) is the score for the soil's ability to provide habitat for soil biota functional groups, and s(C stab) is the score for the soil's ability to encapsulate and stabilize physicochemically soil C in microaggregates. The highest numerical weights were assigned for C stab, PSS and NC functions. This is because of the importance of these soil functions in fulfilling the overall goals of maintaining a healthy soil within the challenges of sustainable intensification strategy. Associated with each soil function were soil health indicators which influenced the related function. A weight that reflected its relevance in fulfilling that soil function was assigned for each indicator. Numerical weights assigned must sum up to 1.0 at each level. Multiplying each indicator score by,

Table 9.4. Soil health score card for the organic system (OS) obtained from non-linear transformed data of EO-MDS.

	Soil ecological function			Indicator Level 1			Indicator Level 2		
	Weight	Score	Value	Weight	Score	Value	Weight	Score	Value
Physical stability and structural support	0.2	0.85	0.17						
macroaggregate stability				0.4	1	0.4			
organic carbon				0.3	1	0.3			
mean weighted diameter				0.1	0.44	0.04			
respiration				0.1	0.68	0.07			
earthworms				0.1	0.39	0.04			
Water movement and avalaibility	0.15	0.43	0.06						
Accommodate water entry				0.4	0.36	0.14			
stability aggregate index							0.5	0.03	0.01
bulk density							0.25	0.99	0.25
earthworms							0.25	0.39	0.1
porosity				0.2	0.96	0.19			
water-filled pore space				0.2	0	0			
electrical conductivity				0.1	0	0			
pH				0.1	0.93	0.09			
Nutrient cycling	0.2	0.75	0.15						
C cycle				0.3	0.86	0.26			
SOC							0.3	1	0.3
active Carbon							0.3	0.99	0.3
β-glucosidase							0.2	0.64	0.13
respiration							0.2	0.68	0.14
N cycle				0.25	0.72	0.18			
total N							0.3	0.22	0.07
P.M.N.							0.3	0.96	0.29
Leuc-amminopeptidase							0.3	0.93	0.28
chitinase							0.1	0.82	0.08
P cycle				0.15	0.98	0.15			
P available							0.6	1	0.6
alkaline phospatase							0.4	0.94	0.38
pH				0.1	0.93	0.09			
CEC				0.1	0.55	0.06			

Table 9.4. Continued.

	Soil ecological function			Indicator Level 1			Indicator Level 2		
	Weight	Score	Value	Weight	Score	Value	Weight	Score	Value
water-filled pore space				0.05	0	0			
arylsulfatase (S cycle)				0.05	0.48	0.02			
Filtering and buffering	0.1	0.96	0.1						
SOC				0.5	1	0.5			
acetate esterase				0.25	0.93	0.23			
bulk density				0.2	0.99	0.2			
respiration				0.05	0.68	0.03			
Biodiversity and habitat	0.1	0.56	0.06						
DNA content				0.33	0.36	0.12			
QBS-ar				0.33	0.94	0.31			
earthworms density				0.33	0.39	0.13			
Long-term C stabilization	0.25	0.78	0.2						
mM-C stock				1	0.78	0.78			
Soil health			0.73						

and summing up these values, we obtained the score for a specific soil function. The soil's function score represented the ability of that soil in supplying that particular ecological function (Hussain *et al.*, 1999). The soil functions values were summed to obtain the overall soil health value, reflecting the whole ecological soil functionality.

Statistical analysis

Soil health index outcomes of the EO-MDS and PCA-MDS were compared with experimental data to understand the dominant mechanisms during the indexing steps. We used SigmaStat 3.5 for all statistical analyses. Analysis of variance was performed to describe the significant differences between production systems using Tukey's multiple comparison test at P=0.05. In all cases, data were tested for normality and homoscedasticity. We performed a stepwise regression to examine the ability of the MDSs in explaining variation in several sustainability endpoints. This step was carried out for all indicators (independent variables) and endpoints (as iterative dependent variables). This allowed us to compare the adjusted R^2 values from regressions using scores versus observed data. The Pearson correlation coefficient was used to test the level of correspondence between each soil function and the final SHI. The prcomp command in R software was used to perform the principle components analysis. We used CurveExpert, version 1.6 to obtain the best fitted scoring algorithms, to plot the scores versus observed data, and to model the scoring curves in response to site-specific factors changes.

9.3 Results and discussion

Indicator selection

The observed values (Table 9.5a) showed that soil in the OS had a significantly greater number of stable aggregates, especially in the small macroaggregate fraction (SmAI), than CS soils. In the OS, reduced tillage intensity, compost application and green manure likely lead to significantly higher levels of SOC, total N content, and mean weighted diameter (MWD). The active-C content of the total SOC content showed in percentage, lower values (3.84±0.21%) in the OS than in the CS (7.41±0.26%). This suggested that a higher percentage of SOC fractions were involved in the process of humification in the OS. Related to C cycling in the OS, there were also significantly higher values of C and N mineralization (as expressed by respiration and PMN indicators), as well as higher values for dsDNA. Our results also showed that five out of six soil enzymatic activities, involved in the cycling of C, N, P and S, were greater in OS than CS. Organically managed soil showed a higher microbial biomass and a more active soil microbial community. Furthermore, the indicators of biodiversity in the OS showed a higher level of adaptation of microarthropods to edaphic life (QBS-ar) and for earthworm density. The chemical indicators showed, instead, higher values of nutrient content in CS compared to OS.

Applying the PCA selection technique to MDS data, we observed that the first four PCAs had Eigenvalues >1, representing 84.96% of cumulative EO-MDS variance (Table 9.1). Under PC1, the highly weighted variables, were not correlated to each other, were SOC (highest factor loading) and DNA content were strongly correlated. Under PC2, extractable-P had the highest factor loading, and thus, was retained for the PCA-MDS, together with active-C, EC, BD, porosity and WFPS. Arysulfatase and chitinase were selected in PC3. Earthworm density, respiration, alkaline phosphatase, β-glucosidase, CEC and SAI were retained in PC4. The final PCA-MDS was composed of 16 indicators, whereas there were 24 with EO-MDS.

Table 9.5. Comparison of means and standard deviations (in brackets) of measured indicator values and with non-linear transformed scores. The latter data set was used for the EO-MDS and PCA-chosen MDS.[1]

	(a) Measured indicator values (n=8)		(b) Non-linear transformed scores	
	OS	CS	OS	CS
BD (g/cm^3)	1.19 (±0.05)a [2]	1.21 (±0.03)a	0.99 (±0.01)a	0.94 (±0.05)a
SAI (%)	36.4 (±0.9)a	22.6 (±0.6)b	0.04 (±0.02)a	0.00 (±0)a
SmAI (%)	19.1 (±2.2)a	11.8 (±3.2)b	1.00 (±0)a	0.73 (±0.38)b
MWD (mm)	0.80 (±0.01)a	0.69 (±0.04)b	0.45 (±0.03)a	0.25 (±0.06)b
Por (%)	45 (±2)a	46 (±1)a	0.92 (±0.06)a	0.95 (±0.02)a
WFPS (%)	19.9 (±4.5) a	45.0(±8.1) b	0.07 (±0.09)a	0.94 (±0.09)b
Active-C (mg/kg)	679 (±12)a	676 (±21)a	0.99 (±0)a	0.93 (±0.02)b
SOC (g/kg)	17.7 (±0.6)a	9.2 (±1.2)b	1.00 (±0)a	1.00 (±0)a
Resp (g C-CO$_2$/g)	6.43 (±0.3)a	4.7 (±0.2)b	0.75 (±0.09)a	0.11 (±0.03)b
Earthworm (#/m^2)	62 (±3)a	3 (±1)b	0.60 (±0.02)a	0.02 (±0.01)b
QBS (value)	93.3(±3.8) a	62.3 (±5.2)b	0.95 (±0.01)a	0.70 (±0.07)b
DNA (ng/g$_{ds}$)	1,863 (±621)a	469 (±118)b	0.67 (±0.28)a	0.07 (±0.02)b
alkP [3] (nM /g$_{ds}$)	75.4 (±13.8)a	58.1 (±1.5)a	0.94 (±0.01a)	0.16 (±0.07)b
aryS (nM/g$_{ds}$)	18.4 (±1.1)a	3.6 (±0.1)b	0.46 (±0.03a)	0.77 (±0.04)b
chit (nM/g$_{ds}$)	12.2 (±1.5)a	4.6 (±0.5)b	0.82 (±0.01a)	0.10 (±0.06)b
β-gluc (nM/g$_{ds}$)	19.6 (±7.7)a	5.7 (±0.6)b	0.74 (±0.13a)	0.11 (±0.05)b
leuc (nM/g$_{ds}$)	52.3 (±0.2)a	15.0 (±0.6)b	0.92 (±0)a	0.58 (±0.09)b
ester (nM/g$_{ds}$)	1,235 (±178)a	805 (±299)b	0.91 (±0.03)a	0.19 (±0.8)b
N tot (mg/kg)	1.44 (±0.02)a	1.44 (±0.09)b	0.23 (±0.01)a	0.25 (±0.02)a
PMN mg N/g$_{ds}$ week	4.78 (±1.73)a	0.66 (±0.4)b	0.78 (±0.27)a	0.04 (±0.02)b
pH	6.9 (±0.3)a	7.3 (±0.2)a	0.61 (±0.45)a	0.14 (±0.19)b
EC (mS/cm^2)	0.24 (±0.01)a	0.27 (±0.02)a	0.00 (±0)a	0.00 (±0)a
CEC (cmol(+)/kg)	33.3 (±0.1) a	40.9 (±2.8)b	0.55 (±0.01)a	0.85(±0.08) b
extr-P (mg/kg)	49.1 (±1.8)a	102.4 (±17.1)b	1.00 (±0)a	0.88 (±0.08)b

[1] Abbreviations used: BD = bulk density; SAI= stable aggregate index; SmAI=stable macro aggregate index; MWD = mean weight diameter with slaking; Por = porosity; WFPS= water-filled pore space; SOC = soil organic carbon; Resp.= soil basal respiration; QBS = biological soil quality index; DNA = soil double-helix DNA content; alkP = alkaline phosphatase; aryS = aryslsulfatase; chit = chitinase; β-gluc = β-glucosidase; leuc = leucine aminopeptidase; ester = acetate esterase; PMN = potentially mineralizable N; EC = electrical conductivity; CEC = cation exchange capacity; extr-P = extractable P.

[2] For a given indicator, mowing treatment means within management systems followed by different letters differ significantly ($P<0.05$ by Tukey test, n=8).

[3] Unit measure for enzymatic activities: nM 4-Methylumbelliferone/g$_{dried\ soil}$.

Soil carbon fractions

Significantly different SOC stocks were observed for small macroaggregates (sMA) in the OS (2,147±105 g C/m^2 sand-free) compared to the CS (1,900±101 g C/m^2 sand-free). Similar results were found by Chen *et al.* (2009). Total mM-C stock in sMA was significantly higher in the OS (413±5 g C/m^2) than in the CS (317±28 g C/m^2). In the OS, the main mechanism involved in long-term C stabilization was likely the physical protection of microaggregates inside the macroaggregates. Denef *et al.* (2007) found that the chemically stabilized mineral fraction was responsible for changes in SOC with time as a result of conservation soil tillage practices. In our research, instead, physically occluded POM inside mM fraction (intra-mM-POM) was in the OS (119±7 g C/m^2) compared to the CS (53±5 g C/m^2). Consistently, Gulde *et al.* (2008) concluded that as C input increased, this fraction is able to capture and stabilize C. The slower macroaggregate turnover due to reduced tillage intensity (Six *et al.*, 2000) and the addition of organic amendments (Gulde *et al.*, 2008) may have resulted in the production of more stable microaggregates that could protect POM. For these reasons we used mM-C stock as an indicator of long-term C stabilization in the SHI.

Indicators of transformation

The non-linear scoring procedure showed that rating soil health according to this procedure resulted in ecologically-sound soil data scores (Table 9.5b). Also, in Andrews *et al.* (2002) the non-linear scores, even if they were obtained from a simple MDS, gave better results than linearly transformed scores. In our study SAI, SOC and Ntot indicators scores supported the results found by Karlen *et al.* (2008) and Lima *et al.* (2013). Between the two management systems, we did not find significant differences for these indicators in non-linearly scored treatment means (Table 9.5b). The reason for the lack of differences in scores for SOC, SAI, Ntot were either all within, or completely outside, the optimum range of productivity (crop specific) and environmental goals. For all indicators, also for the optimum and less-is-better curve, the non-linear scoring method performed well, which was also observed for the same indicators in Andrews *et al.* (2004) and Karlen *et al.* (2008).

The R^2 results indicated that EO-MDS was a better selection method in comparison to the multivariate method (PCA-MDS) for all end-points (Table 9.6). This was because PCA works better with a high and well-balanced number of observations and variables (Andrews *et al.*, 2002; Shukla *et al.*, 2006). Both observed values and transformed scores explained the variability when sustainability goals such as pesticides use, tillage intensity and the EU agri-environmental payments were used.

Table 9.6. Adjusted R^2 coefficients of determination for multiple linear regression for observed and transformed data sets (as independent variables) chosen on the basis of expert opinion (EO) and PCA analysis. The iterative dependent variables are several end-points, considered as sustainability goals.

End point (sustainability goals)	Observed values		Transformed values	
	EO	PCA	EO	PCA
Yield (t/ha)	0.79	0.76	0.74	0.69
Tillage (no. field passes for tillage operations)	0.98	0.92	0.99	0.95
Pesticide use (t/ha)	0.88	0.76	0.90	0.79
SOC change (%) [1]	0.73	0.62	0.86	0.85
Seasonal Weed cover (%)[2]	0.79	0.75	0.82	0.79
EU agri-environmental payments (€/ha)[3]	0.87	0.85	0.92	0.90

[1] % SOC difference in observed TOC and mean of soil map unit (ARPAV, 2004).

[2] Estimated % of weed cover averaged over the growing season of farm plots.

[3] Agri-environmental payments are 5y-contracts, within EU Common Agricultural Policy which offer to farmers a remuneration for voluntary based adoption of environmentally-friendly farming practices.

Soil health index outcomes comparison

For both MDSs, the OS always had a significantly greater SHI score compared to the CS (Figure 9.2). Organically managed soils showed a medium to high soil health status (0.72±0.01 for EO-MDS and 0.75±0.01 for PCA-MDS) compared to CS soils (0.47±0.07 for EO-MDS and 0.55±0.08 for PCA-MDS). For vegetable production, slightly lower soil quality values for both OS and CS management practices were also observed in a 4 year tomato (*Solanum lycopersicum* L.) – safflower (*Carthamus tinctorius* L.) – maize (*Zea mays* L.) rotation by Andrews *et al.* (2002).

Eliminating the redundant indicators under different PCs, we found little differences in the sensitivity of the SHI. The exclusion of leucine-amminopeptidase and QBS-ar (biological indicators), N content and pH (chemical indicators), SmAI and MWD (physical indicators) slightly decreased the SHI sensitivity. The same changes in terms of index sensitivity were found by Romaniuk *et al.* (2011) when some of the biochemical parameters were excluded.

Soil of the OS showed a greater ecosystem performance (0.86±0.01 PSS value) for physical stability and structural support (Figure 9.3) compared to the CS (0.63±0.04

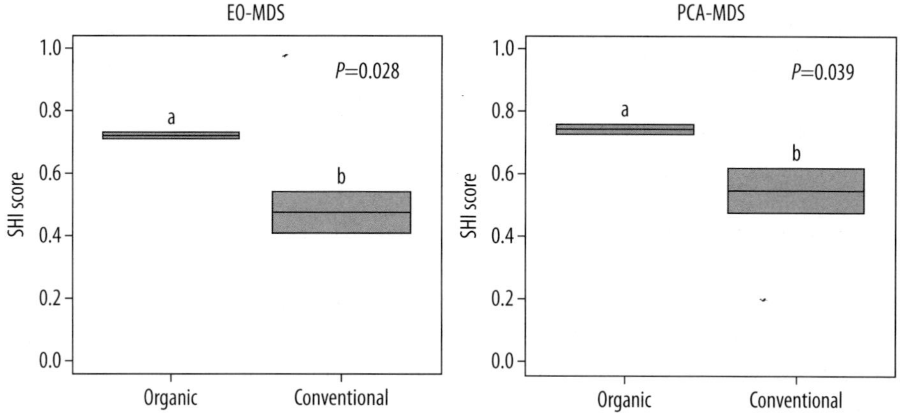

Figure 9.2. Soil health index (SHI) outcomes for organic and conventional systems, including ANOVA *P*-values and Tuckey test results. The mid-point, high, and low bars represent the median, maximum, and minimum of SHI values. Management system labelled with different letters are significantly different at *P*=0.05.

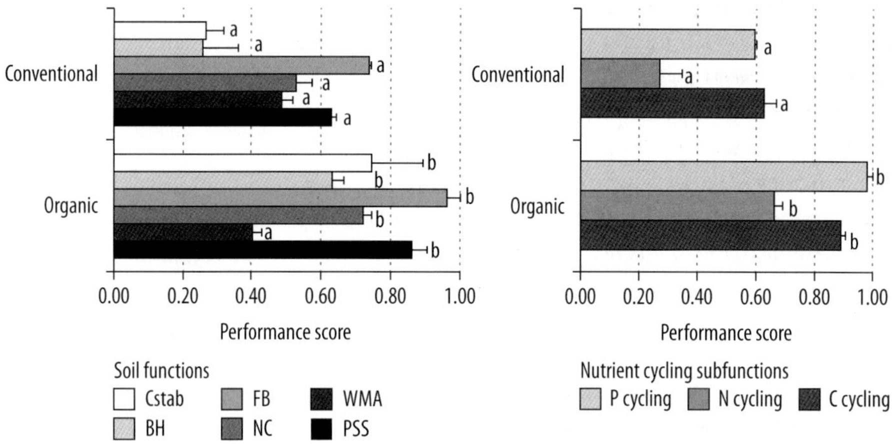

Figure 9.3. Performance scores obtained for (a) the six soil functions of soil height index and (b) for nutrient cycling subfunctions. The data shown are obtained from EO-MDS. Error bars represent standard deviation. Each soil function is affected by the indicators listed in Table 9.4.
Cstab = long term C stabilization; BH = biodiversity and habitat; FB = filtering and buffering; NC = nutrient cycling; WMA = water movement and availability; PSS = physical stability and structural support.

PSS value). For nutrient cycling, soil in the OS performed significantly better for C, N and P. The lowest scores were found in the CS for N cycling (0.27±0.03), which was likely due to the low scores obtained for the PMN and leucine-amminopeptidase activity. The greatest differences in soil performance were observed for biodiversity, habitat, and long-term C stabilization. In the OS, the soil microenvironment, having received less chemical and mechanical disturbance over the years, was more suitable as habitat for soil fauna (0.63±0.01 for OS and 0.24±0.04 for CS). Furthermore, having a greater amount of physically protected C in microaggregates (mM-C fraction) may result in a greater resistance to degradation and extreme events under future climate change scenarios in the OS (0.75±0.05 Cstab value) compared to the CS (0.27±0.15 Cstab value) soil.

Overall soil health status was mostly influenced by SOC, SmAI, MWD, earthworm density, QBS-ar, PMN and all enzymatic activities (correlation analysis, $P<0.001$)). These parameters greatly influenced soil ecological function (SF) and behaviour between the two different management systems. For all SF, these indicators had high correlation coefficient values (always >0.9). The final SHI outcomes, as discussed above, are well correlated with sustainability management goals. SHI proved to be well correlated for EO-MDS with tillage intensity (-0.89) and the percent weed cover (0.96), which was also by Andrews *et al.* (2004), and also in our study with EU-payments (0.95).

9.4 Conclusions

The indicator selection technique based on expert opinion combined with a site-specific non-linear transformation phase proved to be an efficient way to rate soil health between organic and conventionally managed agroecosystems for vegetable production. For both MDSs, the site under organic agroecosystem management practices always had significantly greater SHI scores compared to the conventional agroecosystem. The developed indexing technique demonstrated that organic farming supports a more healthy soil. Both indicator scores and overall SHI scores were most effective in representing specific agroecosystem management practices. Considering the site-specific conditions, the SHI helped to better interpret indicator data for productivity and environmental protection. The SHI proved a powerful tool in representing the different aspects of ecological soil functionality. Our results showed that organically managed soils had a good capability to resist to extreme climatic events, to efficiently provide nutrients for plants, to sustain an active soil food web, and to stabilize C in the long-term. Our findings suggested that using the SHI was an adequate approach to determine the most sustainable soil management program, based on soil C sequestration, in order to mitigate the effects of climate change.

References

Andrews, S.S. and Carroll, C.R., 2001. Designing a soil quality assessment tool for sustainable agroecosystem management. Ecological Applications 11: 1573-1585.

Andrews, S.S., Karlen, D.L. and Cambardella, C.A., 2004. The soil management assessment framework: a quantitative soil quality evaluation method. Soil Science Society of America Journal 68: 1945-1962.

Andrews, S.S., Karlen, D.L. and Mitchell, J.P., 2002. A comparison of soil quality indexing methods for vegetable production systems in Northern California. Agriculture, Ecosystems and Environment 90: 25-45.

ARPAV, 2004. Carta dei suoli del bacino scolante in laguna di Venezia. Servizio Osservatorio Suolo e Rifiuti di Castelfranco Veneto, Dipartimento Provinciale di Treviso, Venice, Italy.

Ascher, J., Ceccherini, M.T., Pantani, O.L., Agnelli, A., Borgogni, F., Guerri, G., Nannipieri, P. and Pietramellara, G., 2009. Sequential extraction and genetic fingerprinting of a forest soil metagenome. Applied Soil Ecology 42: 176-181.

Bastida, F., Zsolnay, A., Hernández, T. and García, C., 2008. Past, present and future of soil quality indices: a biological perspective. Geoderma 147: 159-171.

Bloem, J., Hopkins, D.W. and Benedetti, A., 2006. Microbiological methods for assessing soil quality. CABI, Wallingford, UK.

Carter, M.R., 2002. Soil quality for sustainable land management: organic matter and aggregation interactions that maintain soil functions. Agronomy Journal 94: 3-47.

Chen, H., Hou, R., Gong, Y., Li, H., Fan, M. and Kuzyakov, F. 2009. Effects of 11 years of conservation tillage on soil organic matter fractions in wheat monoculture in Loess Plateau of China. Soil and Tillage Research 106: 85-94.

Culman, S.W., Snapp, S.S., Freeman, M.A., Schipanski, M.E., Beniston, J., Lal, R., Drinkwater, L.E., Franzluebbers, A.J., Glover, J.D., Grandy, A.S., Lee, J., Six, J., Maul, J.E., Mirsky, S.B., Spargo, J.T. and Wander, M.M., 2012. Permanganate oxidizable carbon reflects a processed soil fraction that is sensitive to management. Soil Science Society of America Journal 76: 494-504.

Denef, K., Zotarelli, L., Boddey, R.M. and Six, J., 2007. Microaggregate-associated carbon as a diagnostic fraction for management induced changes in soil organic carbon in two Oxisols. Soil Biology and Biochemistry 39: 1165-1172.

Dominati, E., Patterson, M. and Mackay, A., 2010. A framework for classifying and quantifying the natural capital and ecosystem services of soils. Ecological Economics 69: 1858-1868.

Doran, J.W. and Zeiss, M.R., 2000. Soil health and sustainability: managing the biotic component of soil quality. Applied Soil Ecology 15: 3-11.

Dungait, J.A.J., Hopkins, D.W., Gregory, A.S. and Whitmore, A.P., 2012. Soil organic matter turnover is governed by accessibility not recalcitrance. Global Change Biology 18: 1781-1796.

Faber, J.H. and Van Wensem, J., 2012. Elaborations on the use of the ecosystem services concept for application in ecological risk assessment for soils. Science of the Total Environment 415: 3-8.

Fornasier, F. and Margon, A., 2007. Bovine serum albumin and triton X-100 greatly increase posphomonoesterases and arylsulfatase extraction yield from soil. Soil Biology and Biochemistry 39: 2682-2684.

Gugino, B.K., Idowu, O.J., Schindelbeck, R.R., Van Es, H.M., Wolfe, D.W., Moebius-Clune, B.N., Thies, J.E. and Abawi, G.S., 2009. Cornell soil health assessment training manual. 2nd ed. Cornell University, Geneva, NY, USA.

Gulde S., Chung H., Amelung W., Chang C. and Six J., 2008. Soil carbon saturation controls labile and stable carbon pool dynamics. Soil Science Society of America Journal 72: 605-612.

Hussain, I., Olson, K.R., Wander, M.M. and Karlen, D.L., 1999. Adaptation of soil quality indices and application to three tillage systems in southern Illinois. Soil and Tillage Research 50: 237-249.

Karlen, D.L. and Stott, D.E., 1994. A framework for evaluating physical and chemical indicators of soil quality. In: Doran, J.W., Coleman, D.C., Bezdicek, D.F. and Stewart, B.A. (eds.) Defining soil quality for a sustainable environment. Soil Science Society of America Journal 35: 53-72.

Karlen, D.L., Ditzler, C.A. and Andrews, S.S., 2003. Soil quality: why and how? Geoderma 114: 145-156.

Karlen, D.L., Tomer, M.D., Neppel, J. and Cambardella, C.A., 2008. A preliminary watershed scale soil quality assessment in north central Iowa, USA. Soil and Tillage Research 99: 291-299.

Kibblewhite, M.G., Ritz, K. and Swift, M.J., 2008. Soil health in agricultural systems. Philosophical Transmission of the Royal Society B 363: 685-701.

Lal, R., 2011a. Sequestering carbon in soils of agro-ecosystems. Food Policy 36: S33-S39.

Lal, R., 2011b. Soil health and climate change: an overview. In: Singh, B.P. (ed.) Soil health and climate change. Soil Biology vol. 29. Springer-Verlag, Berlin, Germany, pp. 3-24.

Liebig, M.A., Miller, M.E., Varvel, G.E., Doran, J.W. and Hanson, J.D., 2004. AEPAT: a computer program to assess agronomic and environmental performance of management practices in long-term agroecosystem experiments. Agronomy Journal 95: 316-322.

Lima, A.C.R., Brussaard, L., Totola, M.R., Hoogmoed, W.B. and De Goede, R.G.M., 2013. A functional evaluation of three indicator sets for assessing soil quality. Applied Soil Ecology 64: 194-200.

Luo, Y. and Zhou, X., 2006. Soil respiration and the environment. Academic Press, Burlington, MA, USA.

Màrquez, C.O., Garcia, V.J., Cambardella, C.A., Schultz, R.C. and Isenhart, T.M., 2004. Aggregate-size stability distribution and soil stability. Soil Science Society of America Journal 68: 725-735.

Ministero delle Politiche Agricole, Alimentari e Forestali (MIPAAF), 2000. Metodi di analisi fisica, chimica e biochimica del suolo. Osservatorio nazionale pedologico per la conservazione del suolo, Rome, Italy.

Morari, F., Lugato, E. and Giardini, L., 2008. Olsen phosphorus, exchangeable cations and salinity in two long-term experiments of north-eastern Italy and assessment of soil quality evolution. Agriculture, Ecosystems and Environment 124: 85-96.

Parisi, V., Menta, C., Gardi, C., Jacomini, C. and Mozzanica, E., 2005. Microarthropod communities as a tool to assess soil quality and biodiversity: a new approach in Italy. Agriculture, Ecosystem and Environment 105: 323-333.

Peigné, J., Cannavaciuolo, M., Gautronneau, Y., Aveline, A., Giteau, J.L. and Cluzeau, D., 2009. Earthworm populations under different tillage systems in organic farming. Soil and Tillage Research 104: 207-214.

Pelosi, C., Bertrand, M., Capowiez, Y., Boizard, H. and Roger-Estrade, J., 2009. Earthworm collection from agricultural fields: comparisons of selected expellants in presence/absence of hand-sorting. European Journal of Soil Biology 45: 176-183.

Potter, S.R., Andrews, S., Atwood, J.D., Kellogg, R.L., Lemunyon, J., Norfleet, L. and Oman, D., 2006. Model simulation of soil loss, nutrient loss, and change in soil organic carbon associated with crop production. Conservation Effects Assessment Project, NRCS, Washington, DC, USA.

Powlson, D.S., Gregory, P.J., Whalley, W.R., Quinton, J.N., Hopkins, D.W., Whitmore, A.P., Hirsch, P.R. and Goulding, K.W.T., 2011b. Soil management in relation to sustainable agriculture and ecosystem services. Food Policy 36: S72-S87.

Powlson, D.S., Whitmore, A.P. and Goulding K.W.T., 2011a. Soil carbon sequestration to mitigate climate change: a critical re-examination to identify the true and the false. European Journal of Soil Science 62: 42-55.

Romaniuk, R., Giuffré, L., Costantini, A., Bartoloni, N. and Nannipieri, P., 2011. A comparison of indexing methods to evaluate quality of soils: the role of soil microbiological properties. Soil Research 49: 733-741.

Rutgers, M., Van Wijnen, H.J., Schouten, A.J., Mulder, C., Kuiten, A.M.P., Brussaard, L. and Breure, A.M., 2012. A method to assess ecosystem services developed from soil attributes with stakeholders and data of four arable farms. Science of the Total Environment 415: 39-48.

Schils, R., Kuikman, P., Liski, J., Van Oijen, M., Smith, P., Webb, J., Alm, J., Somogyi, Z., Van den Akker, J., Billett, M., Emmett, B., Evans, C., Lindner, M., Palosuo, T., Bellamy, P., Alm, J., Jandl, R. and Hiederer R., 2008. Review of existing information on the interrelations between soil and climate change. Final report of CLIMSOIL Project. European Community, Brussels, Belgium.

Schindelbeck, R.R., Van Es, H.M., Abawi, G.S., Wolfe, D.W., Whitlow, T.L., Gugino, B.K., Idowu, O.J. and Moebius-Clune, B.N., 2008. Comprehensive assessment of soil quality for landscape and urban management. Landscape and Urban Planning 88: 73-80.

Schmidt, M.W.I., Torn, M.S., Abiven, S., Dittmar, T., Guggenberger, G., Janssens, I.A., Kleber, M., Kögel-Knabner, I., Lehmann, J. and Manning, D.A.C., 2011. Persistence of soil organic matter as an ecosystem property. Nature 478: 49-56.

Shukla, M.K., Lal, R. and Ebinger, M., 2006. Determining soil quality indicators by factor analysis. Soil and Tillage Research 87: 194-204.

Six, J., Conant, R.T., Paul, E.A. and Paustian, K., 2002. Stabilization mechanisms of soil organic matter: implications for C-saturation of soils. Plant and Soil 241: 155-176.

Six, J., Elliott, E.T. and Paustian, K., 2000. Soil macroaggregate turnover and microaggregate formation: a mechanism for C sequestration under no-tillage agriculture. Soil Biology and Biochemistry 32: 2099-2103.

Stott, D.E., Andrews, S.S., Liebig, M.A., Wienhold, B.J. and Karlen, D.L. 2010. Evaluation of β-glucosidase activity as a soil quality indicator for the soil management assessment framework. Soil Science Society of America Journal 74: 107-119.

Wienhold, B.J., Karlen, D.L., Andrews, S.S. and Stott, D.E., 2009. Protocol for indicator scoring in the soil management assessment framework (SMAF). Renewable Agriculture and Food Systems 24: 260-266.

Zobeck, T.M., Halvorson, A.D., Wienhold, B., Acosta-Martínez, V. and Karlen, D.L., 2008. Comparison of two soil quality indexes to evaluate cropping systems in northern Colorado. Journal of Soil and Water Conservation 63: 329-338.

Chapter 10.

Using microbial community interactions within plant microbiomes to advance an evergreen agricultural revolution

M.E. Lucero[1], S. DeBolt[2], A. Unc[3], A. Ruiz-Font[4], L.V. Reyes[5], R.L. McCulley[6], S.C. Alderman[7], R.D. Dinkins[8], J.R. Barrow[9] and D.A. Samac[10]

End-O-Fite Enterprises LLC, 229 Vista Del Rey Drive, Vado, NM 88072, USA; [2]Department of Horticulture, University of Kentucky, N-318 Ag Sciences Center, Lexington, KY 40546, USA; [3]Boreal Ecosystems Research Initiative, Memorial University of Newfoundland, Grenfell Campus, Corner Brook, NL A2H 6P9, Canada; [4]Center for Applied Biotechnology Investigations, National Polytechnic Institute, Ex-Hacienda San Juan Molino, Carretera Tecuexcomac-Tepetitla, Lardizabal, Tlaxcala 90700, Mexico; [5]Graduate School, New Mexico State University, P.O. Box 30001 MSC 3-RRGS, Las Cruces, NM 88003, USA; [6]Department of Plant and Soil Sciences, University of Kentucky, 1405 Veterans Drive, Lexington, KY 40546, USA; [7]USDA-ARS, Forage, Seed, and Cereal Research Laboratory, 3450 Southwest Campus Way, Corvallis, OR 97331, USA; [8]USDA-ARS, Forage Animal Production Research Unit, 1100 S. Limestone, Lexington, KY 40503, USA; [9]USDA-ARS, Jornada Experimental Range, 2995 Knox Street, Las Cruces, NM 88003, USA; [10]USDA-ARS, Plant Science Research Unit, 1991 Upper Buford Circle, St. Paul, MN 55108, USA; malucero64@gmail.com

Abstract

Innovative plant breeding and technology transfer fostered the Green Revolution (GR), which transformed agriculture worldwide by increasing grain yields in developing countries. The GR temporarily alleviated world hunger, but also reduced biodiversity, nutrient cycling, and carbon (C) sequestration that agricultural lands can provide. Meanwhile, economic disparity and food insecurity within and among countries continues. Subsequent agricultural advances, focused on objectives such as increasing crop yields or reducing the risk of a specific pest, have failed to meet food demands at the local scale or to restore lost ecosystem services. An increasing human population, climate change, growing per capita food and energy demands, and reduced ecosystem potential to provide agriculturally relevant services have created an unrelenting need for improved crop production practices. Meeting this need in a sustainable fashion will require interdisciplinary approaches that integrate plant and microbial ecology with efforts to advance crop production while mitigating effects of a changing climate. Metagenomic advances are revealing microbial dynamics that can simultaneously improve crop production and soil restoration while enhancing crop resistance to

M. Oelbermann (ed.) **Sustainable agroecosystems in climate change mitigation**
DOI 10.3920/978-90-8686-788-2_10, © Wageningen Academic Publishers 2014

environmental change. Restoring microbial diversity to contemporary agroecosystems could establish ecosystem services while reducing production costs for agricultural producers. Our framework for examining plant-microbial interactions at multiple scales, modeling outcomes to broadly explore potential impacts, and interacting with extension and training networks to transfer microbial based agricultural technologies across socioeconomic scales, offers an integrated strategy for advancing agroecosystem sustainability while minimizing potential for the kind of negative ecological and socioeconomic feedbacks that have resulted from many widely adopted agricultural technologies.

Keywords: appropriate technology, Green Revolution, metagenomics, microbial diversity, sustainable agriculture

10.1 Introduction

Innovations in plant genetics and agronomic practices from 1940 to 1970 launched the Green Revolution (GR), which increased crop production worldwide and accelerated development of industrialized agriculture (Ortiz *et al.*, 2007; Stanger and Lauer, 2008). The overall benefit of resulting agricultural practices and their impacts on both food security and environmental health have been debated for decades. The loss of microbial diversity associated with agroecosystems as a result of industrialized agricultural practices has not been a primary focus within this debate. Recently, advances in DNA sequencing of uncultured environmental samples (i.e. metagenomics) amplified awareness that complex microbial communities interact with plants to promote growth. New understanding of this interdependence between plants and microbiomes provides evidence that restoring microbial diversity to agroecosystems is crucial for mitigating impacts of climate change to achieve agricultural sustainability and food security.

In the discussion that follows, we will review GR outcomes and other components of industrialized agriculture that have reduced agroecosystem sustainability. We will highlight the complexities of agroecosystems which challenge contemporary efforts to increase food security and sustainability through improved crop production technologies. We will discuss the need to address this complexity with multidisciplinary efforts that consider impacts across varied spatial, temporal, and socioeconomic scales in order to develop diverse, appropriate technologies that make agriculture socially, economically, and environmentally sustainable. Finally, we will highlight the potential for microbe based technologies to contribute to such efforts in a manner that beneficially impacts both climate change mitigation and agroecosystem sustainability.

10.2 Green Revolution advancements decreased agroecosystem sustainability

Green Revolution technological advances resulted in plant genetic combinations that responded to chemical inputs with high yields. Implementation in Mexico transformed the country from a wheat-importing to a wheat-exporting nation. This prompted India, Pakistan, and Turkey to import Mexican wheat germplasms and technologies. Revolutionary success in these countries was attributed not only to the technical breakthroughs in plant nutrition and genetics, but also to the strategic coordination of social and economic factors. This coordinated effort, described as the kick-off strategy (Borlaug and Aresvik, 1973), was implemented in famine-threatened nations in the 1960's. Three components were involved:

1. distribution of selected wheat germplasms;
2. technology transfer via demonstration plots staged to make yield increases resulting from chemical fertilizers and hybrid grain cultivars visible to local producers, planners, government officials, and scientists;
3. government price supports that ensured profitability of the new practices.

India, Pakistan, and Turkey each saw wheat production double within five years after adapting Mexican wheat varieties (Ortiz *et al.*, 2007). The powerful beneficial impacts that ensued supported Norman Borlaug's Nobel Peace Prize in 1970, and contributed to a global transformation in agricultural practices.

Gains achieved through the GR inspired hope that technologies to increase crop yields would end, or at least reduce, world hunger. Unfortunately, this GR promise was not realized in a uniform or sustainable fashion. The failure to sustainably reduce hunger, despite increased production and continual technological advancement, can be attributed to a multitude of complex socioeconomic, political, and environmental factors that interact across scales and disciplines to negatively influence the geographic distribution of food and its utilization (Figure 10.1). Borlaug cited population growth and insufficient global understanding of crop production as factors that limited benefits through improved plant genetics. He lamented the limited disciplinary scope of agricultural research efforts in his time (Borlaug, 1977). M.S. Swaminathan, who was instrumental in bringing Borlaug's high yielding varieties to India, spent subsequent decades articulating the diverse ecosystem threats posed by narrowly focused efforts to increase crop yields.

The potential for temporal and spatial variation influencing food security, including economic disparity, human population growth, market demands, infrastructure, governmental policies, weather, climate, terrain, soil quality, genetic drift and pests, is high. This makes efforts to understand the consequences of technological changes

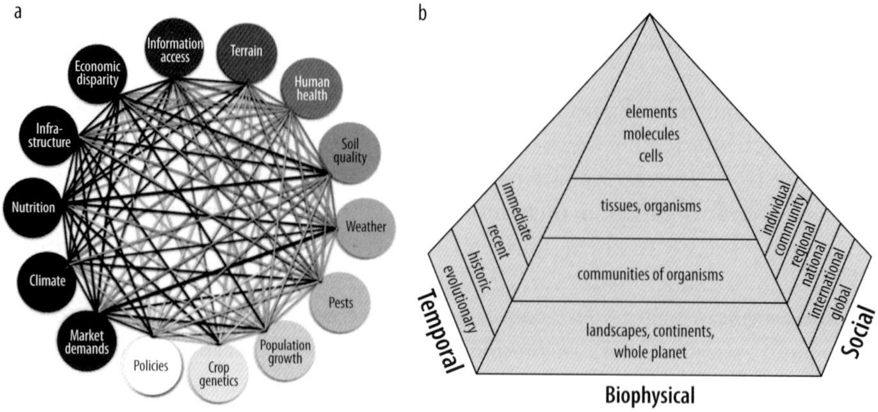

Figure 10.1. (a) Variables that interact to influence food security span infinitely complex disciplines and scales. Variables are selected from an infinite spectrum of interdependent factors that influence food security. Lines between variables illustrate potential for interactions that span spatial and temporal scales. (b) This subset of variables is broadly grouped into scaled temporal, biophysical, and social categories. Lines separating variables or groups of variables at each scale represent boundaries that can limit an individual's area of expertise.

tedious and costly to address (Figure 10.1a). Concerns that agricultural research is threatened by insufficient support for interdisciplinary research were articulated by agricultural scientists as early as the nineteenth century, when W.J. Spillman experimented with wheat genetics (Carlson, 2005). In order to document progress despite limited resources, it becomes tempting to achieve advancement or impact in narrowly focused areas deemed important by many (e.g. crop yields). However, this approach fails to address the significance of those interacting factors that are more difficult to control, such as climate change, insect and microbial populations, human understanding of ecology and economics. Hence, scientists frequently specialize in relatively narrow areas of expertise that span a minimal number of disciplines and scales (Figure 10.1b).

The GR provides an illustration of the potential for interacting factors to create long-term problems when technological breakthroughs in a few areas, including plant genetics, mechanization, and the development of chemical fertilizers, are broadly implemented. Green Revolution technologies clearly allowed growers to temporarily increase crop yields on depleted soils. The dire threat of famine at the time prompted complementary efforts in international diplomacy, extension, and government intervention that evaded famine and expanded the global reach of industrialized agriculture. The need for rapid decision-making outweighed the need for careful, long-term planning. Unfortunately, after famine was evaded, GR practices were widely adopted at the expense of developing long-term soil building practices. This occurred despite concerns expressed by GR leaders that widespread adoption could lead to an era of agricultural disaster (Swaminathan, 1968).

Today, our global population includes nearly 1 billion humans who suffer from chronic hunger (FAO, 2008), and others who suffer from malnutrition. Meanwhile, soil erosion, reduced soil and water quality, and lost biodiversity associated with contemporary agricultural practices is high (Gunningham, 2007; Kiers *et al.*, 2008). The challenge we face is to devise new agricultural approaches that consider not only improved crop yield and hardiness under controlled management conditions, but also the broader, less manageable ecological and socioeconomic factors that interact across scales to influence food security and environmental quality. To meet this challenge, we must recognize the full complexity of the problem.

Figure 10.1a illustrates the arguably infinite spectrum of interdependent variables that interact to influence food security. Due to their intrinsic interconnectedness, manipulation of any set of variables will have direct or indirect, immediate or deferred influences on other variables. For example, a manager's access to information about pest populations may prompt decisions to treat crops with insecticide. The decision may be driven by the real need to manage income. However, the action will also impact soil and water quality, biodiversity, human health, nutrition, and other variables in ways that are more difficult to evaluate.

Scaling of variables is illustrated in Figure 10.1b, and interactions across scales must also be considered in order to better understand the potential for decisions to manage a set of variables at one scale, such as altered amino acid biosynthetic pathways engineered into a plant genome at the molecular scale, to influence other variables including microbial interactions, crop production costs, human nutrition, public perception, at other scales. Hence, managers, researchers and policy makers addressing food security are faced with the task of assessing limitless potential outcomes with finite resources.

10.3 Multi-scaled and multidisciplinary efforts help ensure sustainability

To truly make agricultural advances that benefit society as a whole, it is important to integrate an understanding of those technologies developed within one discipline, and applied at a single scale. This includes advances that enhance cellular communication between plant roots and rhizobia, with the broader, less tangible effects that these variables may have on related factors at broader scales such as pest populations, human nutrition, and environmental quality. In addition to these broader impacts, we must also acknowledge that GR technologies, which were implemented to alleviate hunger by increasing crop production, are recognized by many to have confounded issues by supporting population growth, loss of small farms, environmental degradation, and calorie rich, nutrient poor dietary practices (Das, 2002; Lairon, 2010; Pinstrup-Anderson and Hazell, 1985).

With an increasing human population, industrial development also progresses. Climate change and environmental degradation ensue, which challenge sustainable crop production (FAO, 2008; Sanchez and Swaminathan, 2005; Swaminathan, 2010; UN Millennium Project, 2005). For example, climate change has the potential to increase the abiotic stress on crops while altering the environmental niches for pests, pathogens, and weeds. Alternately, enhancing food production using current technologies increases fuel requirements, but depleted fuel reserves drive demand for biofuels, making crop production for food and fuel directly competitive. Each of these concerns is closely tied to parallel concerns with environmental quality.

Numerous local, regional, and global efforts are currently addressing food security with innovative information and technology transfer efforts (Bill and Melinda Gates Foundation, 2009; FAO, 2011; Lombard *et al.*, 2006; Ortiz, 2006; Ortiz *et al.*, 2007). The term Evergreen Revolution (ER) was coined to articulate the need for making the outcomes of these efforts sustainable (Swaminathan, 2010). It is uncertain whether contemporary approaches, which are all important ER building blocks, will be adequate to address current and future food demands (Sanchez and Swaminathan, 2005). Notably, the majority of these efforts remain phytocentric, emphasizing plant-based technologies for enhancing crop production (Lynch, 2007; Ortiz *et al.*, 2007).

Recent advances in understanding plant-microbe interactions, driven by new technologies that allow the rapid assessment of microbial communities, hold significant promise for meeting global needs to accelerate the ER. In this chapter, we highlight the untapped potential of microbial communities to improve crop yields while restoring soil building, nutrient cycling, and pest reduction properties to agroecosystems. We note features of microbe assisted plant production that offer socioeconomic benefits for small farmers and carbon (C) sequestration features that offer relevance for mitigating the impacts of climate change. We also describe a fundamental strategy that can accelerate understanding of plant-microbe interactions within the multi-scaled context of entire agroecosystems. We outline existing and developing networks designed to promote information sharing and participatory research to extend new knowledge across socioeconomic and geographic scales. We end by summarizing the critical need for policies that promote this and other broad, integrated, and long term efforts to advance sustainable agriculture and related food security.

10.4 Microbial interactions across plant, soil, and environmental interfaces

In undisturbed ecosystems, plants are associated with a continuum of other organisms, the majority of which are microbial. These include epiphytes on plant surfaces, endophytes within plants, and rhizosphere and soil microbes associated with subsurface plant organs and soil interfaces (Figure 10.2). These microbiomes sequester significant amounts of atmospheric C, thus aiding in mitigation of climate change (Müller-Stöver *et al.,* 2012). Furthermore, microbiomes connect plants to surrounding substrates (Green *et al.*, 2008; Khidir *et al.*, 2008), cycle nutrients (Barrow *et al.*, 2008; Green *et al.*, 2008; Rodriguez and Fraga, 1999) stabilize soil (Gale *et al.*, 2000); enhance tolerance to biotic and abiotic stress (Barrow *et al.*, 2008), and buffer the impact of environmental factors on plants (Gale *et al.*, 2000; Lee *et al.*, 2009; Okon, 1985; Plett and Martin, 2011; Rodriguez and Fraga, 1999; Rodriguez and Redman, 2008). Remarkably, these microbial functions parallel many of the management actions humans perform to enhance crop growth in agroecosystems (Barrow *et al.*, 2008). For example, microbes can produce alkaloids that deter insect pests, reducing the need for applied pesticides (Kuldau and Bacon, 2008; Schardl *et al.*, 2004). Microbes fix atmospheric nitrogen, and solubilize phosphorus, which can replace the need for mineral fertilizer (Richardson, 2001; Rodriguez and Fraga, 1999; Stanger and Lauer, 2008). Microbes improve soil aggregation, increasing both soil aeration and soil water holding capacity (Gale *et al.*, 2000) in ways that may reduce the need for intensive irrigation management or deep ploughing.

Notably, while much crop research is devoted to the study of mycorrhizal fungi, rhizobia, and other root-associated microbes that fix nitrogen, endophytes (Barrow *et al.*, 2008) and soil crust microbes (Briggs and Morgan, 2012; Evans and Belnap, 1999; Green *et al.*, 2008) have been most heavily studied in pasture grasses and natural ecosystems. Endophytes are of growing interest to agriculture because the benefits they confer to plants may be transferred across generations, offering a rapid, inexpensive alternative to genetic engineering (Barrow *et al.*, 2008). Soil crust microbes, which have the capacity to fix nitrogen and promote soil stability (Briggs and Morgan, 2012; Evans and Belnap, 1999), are largely ignored in agroecosystems. Industrial agricultural practices such as mechanical planting, tilling and cultivation, application of synthetic fertilizers and pesticides, and harvesting disturb structural integrity and functional diversity of associated microbial communities, forcing increased reliance on artificial inputs to maintain crop yields.

Historically, advances that increased agricultural production were achieved largely through bypassing plant-microbiome associations, relying instead on tillage to manage weeds and soil compaction, and agrochemicals to provide nutrients and to remove

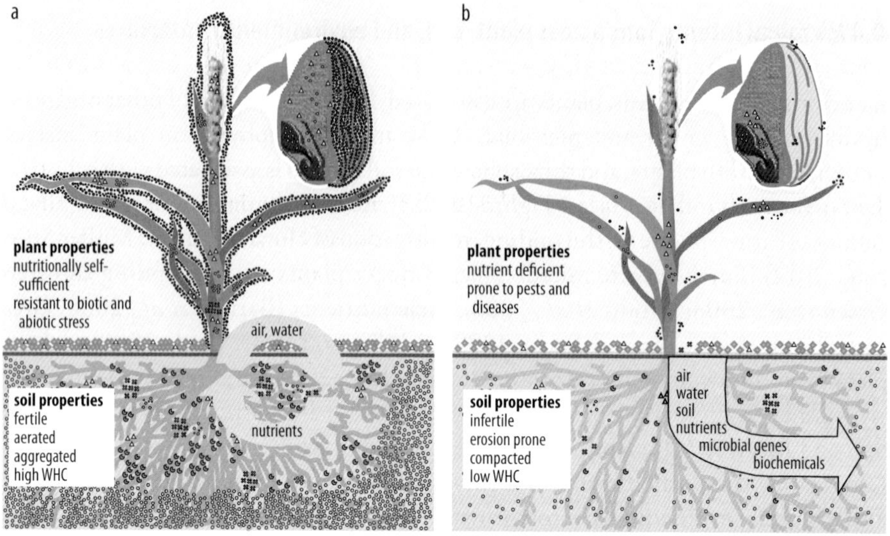

Figure 10.2. (a) Plants in undisturbed ecosystems interact with a continuum of microbial endophytes (light grey circles illustrated in plant tissues) and epiphytes (dark grey circles illustrated on plant surfaces), rhizosphere microbes (black outlined grey crescents illustrated on and near plant roots), soil crust microbes (dark grey outlined grey diamonds illustrated on soil surface), facultative (grey outlined white triangles illustrated in and on plants and soil), and free living soil microbes (black outlined light grey circles illustrated in soil) that interact to cycle basic nutrients and/or respond to biotic and abiotic stress. Microbes build valuable soil properties including aggregation, aeration, and water holding capacity (WHC). (b) In conventional agroecosystems, microbial communities may be altered and/or reduced by fertilization, fumigation or application of broad spectrum pesticides. These practices can deplete nutrients, decrease stress tolerance, and leave soil erosion-prone and compacted, reducing ecosystem capacity to sequester atmospheric carbon. Resulting crops may also differ in genetic, biochemical, and/or nutritional quality.

biotic stressors. Such an approach was unavoidable, because until recently, microbial communities and their associated effects on plant physiology could not be efficiently detected or studied. Bypassing the role of microbes and other plant-supporting biota resulted in the implementation of practices, such as application of non-selective pesticides to reduce pathogens, which unintentionally decoupled plants from their supporting organisms, including those vast microbial communities whose co-evolutionary success mandated efficient synchronization of plant primary production with decomposition, nutrient solubilization, and soil formation processes.

Many contemporary agronomic practices and agrochemicals negatively influence microbial community structures, adversely impacting soil fertility (Bueno and Ladha,

2009; Fox *et al.*, 2007) and C emissions (Dubey and Lal, 2009), while contributing to environmental degradation (Kiers *et al.*, 2008). As environmental impacts associated with these actions, including reduced C sequestration, loss of beneficial insect populations, increased soil erosion, and groundwater contamination by agrochemicals are recognized, the actions themselves have been justified by our critical need to feed the expanding human population. This justification persists despite awareness that the practices themselves negatively impact the input costs associated with crop production, making food security increasingly difficult for low income communities to attain (Kiers *et al.*, 2008).

10.5 Evaluating microbial potential to revolutionize agroecosystem sustainability

An alternative to advancing food security and mitigating climate change through phytocentric agricultural technologies is to leverage entire microbial communities to cycle nutrients, build soil aggregates, and boost plant stress tolerance (Barrow *et al.*, 2008; Dubey and Lal, 2009; Rodriguez and Redman, 2008). Potential benefits of exploiting microbial interactions with plants have been recognized for well over a century (Frank, 2005). Indeed, some of these benefits have already been adopted through improved crop rotation, precision farming, and organic farming, reduced tillage, and integrated pest management (Dubey and Lal, 2009; Stanger and Lauer, 2008; Swaminathan, 2010). Introducing nitrogen-fixing or phosphorus-mobilizing microbes into crop production systems can increase nutrient availability (Rodriguez and Fraga, 1999; Stanger and Lauer, 2008). For example, transfer of seed-borne endophytic microbes may create new germplasms with improved stress tolerance (Barrow *et al.*, 2008).

To date, outcomes to harness microbial services vary. Such variation is attributed to insufficient understanding of the complexity of microbial communities, lack of genetic compatibility between tested hosts and microbes, and poor microbial detection and monitoring technologies (Barrow *et al.*, 2008; Richardson, 2001). New high throughput DNA sequencing approaches offer the first significantly feasible opportunities to broadly examine those microbial interactions important for increasing plant production (Committee on Metagenomics, 2007). These techniques, including metagenomics to describe microbial community genetic diversity, and transcriptomics, proteomics or metabolomics to describe molecular scale community interactions are now applied to plant and soil microbial systems to explore microbial influences on sustainable plant productivity.

Even the most advanced, high throughput methods available today fail to detect large portions of the microbial communities present (Lucero *et al.*, 2011), and it is unlikely that the incredible diversity of microbial kingdoms will never be entirely understood. This is one reason why an approach that utilizes knowledge of biotic patterns and processes observed at larger scales may aid in understanding microbial process at fine scales which may leverage crop production.

While it will never be possible to assess all the variables that could interact within an agroecosystem, these complex and cross-scale interactions can be better understood through networked efforts that combine data from diverse ecosystem components, and simulation models for multiple, linked scales (Peters *et al.*, 2007). Such networked efforts enable specialists within highly focused areas of expertise such as genomics, microbial ecology, soil science and economics, to consider broader suites of drivers and responses, including immediate and latent responses to management actions. Such a networked approach to address complex ecological questions has already been established (Moran *et al.*, 2008; Peters *et al.*, 2004, 2007).

A conceptual framework relevant for advancing sustainable, microbial based agroecosystems to enhance food security and mitigate climate change could include a synthesis of ecological data collected from fine to broad spatial and temporal scales, and socioeconomic data collected from diverse populations, as detailed below. Such data, much of which is already available through unrelated analyses carried out across varied natural science disciplines, could be linked through simulation models that illustrate immediate and long term effects of management actions on microbial community assemblages and agroecosystem function. These linked models may be useful for predicting immediate and long-term consequences of management actions with varied impacts on microbial community structures, allowing the exploration of risks and benefits of new technologies before their implementation.

Elemental, cellular, and molecular scale

Experts in chemistry, molecular biology, and biochemistry could further contribute to our understanding of interactions that occur at very fine scales, including nutrient cycling and microbial community structure and function. Genetic and metabolic differences between microbial communities can be revealed through nucleic acid sequencing or metabolic profiling (Committee on Metagenomics, 2007; Lucero *et al.*, 2011). Whole genome sequencing, proteomics, and/or metabolomics of bacteria and fungi associated with plants may advance our understanding of microbial functions that influence plant production, drought and salinity tolerance, chemical defense, growth, pathogenesis, nutrient cycling, primary and secondary metabolism, and nutrition. Microbial genomes deemed valuable for germplasm development may be sequenced to evaluate genes regulating host specificity, biotic and abiotic stress tolerance, pathogenesis, and secondary metabolism (Kuldau and Bacon, 2008; Schardl *et al.*, 2004). Metagenomes (genomic sequences of entire microbial communities) associated with a plant host in a selected habitat may reveal the complete metabolic potential of microbiomes to influence crop production and other kinds of C sequestration, including the rates of C fixation and biomineralization catalysed by soil microbes. Databases such as the National Center for Biotechnology Information (Genbank) or Cyberinfrasturcture for Advanced Microbial Ecology Research and Analysis (CAMERA) already provide publicly available collections of ecological and metagenomic data that, in combination with related research papers can provide foundational information for fine scale simulation models. When linked to simulations carried out at larger scales, these models can be used to forecast how microbial communities might respond to selected management actions such as soil tillage or irrigation, and to understand how these microbial responses might influence factors observed at larger scales including global C sequestration.

Tissue and whole plant scale

Plant physiologists, plant breeders, and microbial ecologists may interact to explore variations in microbiomes associated with native plants and crop plants of varied domestication histories to further understand how management practices impact plant associated microbial communities. In addition, microbial interactions that influence plant productivity and nutrient content may be evaluated. Tag encoded pyrosequencing of microbial communities associated with host plants and surrounding soils can be utilized in conjunction with plant performance bioassays to assess microbial community composition associated with variations in plant production parameters. Laboratory, greenhouse, and field bioassays in which microbial communities are manipulated by adding or removing specific groups of microbes can reveal conditions wherein microbes interact with host plants and/or with other plant associated microbes. Experiments across drought or other environmental stress gradients can reveal mechanisms by which microbes relate to whole plant stress responses that influence crop production. Simulated crop production forecasts may be fed into fine scale models to demonstrate how these vegetative changes might influence C and nitrogen cycling within the soil microbial community. Outputs from both linked simulations can be used as variables in global scale simulations.

Community or biome scale

Microbial communities associated with seeds, leaves, roots, and soil associated with specific cropping systems including conventional monocultures, rotated crops and no-till systems, can be examined to understand how microbial dynamics are influenced by plant community composition. Combining such analyses with comparisons of microbial persistence across trophic levels could reveal microbe-plant interactions with the potential to propagate across food chains. Analyses of microbes associated with neighboring and invasive plants may provide insights into community interactions important for resisting pest invasion, with implications for both crop production and environmental restoration. The resulting ability to predict which cropping systems will be most resistant to invasions or other environmental threats will facilitate planning that ensures productivity even when climatic change makes historic practices obsolete.

Parameters influencing patch scale soil characteristics are already extensively documented at many national scales. For example, soil map units and related ecological site descriptions are publicly available through the USDA Natural Resource Conservation Service (USDA-NRCS). The ambitious, global scale ISRIC World Soil Information Database strives to meet growing global demands by making spatially related data describing soil, climate, geomorphology, vegetation, land use, and other variables available worldwide. By documenting spatial coordinates and describing the ecological parameters associated with field sample locations, these existing, detailed, and publicly available long-term datasets may be leveraged to understand habitat parameters that influence crop production.

Landscape to global scale

Geospatial coordinates and ecological site descriptions associated with each environmental sample could facilitate cross-site comparisons and reveal distinct habitat and management variables that influence microbiome compositions across geographic scales. Such analyses can be powerful for understanding the soil, landscape, and climatic factors influencing crop responses to microbial modifications, and for understanding how these responses might differ as habitats change. Phylogeographic analyses across regional to global scales can be applied to exploring the role of temporal, climatic, and edaphic factors on plant and microbial adaptation, community structure, and co-evolution. Disentangling effects of different factors can best be achieved through these geographic scale analyses where multiple factors can be examined separately and in combination. Linking these broad scale factors, like climate variables or landscape effects, to simulation models predicting crop productivity can aid land managers in the development of locally relevant management practices.

Dynamics that vary across temporal scales

The enormous genetic diversity of microbiomes and their short generation times allows rapid and dynamic response to environmental change. Changes in microbial community structure following seasonal changes or management activities are expected. Using these changes to benefit production or buffer agroecosystem responses to climate change will require further understanding of the temporal dynamics of microbial responses to environmental drivers. Endophyte associations, which may transcend plant generations or influence host ecology, can best be understood through evolutionary analyses of host plants and associated microbes. Such studies, carried out in conjunction with phylogenetic analyses of genetically related plants, could provide transformational insights into how microbes influence plant adaptation. Phylogenetic comparisons that evaluate microbes associated with native, heirloom, and commercial plant germplasms will be uniquely valuable for illustrating important associations that have been lost as a result of various domestication, breeding, and selection practices (Couch *et al.*, 2005; Doebley *et al.*, 2006; Eyre-Walker *et al.*, 1998).

A hierarchy in temporal scales roughly corresponds to the spatial scales described above. While microbial communities can respond rapidly to climate drivers or management, with dominant taxa shifting over periods of minutes to days, dominant taxa in plant communities may take weeks, months, or even years to change. Insights from these spatial and temporal interactions suggested that rapidly responding microbial communities may assist plants in adapting to climate change more rapidly than may be expected through plant genetic change alone. Experiments that test these insights will be valuable for devising strategies that enhance plant production or that make plant communities resilient to climate change.

Management across socioeconomic scales

In agroecosystems, crops are influenced not only by standard biophysical parameters such as climate, soil, and wildlife, but also by human management decisions and actions. Humans decide what crop species to plant, where to plant them, and how to alter the natural environment in order to support crop plants. Such decisions are largely information driven, because they are based on our understanding of how biophysical parameters influence plant productivity and interact over time with various factors including production costs, market prices, labor, consumer tastes and preferences. Like biophysical and temporal parameters, socioeconomic parameters interact across scales. Hence, an individual crop producer's decision to grow a specific crop might be influenced at the national level by information about government price incentives and at the local or regional level by knowledge of consumer demand, production capacity, and marketing networks. Such information driven human management decisions are among the most powerful determinants of food security, climate change mitigation, and agroecosystem sustainability. For this reason, it is important that managers have sufficient information, resources, and decision making authority to implement practices that are adapted to the conditions influencing their local environment.

Models and experiences that illustrate the complex, cross-scale interactions impacting people and environments whenever new management approaches are introduced will be helpful to evaluate management decisions prior to their implementation. Making such illustrations widely available, while reducing broad policies that limit management and purchasing options, will empower individuals to explore immediate and long-term consequences of varied actions, thus learning from a combination of real and simulated experiences. This will allow producers and consumers to make informed decisions that effectively balance short-term financial and caloric needs with locally adapted strategies that promote long-term human health and environmental integrity.

Global advances in educational and information technology, including the widespread availability of internet services and cell phone applications promise opportunity to make such models and tools broadly available to diverse socioeconomic strata (Bill and Melinda Gates Foundation, 2009; Sanchez and Swaminathan, 2005; Swaminathan, 2010). Modernized kick-off strategies will need to fully utilize these advancements while simultaneously increasing time honoured extension and education efforts, in order to promote effective information transfer to policy makers, crop producers, food processors, distributors, and consumers.

Information access across socioeconomic scales

The need for data accessibility to broad sectors of public users cannot be understated. Insufficient community level understanding of environmental and nutritional dynamics behind food production technologies could lead to an exacerbation of existing socioeconomic disparity (FAO, 2011; Sanchez and Swaminathan, 2005; Swaminathan, 2010). If the goal of an ER is to achieve sustainable agroecosystems, it is critical that research and the development of new agricultural and food technologies be accompanied by aggressive, innovative efforts to share our understanding of food production technologies and their potential environmental effects. The collective understanding of such efforts must include not only understanding of how to increase crop yields or nutritional values, but also understanding of how different management practices and consumer choices influence ecosystem services that are crucial for the long-term sustainability of agroecosystem management practices. This task can be made manageable by including producers, educators, market developers, policy makers, and consumers in various aspects of planning and implementing agricultural, research and technology transfer efforts. Models such as those embraced by non-profit groups like the International Centers for Appropriate Technology and Indigenous Sustainability (iCATIS), illustrate how these difficult efforts may be advanced through networked partnerships.

10.6 Microbial potential to advance food security and mitigate climate change

Agricultural practices that promote integrated, functional plant and soil microbiomes are expected to promote plant and soil quality in manners that sustain, or even increase crop yields while mitigating the impacts of climate change. Organic farming protocols represent steps in the right direction, because they reduce the chemical disruption of plant and soil microbiomes. However, such protocols do not necessarily reduce mechanical disruption or promote microbial community development sufficiently to foster microbial communities that effectively manage soil moisture and cycle C, nitrogen, and other nutrients. Efforts that evaluate crop yields in conjunction with management practices and microbial community structure are needed. Such efforts are expected to reveal existing practices that simultaneously foster high crop yields, low input costs, and improved ecosystem services. Such practices are likely to vary across agroecosystems, because microbial communities that thrive in one habitat will not be optimal in other habitats.

This natural variation in adapted soil microbial communities and the resulting need for variation in management practices could theoretically provide positive socioeconomic feedbacks by increasing demand for local agricultural experts and improve opportunities for small, diversified farms. The resulting shift from agricultural leadership and expertise provided by broad scale government and corporate entities towards leadership within local communities may foster more direct interactions between growers and consumers. These interactions can improve food security at the community level by promoting diversified food systems that are more responsive to culturally and environmentally diverse demands. Benefits may be particularly significant within indigenous communities, where ancient and locally preserved cultural practices may hold critical clues to develop sustainable technologies that restore microbial diversity to local and regional agroecosystems. Communities that have retained these geographically specific knowledge bases will be well-positioned to initiate new commercial practices that involve training local growers to produce food without agrochemicals, sell climatically adapted heirloom seeds, and develop smallscale technologies for processing regionally unique food types. Such foods are already sought by a new generation of consumers that is increasingly interested in organic and locally grown foods.

10.7 Conclusions

Historic efforts to promote sustainable agroecosystem management practices have demonstrated that increasing plant production without attention to ecological and social impacts is devastating to long-term agroecosystem stability and food security. New breakthroughs revealing the complexity and importance of plant microbiomes offer exciting potential for developing management practices that promote sustainable food security and mitigate climate change. Practices that foster quality plant microbiomes could substitute biotic interactions that enhance crop fitness and yields while restoring soil quality and nutrient cycling functions, including C sequestration. Properly developed soil microbial communities may eliminate the need for mineral fertilizers. Plants colonized by drought tolerant, nitrogen fixing and halophytic microbial consortia may produce crops in arid or saline soils. Restoring endophytic microbes may enhance crop nutritional value by adding metabolic pathways that supplement the vitamins, proteins, and antioxidants produced by their host plants. Because microbial community composition can change rapidly in response to environmental change, microbial associations also lend metabolic plasticity to plant hosts, facilitating adaptation to climate change.

Breakthroughs in communication and information transfer will help ensure that new microbial technologies are implemented in ways that minimize undesired environmental impacts. Global databases increasingly facilitate developing models that illustrate the multi-scaled interactions between plants, microbes, humans, and changing environments. Such robust, multi-scaled models will enable users to thoroughly assess benefits and risks associated with new crop technologies. These models should help prevent the kinds of unintended consequences of the GR. For this reason, multi-scaled ecological models, in combination with new microbial technologies, could accelerate global progress towards food security and sustainable agriculture. This is particularly true if both are made accessible through networked, interdisciplinary, and participatory efforts that promote our understanding of biotic interactions across biophysical, temporal, and socioeconomic scales. These endeavours should be synergistic with existing efforts to address food security.

A well-informed public is crucial to success of any such effort. Public participation can minimize the potential for widespread negative ecological and socioeconomic feedbacks that could stem from newly developed agricultural technologies. It is of great importance for agricultural producers, policymakers, researchers, and educators to review time honoured concepts of education and extension as equalizers of socioeconomic disparity and as cornerstones to sustainable development (UN Millennium Project Task Force on Science, Technology, and Innovation, 2005; Mann, 1848; Peters 2006). Policies and practices that support integration of microbial ecology with crop development, agroecology, environmental sciences, nutrition, socioeconomics, and extension will promote robust, adaptable, and sustainable agroecosystems with the improved potential to mitigate impacts of globalization and climate change.

References

Barrow, J.R., Lucero, M.E., Reyes-Vera, I. and Havstad, K., 2008. Do symbiotic microbes have a role in plant evolution, performance and response to stress? Communicative and Integrative Biology 1: 69-93.

Bill and Melinda Gates Foundation, 2009. Global development program fact sheet, Bill & Melinda Gates Foundation. Available at: http://www.gatesfoundation.org/about/Documents/GlobalDevelopmentProgramFactSheet.pdf.

Borlaug, N.E., 1977. The green revolution: can we make it meet expectations? Proceedings of the American Phytopathological Society 3: 6-21.

Borlaug, N.E. and Aresvik, O.H., 1973. The green revolution-an approach to agricultural development and some of its economic implications. International Journal of Agrarian Affairs 4: 385-402.

Briggs, A.L. and Morgan, J.W., 2012. Post-cultivation recovery of biological soil crusts in semi-arid native grasslands, southern Australia. Journal of Arid Environments 77: 84-89.

Bueno, C.S. and Ladha, J.K., 2009. Comparison of soil properties between continuously cultivated and adjacent uncultivated soils in rice-based systems. Biology and Fertility of Soils 45: 499-509.

Carlson, L., 2005. Forging his own path: William Jasper Spillman and progressive era breeding and genetics. Agricultural History 79: 50-73.

Committee on Metagenomics: challenges and functional applications, National Research Council, 2007. The new science of metagenomics: revealing the secrets of our microbial planet. The National Academies Press, Washington, DC, USA. Available at: http://www.nap.edu/openbook.php?record_id=11902.

Couch, B.C., Fudal, I., Lebrun, M.H., Tharreau, D., Valent, B., Van Kim, P., Nottéghem, J.L. and Kohn, L.M., 2005. Origins of host-specific populations of the blast pathogen *Magnaporthe oryzae* in crop domestication with subsequent expansion of pandemic clones on rice and weeds of rice. Genetics 170: 613-630.

Das, R.J., 2002. The green revolution and poverty: a theoretical and empirical examination of the relation between technology and society. Geoforum 33: 55-72.

Doebley, J., Gaut, B. and Smith, B., 2006. The molecular genetics of crop domestication. Cell 127: 1309-1321.

Dubey, A. and Lal, R., 2009. Carbon footprint and sustainability of agricultural production systems in Punjab, India, and Ohio, USA. Journal of Crop Improvement 23: 332-350.

Evans, R.D. and Belnap, J., 1999. Long-term consequences of disturbance on nitrogen dynamics in an arid ecosystem. Ecology 80: 150-160.

Eyre-Walker, A., Gaut, R.L., Hilton, H., Feldman, D.L. and Gaut, B.S., 1998. Investigation of the bottleneck leading to the domestication of maize. Proceedings of the National Academy of Sciences of the USA 95: 4441-4446.

Food and Agriculture Organization (FAO), 2008. High food prices and food security-threats and opportunities, Food and Agriculture Organization of the United Nations, Rome, Italy. Available at: ftp://ftp.fao.org/docrep/fao/011/i0291e/i0291e00.pdf.

Food and Agriculture Organization (FAO), 2011. How does international price volatility affect domestic economies and food security? Food and Agriculture Organization of the United Nations, Rome, Italy. Available at: http://www.fao.org/docrep/013/i2050e/i2050e.pdf.

Fox, J.E., Burow, M.E., McLachlan, J.A, Gulledte, J. and Engelhaupt, E., 2007. Pesticides reduce symbiotic efficiency of nitrogen-fixing rhizobia and host plants. Proceedings of the National Academy of Sciences of the USA 104: 10282-10287.

Frank, B., 2005. On the nutritional dependence of certain trees on root symbiosis with belowground fungi (an English translation of A.B. Frank's classic paper of 1885). Mycorrhiza 15: 267-275.

Gale, W.J., Cambardella, C.A. and Bailey, T.B., 2000. Root-derived carbon and the formation and stabilization of aggregates. Soil Science Society of America Journal 64: 201-207.

Green, L.E., Porras-Alfaro, A., and Sinsabaugh, R.L., 2008. Translocation of nitrogen and carbon integrates biotic crust and grass production in desert grassland. Journal of Ecology 96: 1076-1085.

Gunningham, N., 2007. Incentives to improve farm management: EMS, supply-chains and civil society. Journal of Environmental Management 82: 302-310.

Khidir, H.H., Eudy, D.M., Porras-Alfaro, A., Herrera, J., Natvig, D. O. and Sinsabaugh, R.L., 2008. A general suite of fungal endophytes dominate the roots of two dominant grasses in a semiarid grassland. Journal of Arid Environments 74: 35-42.

Kiers, E.T., Leakey, R.R.B., Izac, A.M., Heinemanen, J.A., Rosenthal, E., Nathan, D. and Jiggins, J., 2008. Ecology: agriculture at a crossroads. Science 320: 320-321.

Kuldau, G. and Bacon, C., 2008. Clavicipitaceous endophytes: their ability to enhance resistance of grasses to multiple stresses. Biological Control 46: 57-71.

Lairon, D., 2010. Nutritional quality and safety of organic food. A review. Agronomy for Sustainable Development 30: 33-41.

Lee, K., Pan, J.J. and May, G., 2009. Endophytic *Fusarium verticillioides* reduces disease severity caused by *Ustilago maydis* on maize. FEMS Microbiology Letters 299: 31-37.

Lombard, K.A., Forster-Cox, S., Smeal, D. and O'Neill, M.K., 2006. Diabetes on the Navajo nation: what role can gardening and agriculture extension play to reduce it? Rural and Remote Health 6: 640.

Lucero, M.E., Unc, A., Cooke, P., Dowd and Sun, S., 2011. Endophyte microbiome diversity in micropropagated *Atriplex canescens* and *Atriplex torreyi* var *griffithsii*. Public Library of Science One 6: e17693.

Lynch, J.P., 2007. Roots of the second green revolution. Australian Journal of Botany 55: 493-412.

Mann, H., 1848. Education and national welfare. Twelfth annual report of horace mann as secretary of Massachusetts State Board of Education. Massachusetts State Board of Education. Available at: http://usa.usembassy.de/etexts/democrac/16.htm.

Moran, M.S., Peters, D.P.C., McClaran, M.P., Nichols, M.H. and Adams, M.B., 2008. Long-term data collection at USDA experimental sites for studies of ecohydrology. Ecohydrology 1: 377-393.

Müller-Stöver, D., Hauggard-Nielsen, H., Eriksen, J., Ambus, P., and Johansen, A., 2012. Microbial biomass, microbial diversity, soil carbon storage, and stability after incubation of soil from grass-clover pastures of different age. Biology and Fertility of Soils 48: 371-383.

Okon, Y., 1985. *Azospirillum* as a potential inoculant for agriculture. Trends in Biotechnology 3: 223-228.

Ortiz, O., 2006. Evolution of agricultural extension and information dissemination in Peru: an historical perspective focusing on potato-related pest control. Agriculture and Human Values 23: 477-489.

Ortiz, R., Crouch, J.H., Crossa, J., Crouch, J.H., Crossa, J., Braun, H.J., Dodds, J.H., Trethowan, R., Ferrara, G.O. and Iwanaga, M., 2007. High yield potential, shuttle breeding, genetic diversity, and a new international wheat improvement strategy. Euphytica 157: 365-384.

Peters, D.P.C., Pielke, Sr. R.A., Bestelmeyer, B.T., Allen, C.D., Munson-McGee, S. and Havstad, K.M., 2004. Cross-scale interactions, nonlinearities, and forecasting catastrophic events. Proceedings of the National Academy of Sciences USA 101: 15130-15135.

Peters, D.P.C., Sala, O.E. and Allen, C.D., 2007. Cascading events in linked ecological and socioeconomic systems. Frontiers in Ecology and the Environment 5: 221-224.

Peters, S.J., 2006. Every farmer should be awakened: Liberty Hyde Bailey's vision of agricultural extension work. Agricultural History 80: 190-219.

Pinstrup-Anderson, P. and Hazell, P.B.R., 1985. The impact of the green revolution and prospects for the future. Food Reviews International 1: 1-25.

Plett, J.M. and Martin, F., 2011. Blurred boundaries: lifestyle lessons from ectomycorrhizal fungal genomes. Trends in Genetics 27: 14-22.

Richardson, A.E., 2001., Prospects for using soil microorganisms to improve the acquisition of phosphorus by plants. Functional Plant Biology 28: 897-906.

Rodriguez, H. and Fraga R., 1999. Phosphate solubilizing bacteria and their role in plant growth promotion. Biotechnology Advances 17: 319-339.

Rodriguez, R. and Redman, R., 2008. More than 400 million years of evolution and some plants still can't make it on their own: plant stress tolerance via fungal symbiosis. Journal of Experimental Botany 59: 1109-1114.

Sanchez, P.A. and Swaminathan, M.S., 2005. Cutting world hunger in half. Science 307: 357-359.

Schardl, C.L., Leuchtmann, A. and Spiering, M.J., 2004. Symbioses of grasses with seedborne fungal endophytes. Annual Review of Plant Biology 55: 315-340.

Stanger, T.F. and Lauer, J.G., 2008. Corn grain yield response to crop rotation and nitrogen over 35 years. Agronomy Journal 100: 643-650.

Swaminathan, M., 1968. Presidential address, agricultural science section the age of algeny, genetic destruction of yield barriers and agricultural transformation, Varanasi, 1968.

Swaminathan, M.S., 2010. From green to an evergreen revolution, In: Science and sustainable food security: selected papers of MS Swaminathan, World Scientific Publishing Company Pte. Ltd., Singapore, pp. 27-40.

UN Millennium Project, 2005. Investing in development: a practical plan to achieve the millennium development goals. Available at: http://www.unmillenniumproject.org/reports/fullreport/htm.
UN Millennium Project, Task Force on Science, Technology, and Innovation, 2005. Innovation: applying knowledge in development. New York. Available at: http://www.unmillenniumproject.org/documents/Science-part1.pdf.

Section 4.

New tools: analytical methods and modeling

Chapter 11.

Using mitigation and adaptation strategies to optimize crop yield and greenhouse gas emissions

L. Brilli[1], R. Ferrise[1], E. Lugato[2], M. Moriondo[3] and M. Bindi[1]
[1]DiSPAA, University of Florence, Piazzale delle Cascine 18, 50144 Florence, Italy; [2]JRC, Via Enrico Fermi 2749, 21027 Ispra Varese, Italy; [3]IBIMET-CNR, Via Caproni 8, 50145 Florence, Italy; lorenzo.brilli@unifi.it

Abstract

The global agricultural sector accounts for ~14% of the total anthropogenic greenhouse gas (GHG) emissions, of which 46% are generated as N_2O emissions from nitrogen fertilizers, 45% is contributed by CH_4 from manure and livestock fermentation, and 9% stems from CO_2 as a result of soil organic matter decomposition. Several mitigation strategies were proposed, with the aim to reduce GHG emissions. However, these practices are generally not evaluated in combination with adaptation strategies that farmers may implement independently to offset climate change impacts. The biogeochemical model DNDC was used to assess the role of sustainable mitigation and adaptation strategies on the optimization on crop yield and GHG emissions (CO_2eq/ha) in an intensively managed agricultural region of Tuscany, Italy, under a predicted warmer and drier climate. The simulated crop was maize (*Zea mays* L.). The model was used to simulate maize yield using different mitigation strategies (type and amount of soil amendments) and adaptation strategies (sowing dates and maize cultivars). The future climate scenario was created by increasing present (1976-2005) air mean temperature by 2 °C and reducing daily rainfall by 10%. Results from the simulation suggested that longer growing season cultivars and earlier sowing dates in association with high N inputs may be adopted as strategies to increase maize yield compared with the standard crop and soil management strategy. Low N inputs with higher residue incorporation may be adopted to mitigate GHG emissions. However, in order to optimize both maize crop yield and GHG emissions only a few strategies were acceptable and included low organic N input with high residue incorporation in conjunction with a longer growing season cultivar and an early sowing date.

Keywords: agroecosystems, DNDC model, global warming potential, Italy, maize, N_2O

DOI 10.3920/978-90-8686-788-2_11,

11.1 Introduction

Agriculture and climate change are characterized by a complex relationship of cause and effect. On the one hand, agriculture is sensitive to climate change with negative impacts on crop yield and increased cultivation costs. On the other hand, agriculture can be a major contributor to global warming through activities that enhance the emission of greenhouse gases (GHG) (Bennetzen *et al.,* 2012; Fisher *et al.,* 2012; IPCC, 2007; Paustian *et al.,* 2004; Smith *et al.,* 2007).

Agricultural lands cover ~37% of the total global land surface (Smith *et al.,* 2008), and contribute 14% of all anthropogenic GHG emissions (Denman *et al.* 2007) and 58% of the world's anthropogenic non-carbon dioxide (CO_2) GHG emissions (Beach *et al.,* 2008). According to Smith *et al.* (2008), agriculture accounts for about 84% and 52% of global anthropogenic nitrous oxide (N_2O) and methane (CH_4) emissions, respectively. The high percentage of these emissions are mainly due to high nitrogen (N) inputs used in agricultural crop production, and from manure and livestock fermentation, while CO_2 emission can be generated by many different sources such as fossil fuels, land use change, fires, and decomposition of soil organic matter (SOM) (Janzen *et al.,* 2004; Smith, 2004c).

The Intergovernmental Panel on Climate Change (IPCC, 2007) indicated an increasing trend in GHG emissions and suggested that global emissions could range from 9.7 to 36.7 $GtCO_2eq/y$ by 2030. According to the US Environmental Protection Agency (US-EPA, 2006), western Europe is the only area where GHG emissions from agricultural sector are projected to decrease by 2020. This countertrend, compared to the rest of the world, is due to the adoption of several policies concerning climate and environment within the European Union (Smith *et al.,* 2007a), such as the European Climate Change Programme (ECCP), European Union Emissions Trading System (EU ETS), and Renewable Energy Directive among others.

Mitigation strategies (MS) are referred to as those practices aimed at reducing and stabilizing GHG emissions to minimize the vulnerability of physical and biophysical systems. Opportunities for mitigating GHG in agriculture include the managements of organic soils, avoiding deep ploughing, maintaining a more shallow water table (Freibauer *et al.,* 2004), the adoption of agronomic practices to sequester carbon (C) by appropriate crop rotations or reduced use of bare fallow (Lal, 2004a; Smith, 2004a,b), and a decrease of N emissions by managing soil chemistry and microbiology or by limiting the dependence on fertilizers through genetic improvement (Thomson *et al.,* 2012).

Concurrent to the implementation of MS, adaptation strategies (AS) were also developed. These are strategies usually adopted by farmers to reduce economic losses due to declining crop yield and associated high input costs from fertilizers and irrigation (Smith and Olesen, 2010). These actions have an effect in the short- to mid-term and involve several agronomic options. More specifically, they concern changes in management of cropping systems by means of changes in cultivar types, sowing dates, and/or type and method of fertilizer application (Bindi and Olesen, 2011). The role of these strategies could be fundamental especially for those intensive farming systems where, besides genetic improvement, the changes in management practices can lead to remarkable increases in crop yield (Duvick, 2005; Evans, 1997).

Currently, many approaches such as direct measurements, modelling and remote sensing have been developed to assess changes in crop yield and GHG emissions under specific agronomic conditions. Modelling approaches have shown to provide a cost effective alternative to direct measurements to quantify the comprehensive effects of management practices on agricultural production and the environment. To date, several researchers have used empirical models such as DNDC and DayCent to evaluate C and N fluxes and emissions from major cropping systems such as rice, grassland, maize, etc. (Del Grosso, 2008; Pathak *et al.,* 2005; Wang *et al.,* 2012). Information gained from these simulations also provided further insight into the effects of limiting and reducing factors such as temperature, water and N supply on crop yield (Bonfante *et al.,* 2010; Confalonieri *et al.,* 2009; Todorovic *et al.,* 2009). Other models, such as the Crop Environment Resource Synthesis (CERES) were developed for assessing the dynamics of specific crop systems such as maize (*Zea mays* L.), wheat (*Triticum aestivum* L.), and barley (*Hordeum vulgare* L.) (Garrison *et al.,* 1999, Hlavinka *et al.,* 2010; Jamieson *et al.,* 1998).

Among the major cropping systems, due to its worldwide cultivation and the great importance regarding food supply, maize is one of the most studied crops. In particular biogeochemical and productivity aspects of this crop have been analysed intensively. For instance Wolf and Van Diepen (1995), using the crop growth model WOFOST to assess the impact of crop management on maize in a future climate scenario, confirmed that an increase in CO_2 concentration did not show any influence on yield, while water availability and changes in management practices were the most important drivers influencing maize yield. Abraha and Savage (2006), using the model CropSyst, showed that, under increased CO_2 concentration regimes, changes in yield of maize in relation to variations of the sowing date, may also cause variation in the mitigation potential. Farahbakhshazad *et al.* (2008), using the model DNDC to analyse the impact of sustainable agroecosystem management practices on crop yield, and C and N dynamics, highlighted that no-tillage significantly increases the soil organic carbon

(SOC) stock but slightly decreases crop yield; however, disadvantages of no-till could be reduced by changing fertilizer application methods.

Up to now, the majority of these studies have focused the effect of MS on GHG emissions or the consequences of AS on crop yield, while the interaction between these two strategies was overlooked. According to Smith *et al.* (2007b), OECD (2010) and FAO (2010), the way to cope with climate change should occur through the interaction of AS and MS, because these strategies should not be considered individually but as complementary. In light of these considerations, in this study the biogeochemical model DNDC (Li *et al.,* 1992a, b) was applied to recognize how MS and AS can be used to optimize maize yields, GHGs emissions and the global warming potential (GWP) in an intensively managed agroecosystem under a warmer climate.

11.2 Materials and methods

Study site

Field data to calibrate the DNDC model was derived from Grosseto (42°46'2 N 11°06' E) in the region of Tuscany in central Italy (Figure 11.1). The most commonly cultivated crop in this region of Italy, because of its favourable geomorphological and pedo-climatic characteristics, is maize. According to the harmonized world soil database, the soils in this region of Italy consist of a sandy clay loam texture with a pH of 7 (FAO, 2009). Average total annual rainfall is 650 mm, with the greatest proportion of precipitation occurring in autumn and spring. According to the Thornthwaite climatic classification, this area was characterized by a subarid-mesothermal climate, with high summer temperatures and mild winters (Thornthwaite, 1948). The coldest month is January with a mean monthly temperature of 7 °C, while the hottest months are July and August (monthly mean maximum temperature of 23 °C).

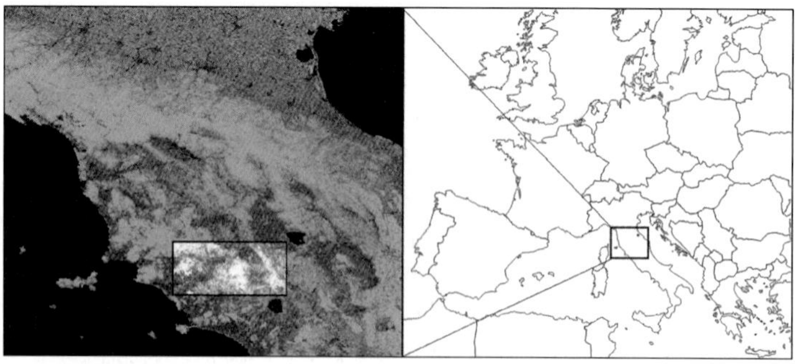

Figure 11.1. Location of the study area in Tuscany, Italy (42°46'2 N 11°06' E).

The soil has a low SOC content ranging from <0.2 to 0.6%. The soils of this region commonly need to be amended with manure or mineral fertilizers to remain productive. Soil is usually cultivated in October, using a plough. Seed-bed preparation occurs just before sowing maize in the early spring (late March-April) and includes cultivating to a 10 cm depth with a disk-harrow. Maize crop residues and N (240 kg N/ha/y) are the only amendments added to the soil. The maize crop is usually irrigated during the summer months at 10 day intervals at a rate of 25 mm/d.

Modelling crop yields and greenhouse gas emissions

The response of maize yield and GHG emissions under future climate scenarios using a variety of maize cultivars and different AS and MS agroecosystem practices were investigated using the DNDC model (Li *et al.,* 1992a, b). DNDC calculates crop growth at a daily time step driven by cumulative temperature, while N stress and water stress may reduce the maximum potential biomass and yield. Moreover, DNDC is able to simulate C and N cycles under both aerobic and anaerobic conditions in agricultural ecosystems at daily or sub-daily time steps (Li *et al.,* 1992a,b; Pathak *et al.,* 2005). The model simulates nitrification, denitrification and fermentation processes, and the resulting C and N emissions. The SOC pool in DNDC is divided in 4 sub-pools (litter, microbe biomass, active and passive humus) with specific decomposition rates. Model simulations were performed using the automatic SOC pool partition without running a previous equilibrium period. CO_2 emissions were estimated by DNDC based on the variation in SOC, which was quantified as the difference in SOC at the end and at the beginning of the simulation period. In this study, the effect of AS and MS on CO_2 emissions were evaluated with respect to changes in in SOC.

To simulate crop growth, SOC dynamics and GHG emissions, the model requires meteorological data, soil properties and agricultural management practices. The minimum level of meteorological data required by DNDC are daily mean temperature and rainfall, however more meteorological information such as daily maximum and minimum temperature and daily radiation can enhance model performances. Therefore, we used daily maximum and minimum temperature, daily precipitation and daily radiation to run the model. DNDC also requires information on land-use management practices, soil texture, pH and levels of SOC at the soil surface. Agricultural management includes the most common practices such as planting and harvest date, crop type, cultivation type and the type and amount of N fertilizer applied.

The model was validated using field data from China, USA, Canada, India, Belgium, Germany (Babu *et al.,* 2006; Beheydt *et al.,* 2007; Giltrap *et al.,* 2010; Grant *et al.,*

2004; Lamers *et al.*, 2007b; Li *et al.*, 2000; Smith *et al.*, 2002) and was used to assess the impact of various management practices on crop productivity (Hastings *et al.*, 2010; Li *et al.*, 2005, 2006). In Italy, a study by Lugato *et al.* (2010) suggested that DNDC can follow the dynamics related to GHG emissions and crop yield in an intensively managed maize agroecosystem.

In DNDC, maximum biomass is an input parameter representing optimum growing conditions and the genetic potential of plant productivity. As such, the three different maize cultivars were used in this study, and their productivity was evaluated using the physiological process-based CropSyst model. CropSyst (Stockle *et al.*, 1994) was widely calibrated for maize cultivation in Mediterranean environments, including northern and central Italy (Donatelli *et al.*, 1996a,b; Morari *et al.*, 2001). More specifically, Dalla Marta *et al.* (2010) calibrated the model using 10 years of yield data (from 1995 to 2004) in Tuscany, confirming the accuracy of CropSyst to simulate maize cultivar, with a medium growing season length, productivity.

Mitigation and adaptation strategies

In DNDC simulations, 5 different crop management practices (representing MS) were used. The MS practices used included: (1) standard practices (STD) using 240 kg N/ha as ammonium nitrate and leaving 15% of crop residue in the field; (2) organic practices (ORG) using 240 kg N/ha as manure and leaving 15% of crop residue in the field; (3) organic and inorganic fertilizer practices (MIX) using 120 kg N/ha as ammonium nitrate, 120 kg N/ha as manure and leaving 15% of crop residue in the field; (4) low organic and crop residue practices (LWO) using 120 kg N/ha as manure and leaving 80% of crop residue in the field; and (5) inorganic fertilizer and crop residue practices (LWC) using 120 kg N/ha as ammonium nitrate and leaving 80% of crop residue in the field.

Additionally, eight different AS were used representing five different sowing dates and three different maize cultivars. Each MS and AS practices was simulated under future climate change scenarios. Standard sowing date was set to 1 April and the remaining sowing dates were simulated corresponding to 15 and 30 days before the standard sowing date (ASD15 and ASD30, respectively) and 15 and 30 days after standard sowing date (DSD15 and DSD30, respectively).

Maize cultivar was the FAO class 500. Cultivars with a shorter growing cycle (SGC) corresponded to the FAO class 300, and cultivars with a longer growing cycle (LGC) corresponded to the FAO class 700. In the simulations with totally organic managements (ORG and LWO), the manure was applied at ploughing (1 October[t]) while, in the other management practices, N fertilizers were distributed at the 6th leaf stage (~30-50 days after the sowing date) and in relation to the length of the growing season of the various maize cultivars used in this study.

Future climate scenarios and soil data

Future climate scenarios were created by increasing the mean air temperature by 2 °C and reducing daily rainfall by 10%. This increase in temperature and reduction in precipitation was based on current (1976-2005) meteorological data obtained from the MARS-JRC archive (http://mars.jrc.ec.europa.eu). The MARS-JRC data-set was obtained by the interpolation of observed daily temperature and rainfall data, and estimated global radiation. Atmospheric CO_2 concentration level was increased to 550 ppm from the current 400 ppm. Future climate scenarios for central Italy were based on variations expected according to A2 SRES scenario provided (IPCC, 2007). Soil input data required by the model (initial SOC concentration, pH, clay content and bulk density) were obtained from the Tuscany soil database (http://sit.lamma.rete.toscana.it/websuoli).

Net global warming potential

The impact of each GHG on atmospheric radiative forcing, in relation to the same mass of CO_2, was defined as its GWP (MacCarty *et al.,* 2008). Concerning a crop production system, the net global warming potential (NGWP) was the overall balance of the net exchange of GHG (Robertson and Grace, 2004). In this context, CH_4, N_2O and CO_2 represented the main GHG emissions from sources such as manure management, fertilization and agricultural practices. Given a 100-year time horizon and GWP values for CH_4 (kg CH_4/ha/y), N_2O (kg N_2O/ha/y) and CO_2 (kg CO_2/ha/y) respectively equal to 25, 298 and 1 (IPCC, 2007). The NGWP (kg O_2eq/ha/y) of each different strategy adopted in this analysis was calculated using the following equation (IPCC, 2007):

$$NGWP = CO_2\ \text{emissions} + 25 \times CH_4\ \text{emissions} + 298 \times N_2O\ \text{emissions} \qquad (1)$$

Data analysis

In order to evaluate the effect of MS and AS on crop yield, GHG emissions, and NGWP under a future climate scenario, data obtained for the STD practices were compared with those obtained from the other selected strategies. More specifically, crop yields, GHG emissions and NGWP simulated by DNDC were averaged for each crop and soil management. Then, the averaged values of crop yields, CO_2, CH_4, N_2O

emissions and NGWP for the STD were compared with those obtained for each of the selected adaptation strategies. Finally, in order to select the adaptation and mitigation strategies that optimized crop yield and minimized greenhouse gas emission, a K-mean cluster analysis with the function *kmeans* of the R/stats package (R Development Core Team, 2006) was performed.

11.3 Results

Maize yield responses

Our results did not show great change in SOC based on 30 years of simulation with CropSyst, indicating that the model was in equilibrium (data not shown). Predicted 30 year crop yields, CO_2, CH_4, N_2O emissions and NGWP for the future scenario were analysed comparing the simulated values obtained for the STD with those obtained from the various MS and AS practices (crop management, cultivar, sowing date) (Table 11.1). Results from this simulation showed that maize yields were strongly influenced by MS and AS (Table 11.1, Figure 11.2). On average, the highest yield reductions (about 3,000 kg/ha or 28%) were obtained by selecting short growth cycle (SGC) cultivars and delaying the sowing dates. Also, a reduction of organic and mineral fertilizer N inputs (LWO and LWC) led to consistent yield reductions of ~2,000 kg/ha or 18%. Whereas, the selection of LGC cultivars and an advance in sowing dates (ASD) showed an increase in maize yield of up to 1,400 kg/ha or 13%. However, such increases in maize yield could only be obtained when N inputs were high. Maize yields, close to those obtained with STD, ranging from a 10% loss and to a 10% gain in crop yield, could be maintained using lower inorganic N inputs and increasing crop residues input when the sowing date was advanced and using cultivars with along growing season (Table 11.1, Figure 11.2).

Impact of greenhouse gas emissions

Results from our simulation showed that CO_2 emissions were strongly affected by N management (Table 11.1, Figure 11.3). The greatest reduction in CO_2 emissions were observed in LWO and ORG N treatments, ranging from 658 to 544 kg C/ha/y. CO_2 emissions from inorganic treatments (LWC and MIX) were higher compared to those from LWO and ORG N treatments. Both treatments showed a considerable reduction in CO_2 emissions (250 kg C/ha/y). The effect on CO_2 emissions was very low, except for SGC and DSD30 which showed an increase of 60 kg C/ha/y, when introducing different maize cultivars and length of growing season (Table 11.1, Figure 11.3).

Table 11.1. Maize yield, greenhouse gas emissions and net global warming potential (NGWP) simulated with DNDC under future climate scenarios for standard (STD) and sustainable (ORG) crop management practices [e.g. STD-ASD30-SGC refers to standard fertilization level and crop residue management (STD), delay of 30 days in sowing date (ASD30), cultivar with short growing season (SGC)], in Tuscany, Italy.

Treatments[1]	Mean values					Differences from STD-STD-STD treatment				
	Yield (kg/ha/y)	CO_2 (kg C/ha/y)	CH_4 (kg C/ha/y)	N_2O (kg N/ha/y)	NGWP (kg CO_2e/ha/y)	Yield (kg/ha/y)	CO_2 (kg C/ha/y)	CH_4 (kg C/ha/y)	N_2O (kg N/ha/y)	NGWP (kg CO_2e/ha/y)
STD-STD-STD	10,647.25	81.17	-0.81	4.63	2,437.36					
STD-STD-SGC	9,090.00	88.63	-0.80	5.80	3,013.62	-1,557.25	7.47	0.01	1.17	576.27
STD-STD-LGC	12,047.83	77.80	-0.82	3.39	1,843.21	1,400.58	-3.37	-0.01	-1.24	-594.15
STD-DSD30-STD	9,802.58	89.77	-0.81	5.46	2,860.57	-844.67	8.60	0.00	0.84	423.21
STD-DSD30-SGC	7,725.83	95.83	-0.81	6.54	3,385.33	-2,921.42	14.67	0.00	1.91	947.97
STD-DSD30-LGC	10,844.58	86.07	-0.82	4.63	2,456.58	197.33	4.90	-0.01	0.00	19.23
STD-DSD15-STD	10,797.83	84.63	-0.81	4.20	2,250.99	150.58	3.47	0.00	-0.43	-186.37
STD-DSD15-SGC	8,560.42	91.63	-0.81	5.68	2,967.77	-2,086.83	10.47	0.01	1.05	530.41
STD-DSD15-LGC	11,509.17	83.37	-0.82	3.46	1,896.97	861.92	2.20	-0.01	-1.17	-540.39
STD-ASD30-STD	10,447.58	82.80	-0.81	5.14	2,683.47	-199.67	1.63	0.00	0.51	246.11
STD-ASD30-SGC	8,977.33	87.90	-0.80	6.43	3,307.92	-1,669.92	6.73	0.01	1.81	870.56
STD-ASD30-LGC	11,385.67	77.87	-0.82	4.19	2,221.36	738.42	-3.30	-0.01	-0.44	-215.99
STD-ASD15-STD	10,769.92	81.17	-0.81	4.83	2,530.27	122.67	0.00	0.00	0.20	92.91
STD-ASD15-SGC	9,257.17	87.33	-0.80	6.15	3,172.64	-1,390.08	6.17	0.01	1.52	735.28
STD-ASD15-LGC	11,495.25	75.00	-0.82	4.07	2,154.66	848.00	-6.17	-0.01	-0.56	-282.70
ORG-STD-STD	10,650.17	-463.73	-1.05	9.04	2,496.23	2.92	-544.90	-0.24	4.41	58.87
ORG-STD-SGC	9,093.00	-418.33	-1.03	10.27	3,241.23	-1,554.25	-499.50	-0.22	5.64	803.87
ORG-STD-LGC	12,051.58	-511.97	-1.07	7.89	1,784.12	1,404.33	-593.13	-0.26	3.27	-653.24
ORG-DSD30-STD	9,797.00	-431.87	-1.05	8.70	2,454.14	-850.25	-513.03	-0.24	4.07	16.78
ORG-DSD30-SGC	7,723.75	-373.73	-1.03	9.71	3,143.82	-2,923.50	-454.90	-0.22	5.09	706.47
ORG-DSD30-LGC	10,849.92	-473.37	-1.07	8.01	1,980.76	202.67	-554.53	-0.26	3.39	-456.60
ORG-DSD15-STD	10,807.08	-463.73	-1.05	8.57	2,278.06	159.83	-544.90	-0.24	3.94	-159.30
ORG-DSD15-SGC	8,561.92	-400.80	-1.03	10.08	3,216.26	-2,085.33	-481.97	-0.22	5.45	778.91

[1] STD = standard practices for crop management, sowing date or growing cycle; MIX = organic and inorganic fertilizer practices; LWO = low organic and crop residue practices; ORG =organic practices; LWC = inorganic fertilizer and crop residue practices; ASD = advanced sowing date (days); DSD = delayed sowing date (days); LGC = longer crop growing cycle; SGC = shorter crop growing cycle.

Table 11.1. Continued.

Treatments[1]	Mean values					Differences from STD-STD-STD treatment				
	Yield (kg/ha/y)	CO_2 (kg C/ha/y)	CH_4 (kg C/ha/y)	N_2O (kg N/ha/y)	NGWP (kg CO_2e/ha/y)	Yield (kg/ha/y)	CO_2 (kg C/ha/y)	CH_4 (kg C/ha/y)	N_2O (kg N/ha/y)	NGWP (kg CO_2e/ha/y)
ORG-DSD15-LGC	11,513.83	-492.77	-1.07	7.96	1,885.03	866.58	-573.93	-0.26	3.33	-552.33
ORG-ASD30-STD	10,446.92	-459.27	-1.05	9.65	2,798.09	-200.33	-540.43	-0.24	5.02	360.73
ORG-ASD30-SGC	8,984.67	-416.67	-1.03	10.97	3,575.62	-1,662.58	-497.83	-0.22	6.34	1,138.26
ORG-ASD30-LGC	11,393.33	-496.63	-1.06	8.73	2,233.02	746.08	-577.80	-0.25	4.11	-204.34
ORG-ASD15-STD	10,770.42	-466.60	-1.05	9.27	2,593.59	123.17	-547.77	-0.24	4.64	156.24
ORG-ASD15-SGC	9,266.00	-423.33	-1.03	10.66	3,405.64	-1,381.25	-504.50	-0.22	6.03	968.28
ORG-ASD15-LGC	11,512.00	-499.93	-1.06	8.56	2,139.29	864.75	-581.10	-0.25	3.93	-298.07
MIX-STD-STD	10,646.67	-169.30	-0.93	6.73	2,501.79	-0.58	-250.47	-0.12	2.11	64.43
MIX-STD-SGC	9,087.50	-119.60	-0.92	7.87	3,214.46	-1,559.75	-200.77	-0.11	3.24	777.10
MIX-STD-LGC	12,054.58	-206.47	-0.95	5.53	1,801.16	1,407.33	-287.63	-0.14	0.90	-636.20
MIX-DSD30-STD	9,803.33	-133.53	-0.93	6.89	2,706.47	-843.92	-214.70	-0.12	2.26	269.11
MIX-DSD30-SGC	7,725.67	-78.97	-0.91	7.84	3,350.66	-2,921.58	-160.13	-0.10	3.21	913.30
MIX-DSD30-LGC	10,848.58	-170.50	-0.95	6.19	2,241.28	201.33	-251.67	-0.14	1.56	-196.08
MIX-DSD15-STD	10,797.42	-163.27	-0.94	6.23	2,289.60	150.17	-244.43	-0.13	1.61	-147.75
MIX-DSD15-SGC	8,562.25	-102.60	-0.91	7.65	3,177.89	-2,085.00	-183.77	-0.10	3.03	740.54
MIX-DSD15-LGC	11,508.75	-187.53	-0.95	5.58	1,893.04	861.50	-268.70	-0.14	0.95	-544.32
MIX-ASD30-STD	10,440.50	-160.77	-0.93	7.26	2,777.23	-206.75	-241.93	-0.12	2.63	339.87
MIX-ASD30-SGC	8,981.75	-119.67	-0.91	8.60	3,558.81	-1,665.50	-200.83	-0.10	3.97	1,121.45
MIX-ASD30-LGC	11,390.83	-193.53	-0.95	6.42	2,266.04	743.58	-274.70	-0.14	1.79	-171.32
MIX-ASD15-STD	10,771.33	-167.80	-0.93	6.90	2,583.57	124.08	-248.97	-0.12	2.27	146.21
MIX-ASD15-SGC	9,262.17	-125.93	-0.92	8.27	3,381.66	-1,385.08	-207.10	-0.11	3.65	944.30
MIX-ASD15-LGC	11,504.42	-196.47	-0.95	6.25	2,174.10	857.17	-277.63	-0.14	1.62	-263.25
LWO-STD-STD	8,915.17	-550.27	-1.04	2.07	-1,082.65	-1,732.08	-631.43	-0.23	-2.56	-3,520.01
LWO-STD-SGC	8,417.75	-497.40	-1.04	2.30	-782.26	-2,229.50	-578.57	-0.23	-2.33	-3,219.62
LWO-STD-LGC	9,032.50	-563.97	-1.05	2.14	-1,100.43	-1,614.75	-645.13	-0.24	-2.49	-3,537.79
LWO-DSD30-STD	8,467.08	-499.90	-1.04	2.14	-867.23	-2,180.17	-581.07	-0.23	-2.49	-3,304.59

[1] STD = standard practices for crop management, sowing date or growing cycle; MIX = organic and inorganic fertilizer practices; LWO = low organic and crop residue practices; ORG =organic practices; LWC = inorganic fertilizer and crop residue practices; ASD = advanced sowing date (days); DSD = delayed sowing date (days); LGC = longer crop growing cycle; SGC = shorter crop growing cycle.

Table 11.1. Continued.

Treatments[1]	Mean values					Differences from STD-STD-STD treatment				
	Yield (kg/ha/y)	CO_2 (kg C/ha/y)	CH_4 (kg C/ha/y)	N_2O (kg N/ha/y)	NGWP (kg CO_2e/ha/y)	Yield (kg/ha/y)	CO_2 (kg C/ha/y)	CH_4 (kg C/ha/y)	N_2O (kg N/ha/y)	NGWP (kg CO_2e/ha/y)
LWO-DSD30-SGC	7,425.58	-403.70	-1.02	2.58	-306.49	-3,221.67	-484.87	-0.21	-2.05	-2,743.85
LWO-DSD30-LGC	8,520.67	-515.33	-1.05	2.27	-862.21	-2,126.58	-596.50	-0.24	-2.36	-3,299.57
LWO-DSD15-STD	8,902.00	-536.27	-1.05	2.02	-1,057.55	-1,745.25	-617.43	-0.23	-2.61	-3,494.91
LWO-DSD15-SGC	8,072.25	-460.87	-1.03	2.45	-574.89	-2,575.00	-542.03	-0.22	-2.17	-3,012.25
LWO-DSD15-LGC	8,852.08	-542.37	-1.05	2.19	-997.97	-1,795.17	-623.53	-0.24	-2.44	-3,435.32
LWO-ASD30-STD	8,963.83	-561.07	-1.05	2.11	-1,105.81	-1,683.42	-642.23	-0.24	-2.52	-3,543.17
LWO-ASD30-SGC	8,407.00	-502.23	-1.04	2.44	-734.24	-2,240.25	-583.40	-0.23	-2.19	-3,171.60
LWO-ASD30-LGC	9,051.75	-577.40	-1.06	2.17	-1,136.90	-1,595.50	-658.57	-0.24	-2.46	-3,574.26
LWO-ASD15-STD	8,997.50	-563.60	-1.05	2.09	-1,122.99	-1,649.75	-644.77	-0.24	-2.54	-3,560.34
LWO-ASD15-SGC	8,548.75	-511.47	-1.04	2.31	-828.53	-2,098.50	-592.63	-0.23	-2.32	-3,265.89
LWO-ASD15-LGC	8,977.17	-568.60	-1.05	2.20	-1,089.66	-1,670.08	-649.77	-0.24	-2.43	-3,527.02
LWC-STD-STD	9,698.33	-209.13	-0.95	0.25	-681.42	-948.92	-290.30	-0.14	-4.38	-3,118.77
LWC-STD-SGC	9,043.75	-153.00	-0.94	0.41	-401.83	-1,603.50	-234.17	-0.13	-4.22	-2,839.19
LWC-STD-LGC	9,818.17	-195.13	-0.95	0.23	-637.72	-829.08	-276.30	-0.14	-4.39	-3,075.08
LWC-DSD30-STD	9,224.67	-136.27	-0.93	0.77	-171.87	-1,422.58	-217.43	-0.12	-3.86	-2,609.23
LWC-DSD30-SGC	7,721.17	-76.53	-0.91	1.15	228.38	-2,926.08	-157.70	-0.10	-3.48	-2,208.98
LWC-DSD30-LGC	9,292.33	-151.30	-0.95	0.79	-218.06	-1,354.92	-232.47	-0.14	-3.84	-2,655.42
LWC-DSD15-STD	9,637.83	-170.60	-0.94	0.18	-572.81	-1,009.42	-251.77	-0.13	-4.45	-3,010.17
LWC-DSD15-SGC	8,569.67	-121.33	-0.93	0.47	-257.34	-2,077.58	-202.50	-0.12	-4.16	-2,694.70
LWC-DSD15-LGC	9,626.92	-173.60	-0.95	0.20	-574.75	-1,020.33	-254.77	-0.14	-4.43	-3,012.10
LWC-ASD30-STD	9,703.58	-201.87	-0.95	0.28	-639.22	-943.67	-283.03	-0.14	-4.34	-3,076.58
LWC-ASD30-SGC	8,988.17	-161.90	-0.94	0.51	-388.32	-1,659.08	-243.07	-0.13	-4.12	-2,825.67
LWC-ASD30-LGC	9,792.00	-208.63	-0.96	0.27	-668.78	-855.25	-289.80	-0.15	-4.35	-3,106.14
LWC-ASD15-STD	9,748.00	-198.47	-0.95	0.28	-629.28	-899.25	-279.63	-0.14	-4.35	-3,066.64
LWC-ASD15-SGC	9,222.50	-165.70	-0.94	0.43	-439.39	-1,424.75	-246.87	-0.13	-4.20	-2,876.75
LWC-ASD15-LGC	9,737.33	-208.27	-0.95	0.29	-658.08	-909.92	-289.43	-0.14	-4.33	-3,095.44

[1] STD = standard practices for crop management, sowing date or growing cycle; MIX = organic and inorganic fertilizer practices; LWO = low organic and crop residue practices; ORG =organic practices; LWC = inorganic fertilizer and crop residue practices; ASD = advanced sowing date (days); DSD = delayed sowing date (days); LGC = longer crop growing cycle; SGC = shorter crop growing cycle.

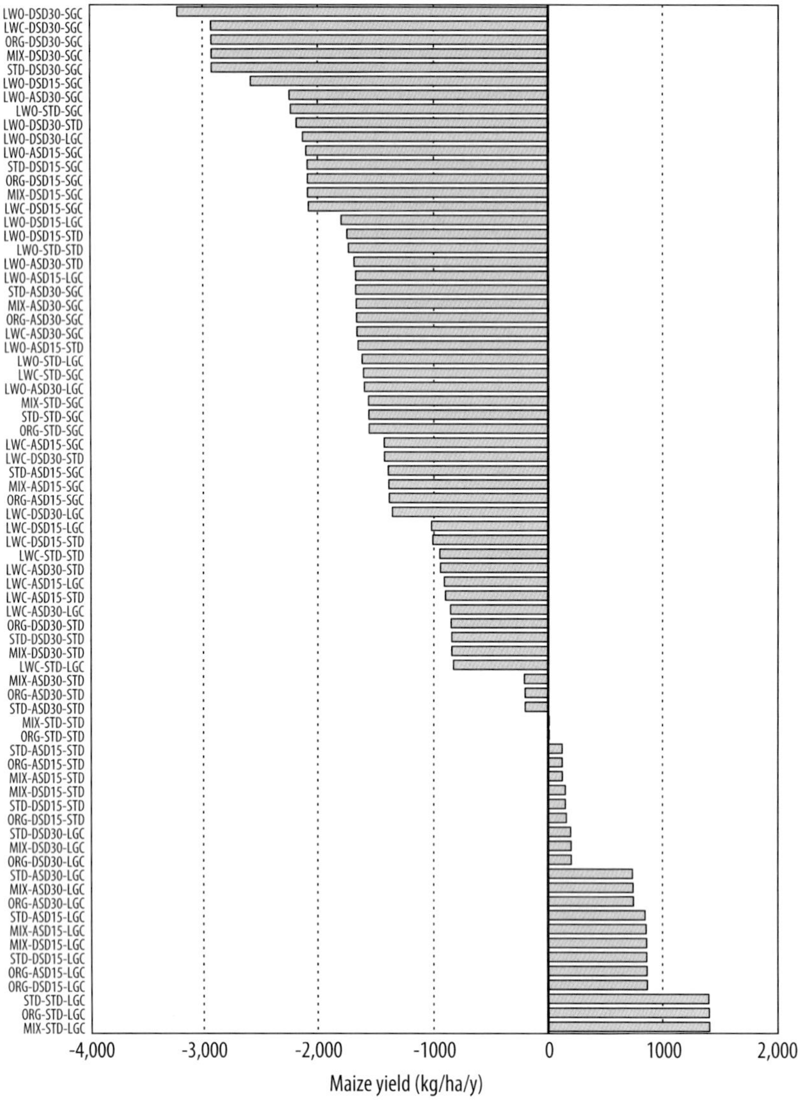

Figure 11.2. Differences in maize yield under a future climate scenario using standard management practices and maize cultivars (STD) compared to those obtained when introducing mitigation and adaptation strategies (crop management, cultivar, sowing date) in Tuscany, Italy.

The balance of the net CH_4 exchange was very similar for standard N inputs, and progressively became negative moving from the treatments with inorganic N inputs (MIX and LWC) to those with higher organic N inputs and crop residues in the soil

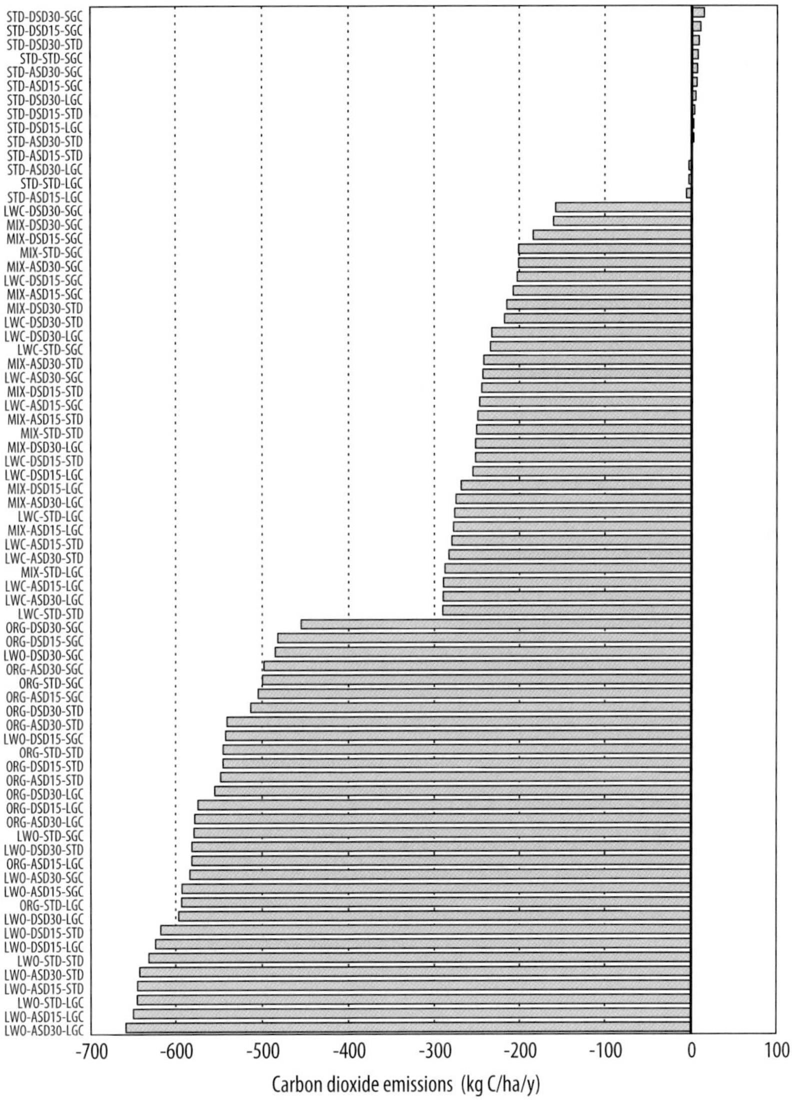

Figure 11.3. Differences in mean CO_2 emission under a future climate scenario using standard management practices and maize cultivars (STD) compared to those obtained when introducing mitigation and adaptation strategies (crop management, cultivar, sowing date), in Tuscany, Italy.

(Table 11.1, Figure 11.4). In particular, ORG and LWO treatments had a decrease in CH_4 emission by 0.24 kg C/ha/y, whereas LWC and MIX had a CH_4 emission of 0.12 kg C/ha/y. The use of cultivars with different growing season lengths and changes in sowing dates did not reduce CH_4 emissions significantly (Table 11.1, Figure 11.4).

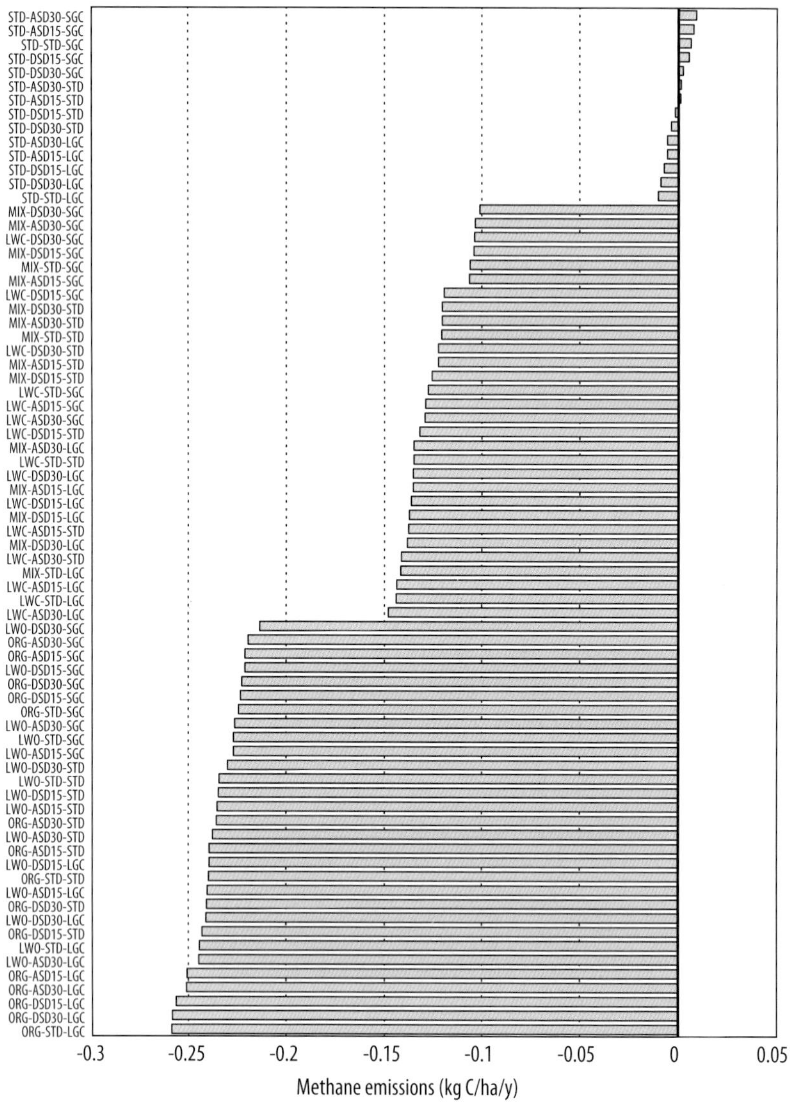

Figure 11.4. Differences in mean CH_4 emission under a future climate scenario using standard management practices and maize cultivars (STD) compared to those obtained when introducing mitigation and adaptation strategies (crop management, cultivar, sowing date), in Tuscany, Italy.

Nitrogen management was the main factor influencing N_2O emissions (Table 11.1, Figure 11.5). In particular, the greatest reductions in N_2O emissions were obtained when N inputs were reduced from 4.20 kg N/ha/y (LWC) to 2.4 kg N/ha/y (LWO).

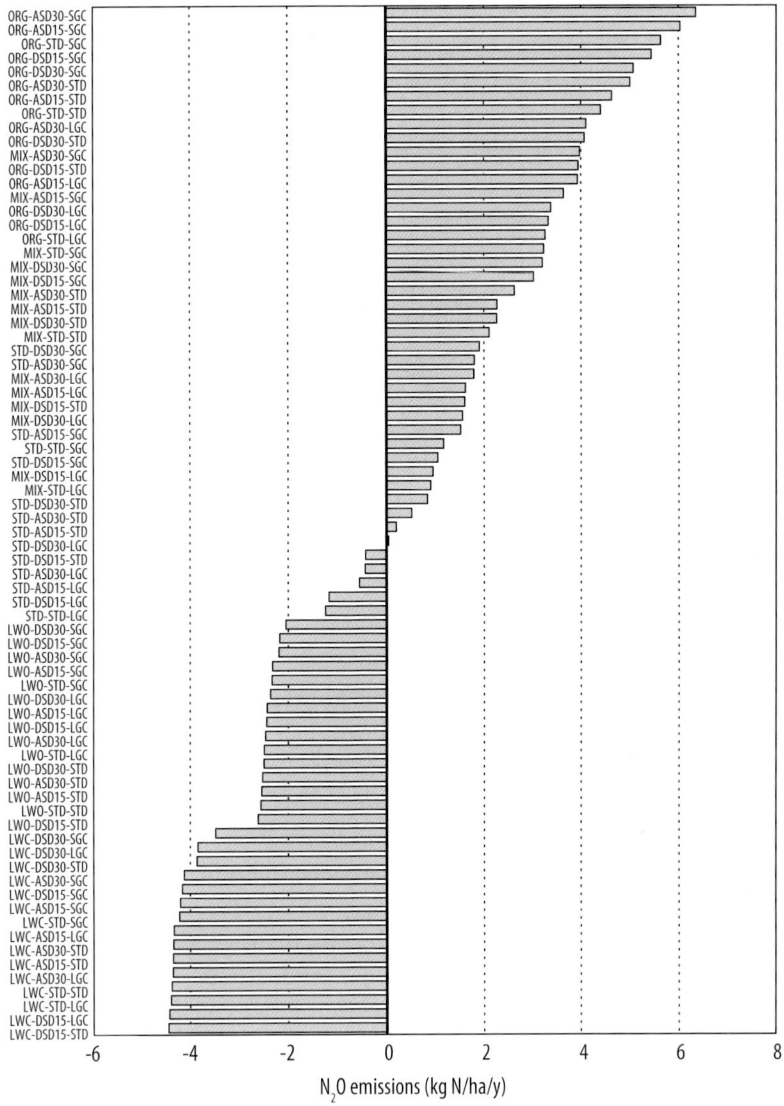

Figure 11.5. Differences in mean N_2O emission emission under a future climate scenario using standard management practices and maize cultivars (STD) compared to those obtained when introducing mitigation and adaptation strategies (crop management, cultivar, sowing date), in Tuscany, Italy.

However, when N inputs were maintained and the organic fraction was increased by 50 to 100%, N_2O emissions increased by 2.3 kg N/ha/y (MIX) and 4.6 kg N/ha/y (ORG). Changes in sowing dates and cultivar type were strongly related to N management with respect to N_2O emissions (Table 11.1, Figure 11.5). However, a longer growing season cultivars and delays in sowing dates reduced N_2O emissions.

Effect on net global warming potential

When GHG emissions were expressed as their NGWP, the interactions between N input, sowing date and cultivar treatments were very complex. The greatest increase in NGWP values were found in treatments with a high input of organic or inorganic N fertilizer (STD, MIX, ORG) (+1000 kg CO_2eq/ha/y) (Table 11.1, Figure 11.6). However, simulations with low input management practices resulted in a decrease in NGWP, especially in LWO and LWC (up to 3,500 kg CO_2eq/ha/y) (Table 11.1, Figure 11.6). In general, the use of long growing season cultivars resulted in a decreased NGWP. However, a delay in the sowing date did not have a consistent effect on NGWP (Table 11.1, Figure 11.6).

11.4 Discussion

Maize yield response

The relationship between climate, maize yield and GHG emissions is difficult to interpret because of a large number of parameters involved including soil, N management, atmospheric CO_2 concentration, and crop characteristics. In this study, a series of adaptation and mitigation strategies were tested, using a maize cropping system under future climatic conditions, to evaluate their effects on crop yield and GHG emissions and identify possible synergies or conflicts between them.

Soil fertility is one of the most important factors for maize production and N is one of the most yield-limiting nutrient factors in maize productivity (Kim and Dale, 2008). Fertilization provides different effects on soil properties which depend on the type and amount of fertilizer used. Manure can lead to favourable changes in soil physical, chemical and biological characteristics, improving soil nutrition (Bronick and Lal, 2005; Marinari *et al.*, 2000). Although mineral fertilizers offer nutrients in forms that are readily available for plant uptake, it is well known that they also contribute to soil structure degradation, soil acidification, changes of soil microbial community structure and a reduction in soil microbial biomass. As expected, maize yield was greatly sensitive to N supply, where the greatest maize productivity occurred in treatments receiving the highest input of either organic or inorganic fertilizer. With respect to the different maize growing cycles, the greatest maize yield occurred when N inputs were high. This may be a response to a longer growing season providing a longer time period in which N is available for plant uptake. This in turn may have resulted in a higher biomass

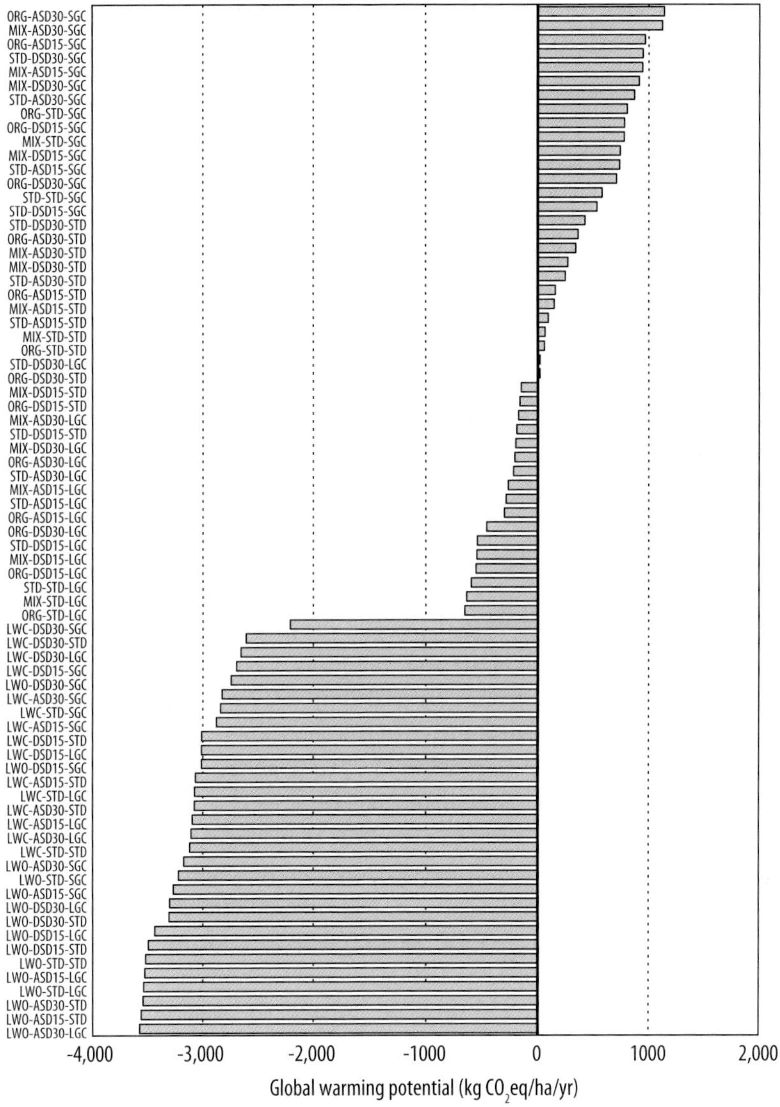

Figure 11.6. Differences in mean net global warming potential under a future climate scenario using standard management practices and maize cultivars (STD) compared to those obtained when introducing mitigation and adaptation strategies (crop management, cultivar, sowing date), in Tuscany, Italy.

production and grain yield. Daynard and Kannenberg (1976) suggested that a longer grain filling period may lead to an increase grain yield.

The selection of the best cultivar and sowing date are usually considered as two fundamental components of maize productivity (Panahi *et al.,* 2010). An optimum sowing date for maize is needed to ensure appropriate root development, leading to an increased tolerance to the water stress (Garcia *et al.,* 2009). In agreement with many studies carried out on maize cultivation (Cantarero *et al.,* 2000; Rizzardi *et al.,* 1994; Saseendran *et al.,* 2005), results from our study showed a yield decrease when the sowing date was postponed. This was likely due to a shorter growing season, which reduced the time available for exploiting the soil N supply. Additionally, a shorter growing season also reduced the time available for the plant to accumulate C, resulting in a lower biomass and yield production. However, cultivars with a longer growing season resulted in an enhanced development of the root system, which would provide a greater opportunity to explore the soil and exploit its N supply. As a consequence, the efficiency of the plant system, such as water stress tolerance, photosynthesis, and C sequestration, increased resulting in a higher biomass and yield production.

Greenhouse gas emissions

Temperature, moisture, redox potential, pH and substrate concentration gradients are the main drivers affecting C and N transport and transformation in the DNDC model (Li *et al.,* 1997). Regarding C dynamics, according to Li *et al.* (1997), the SOC pool in DNDC is determined by the balance between the C input from residues and the C losses from CO_2 emissions. As expected in this study, the results had the lowest CO_2 emissions in treatments with organic managements (LWO and ORG), indicating the soil's capacity to stabilize SOC. These results were in agreement with several field studies (Gomiero *et al.,* 2008; Paustian *et al.,* 2000). For example, Lal (2008) highlighted that soil C sequestration mostly depends on the quantity and quality of biomass returned to the soil. Marriot and Wander (2006) observed an increase in SOC under organic management practices compared to 10 years under conventional management. Lal (2007) confirmed that residue retention can improve soil quality and biological soil properties such as microbial biomass activity and diversity, consequently enhancing the capacity of the soil to sequester C, in turn, reducing CO_2 emissions. Concerning soil amendments, Gattinger *et al.* (2012) and Leifield and Fuhrer (2010) focused on the role of C in soil and suggested that the advantages of organic practices for SOC are mainly due to the higher and often disproportionate use of organic fertilizer compared to those applied for conventional agricultural systems. If the increase in SOC was due to manure, especially at higher rates compared to the incorporation of other residues, then a transfer of C rather than a net gain may have occurred. As such, the positive effects of increasing levels of SOC on GHG emission, through organic amendments, may be lower than those simulated. Therefore, we independently evaluated the effect of each organic treatment (LWO and ORG) on CO_2 emissions, excluding the different contributions from manure or residue retention and C sequestration. Results highlighted that fully organic management practices sequestered more C compared to the other combinations of the various management practices evaluated in our study.

Manure input in association with mineral fertilizers contributed to a decrease in CO_2 emissions (LWC and MIX) compared to STD. When LWC, MIX management and STD were simulated, the balance between C input and C output changed, resulting in an increase in CO_2 emissions compared to the treatment with organic management. As expected, the greatest CO_2 emissions were in the STD treatment, which may be due to the very low capacity of the soil to stabilize SOC. Model simulations agreed with those of Su *et al.* (2006), who highlighted that the long-term application of inorganic fertilizers under conventional management could be inadequate to maintain soil C and nutrients levels, especially without crop residue incorporation. Mann *et al.* (2002) and Yang and Wander (1999) suggested that crop residue removal may lead to a rapid decline in soil C content especially when coupled with conventional tillage.

Although inorganic fertilizers may increase microbial decomposition of plant residues, several studies reported a reduction in SOC over the long-term (Conde *et al.,* 2005; Mack *et al.,* 2004; Poirier *et al.,* 2009). Several experiments carried out in the United States on maize systems confirmed that inorganic N fertilizer depleted SOC and contributed to CO_2 emissions (Black and Tanaka, 1997; Pikul *et al.,* 2001; Reicosky *et al.,* 2002; Russell *et al.,* 2005).

Brown *et al.* (2002) showed that N_2O emissions simulated with DNDC were highly sensitive to agronomic management practices including the type, rate, timing and depth of N application, which influenced soil nitrification and denitrification. In DNDC, nitrification converts ammonium (NH_4^+) to nitrate (NO_3^-) through microbiological oxidation under aerobic conditions, which produces N_2O emissions as a by-product or intermediate (Li *et al.,* 2000). Simulations showed that N_2O emissions increased under MIX, ORG and STD treatments which were characterized by a fertilizer input rate twice that of LWC and LWO. These results were in agreement with Kim and Dale (2008) and Roy *et al.* (2006), who suggested that N_2O emissions in agricultural soils mostly depended on N application rates. Similarly, other studies also reported higher N_2O emissions when N application rates were high (Bouwman *et al.,* 2002; Sehy *et al.,* 2003; Zebarth *et al.,* 2008). In DNDC, the higher N availability due to an increase in fertilizer application also increased the substrate availability for the microbial pool. This can lead to higher nitrifying and denitrifying activity, and hence N_2O emissions (Kroeze *et al.,* 1999). Li *et al.* (1992a,b) indicated that N_2O emissions estimated by DNDC are very sensitive and will respond positively to changes in SOC and clay content. The lowest N emissions were observed only in treatments with low

N inputs. For example, N emissions from LWC were on two times lower compared to that of the LWO treatment. This suggested that among all treatments or management practices, only LWC was within the 1996 and 2000 IPCC N_2O emission factor (EF). The EF quantifies the NO_2 emission rate equal to 1.25% of N input as direct emission from agricultural soil.

Organic amendments can easily lead to the mineralization of the labile C pool, which in turn stimulates microbial activity and consequently increases N_2O emissions (Chadwick *et al.*, 2000). For example, Dambreville *et al.* (2008) reported the highest peak in N_2O emissions from maize after the application of organic fertilizers, while the highest N_2O emissions occurred when ammonium nitrate was applied in the summer. Jones *et al.* (2007) and Scott *et al.* (2000) quantified N_2O from soil treated with manure and inorganic fertilizers. They concluded that N_2O emissions from organic compounds were emitted for a longer time and characterized by higher magnitude compared to inorganic fertilizers.

Results from our study showed that the lowest N_2O emissions occurred in treatments with high inputs when cultivars with a long growing season were used (Table 11.1). This was likely due to a more developed root system in cultivars with a longer growing season. The DNDC simulation showed that cultivars with a shorter growing season produced maize plants with a smaller root system compared to that of the LGC treatment. This suggested that less N was taken up given that root biomass is a model parameter influencing the plant N cycle. These results were similar to that reported by Mackay and Barber (1986). They suggested that N uptake capacity of plants is mainly controlled by N input at the root surface and by the size of the root system (Mackay and Barber, 1986).

No noticeable variability was observed in the LWO treatment with respect to sowing date. However, the ORG treatment showed a slight reduction in N_2O emissions as a result of a delayed sowing date. Treatments or management practices characterized by a partial or full application of inorganic fertilizer (LWC, MIX and STD) showed a wide variability in N_2O emissions, where the greatest N_2O emissions occurred in ASD30 and DSD30 treatments. The partial or full application of inorganic fertilizer resulted in high leaching values. For example, N losses due to leaching in organic fertilizer treatments were 21% lower compared to that using inorganic fertilizers. The effect of inorganic fertilizers on leaching showed that mineral N can be lost readily due to leaching (Beheydt *et al.*, 2007; Tonitto *et al.*, 2007). A large increase in N_2O emission was observed in the DSD30 treatment which was likely due to a delay in the fertilizer application date as a result of a later sowing date. This was likely due to the model's sensitivity to temperature. For example, ambient temperature was greater at the time of fertilizer application in the treatment with a later sowing date compared to that with an earlier sowing date. Li *et al.* (1992a,b) also found that DNDC was highly sensitive to a change in climate and therefore N_2O emission rates. Brown *et al.* (2002) also observed a high sensitivity of the model to changes in temperature and when fertilizer application was moved to a warmer period. Therefore the more the sowing date was advanced (ASD30) or delayed (DSD30), N_2O emissions were high.

Net global warming potential

In this study, the simulation of GHG emissions was used to quantify the net global warming potential (NGWP) of a maize crop in response to a variety of agronomic management practices and strategies. The lowest NGWP was observed in management systems or treatments with low inputs (LWO and LWC). These management practices were characterized by high residue incorporation and a low N supply, leading to an overall lower emission of CO_2 and N_2O. The differences between LWC and LWO were due to differences in the amount of C sequestered and N_2O emitted. More specifically, the LWO management practice contributed to a higher N_2O emission and C sequestration than LWC. This is because in the DNDC model, the GHG emissions are very sensitive to SOC content. Sainju *et al.* (2008) suggested that use of manure may improve the C sequestration effect, off-set atmospheric CO_2 levels, and enhance soil and environmental quality, leading to a lower NGWP. This is consistent with Thelen *et al.* (2010), who determined the influence of manure and compost on short-term C sequestration rates and NGWP in a corn-soybean rotation. They found a decrease in the NGWP when manure was applied (Thelen *et al.*, 2010). The positive effect of lower CO_2 emissions in low input management systems was only partially offset by N_2O emissions, which contributed to the highest GWP in MIX, ORG and STD treatments, which received double the amount of N.

Concerning cultivars the greatest NGWP reduction was observed using cultivars with a longer growing cycle. This is because in DNDC, the SOC content is determined by the balance between the C-input and output. Using LGC management practices, C input as a result of residue incorporation, and C output from CO_2 emissions were higher compared to using cultivars with a shorter growing cycle. More specifically results showed that the CO_2 emissions in LGC management practices were 30% greater compared to those in the SGC management practice. The CO_2 emissions did not show large differences by changing the sowing date, thereby the NGWP variability was mainly due to N_2O emissions.

When fertilizer application was postponed to a period when temperature usually rises (late spring/early summer), a large increase in N_2O emission (DSD30) due to model sensitivity to temperature was observed. Therefore, N_2O emission was the main driver in NGWP trend with respect to a change in the sowing date.

In this study, we calculated the NGWP per unit land-base. However several studies (Gomiero *et al.,* 2008; Kustermann and Hülsbergen, 2008; Lynch *et al.,* 2011) showed that the large benefit in organic management systems compared to conventional systems with respect to CO_2 emissions was lost or, at least, reduced when they were measured per unit product (kg of grain). This was due to differences in yield, especially in those areas characterized by intensive conventional production, where the GWP per product in the organic systems do not strongly reduce GHG emissions. N_2O and CO_2 emissions from organic systems were much lower compared to conventional systems; this was attributed to the higher sequestration of C.

Best combination between mitigation and adaptation strategies to optimize crop yields and reduce GHG emissions

When analysing all management combinations, only a few MS and AS practices successfully reduced GHG emissions while maintaining crop yield. In particular, analysing the 75 combinations of mitigation and adaptation options, on the basis of their capacity to reduce NGWP and to avoid yield reductions, the results showed that these may be clustered in eight groups (Figure 11.7). Strategies within cluster 5 (short growing season cultivar), cluster 8 (short growing season and delay in sowing date), and cluster 1 (delay in sowing date) can be excluded because these combinations led to a decrease in crop yield without a decrease in the NGWP. Similarly, strategies within cluster 6 were not suitable because of the very high loss of crop yield. Among the remaining clusters, cluster 3 (long growing season cultivar) showed an effect on yield only. As such, clusters 4 and 2 may be considered to limit grain yield reductions (less than 20%) and achieve a reduction in the NGWP. These strategies were characterized by the use of low N inputs, preferably in an organic form, in association with high residue incorporation (LWC and LWO), cultivars with a longer growing cycle (LGC), and the earliest sowing date (ASD30) (Table 11.1, Figure 11.7). The best MS practice based on the NGWP was those under organic management (LWC, LWO). This was suggested also by Lynch *et al.* (2011) who analysed the NGWP of many organic and conventional production sectors and practices (tillage, soil C sequestration and energy offsets). They also concluded that organic farming management practices would maintain crop yield and also reduce the NGWP.

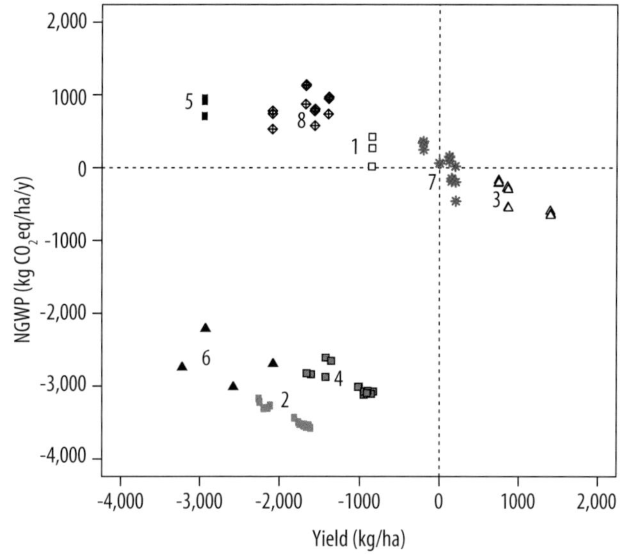

Figure 11.7. Differences in maize yields and net global warming potentials using standard management practices and maize cultivars (STD) compared to those obtained when introducing mitigation and adaptation strategies (crop management, cultivar, sowing date), in Tuscany, Italy. Symbols and numbers indicate the eight clusters obtained using the kmeans function.

11.5 Conclusions

Calibrated crop and biogeochemical models can represent very powerful tools for assessing how mitigation and adaptation strategies can be introduced to optimize crop yield and minimize GHG emissions under future climate scenarios. The sensitivity of a model like DNDC to different agricultural management practices, such as the timing and the type of fertilizer application, length of the growing season and the sowing date, provided an opportunity to enhance our knowledge on the future impacts of climate change on crop yield and GWP. The possibility to evaluate the impact of different agroecosystem mitigation (MS) and adaptation (AS) strategies, allows the detection of which combinations of these strategies may guarantee sustainable crop yields and a reduction in GHG emissions. In this study, results suggested that for the selected future climate scenario (temperature: +2 °C, precipitation: -10%, CO_2: 550 ppm) in the region of Tuscany, Italy, using 75 different crop and soil management options and their effect on maize yield and NGWP, a series of mitigation and adaptation strategies may be available. Our results showed that cultivars with a longer growing season in addition to earlier sowing dates and high N inputs (ORG, MIX, and STD) may be sustainable strategies that help increase maize yield compared to the standard (conventional) crop and soil management strategies. Whereas low N inputs (LWO and LWC) with higher residue incorporation may be adopted to mitigate GHG emissions. However, in order to optimize both maize crop yields and GHGs emission only strategies with low N inputs and high residue incorporation in association with a longer growing season and advanced sowing date may be effective.

Acknowledgements

Special thanks to Nicolina Staglianò and Camilla Dibari, University of Florence, Italy, for the changes proposed and the careful revision of the manuscript.

References

Abraha, M.G. and Savage, M.J., 2006. Potential impacts of climate change on the grain yield of maize for the midlands of KwaZulu-Natal, South Africa. Agriculture, Ecosystems and Environment 115: 150-160.

Babu, Y.J., Li, C., Frolking, S., Nayak, D.R. and Adhya, T.K., 2006. Field validation of DNDC model for methane and nitrous oxide emissions from rice-based production systems of India. Nutrient Cycling in Agroecosystems 74: 157-174.

Beach, R.H., DeAngelo, B.J., Rose, S., Li, C., Salas,W. and Del Grosso, S.J., 2008. Mitigation potential and costs for global agricultural greenhouse gas emissions. Agricultural Economics 38: 109-115.

Beheydt, D., Boeckx, P., Sleutel, S., Li, C. and Van Cleemput, O., 2007. Validation of DNDC for 22 long-term N2O field emissions measurements. Atmospheric Environment 41: 6196-6211.

Bennetzen, H.E., Smith, P., Soussana, J.F. and Porter, J.R., 2012. Identity-based estimation of greenhouse gas emissions from crop production: case study from Denmark. European Journal of Agronomy41: 66-72.

Bindi, M. and Olesen, J.E., 2011. The responses of agriculture in Europe to climate change. Regional Environmental Change 11: S151-S158.

Black, A.L. and Tanaka, D.L., 1997. A conservation tillage-cropping system study in the northern Great Plains of the United States. In: Paul, E.A. (ed.) Soil organic matter in temperate agroecosystems. Boca Raton, Florida, FL, USA, pp. 335-342.

Bonfante, A., Basile, A., Acutis, M., De Mascellis, R., Manna, P., Perego, A. and Terribile, F., 2010. SWAP, CropSyst and MACRO comparison in two contrasting soils cropped with maize in Northern Italy. Agricultural Water Management 7: 1051-1062.

Bouwman, A.F., Boumans, L.J.M. and Batjes, N.H., 2002. Modeling global annual N_2O and NO emissions from fertilized fields. Global Biogeochemical Cycles 16: 1080-1091.

Bronick, C.J. and Lal, R., 2005. Soil structure and management: a review. Geoderma 124: 3-22.

Brown, L., Syed, B., Jarvis, S.C., Sneath, R.W., Phillips, V.R., Goulding, K.W.T. and Li, C., 2002. Development and application of a mechanistic model to estimate emission of nitrous oxide from UK agriculture. Atmospheric Environment 36: 917-928.

Cantarero, M.G., Luque, S.F. and Rubiolo, O.J., 2000. Effects of sowing date and plant density on grain number and yield of a maize hybrid in the central region of Córdoba, Argentina. AgriScientia 17: 3-10.
Chadwick, D.R., Pain, B.F. and Brookman, S.K.E., 2000. Nitrous oxide and methane emissions following application of animal manures to grassland. Journal of Environmental Quality 29: 277-287.
Conde, E., Cardenas, M., Ponce-Mendoza, A., Luna-Guido, M.L., Cruz-Mondragón, C. and Dendooven, L., 2005. The impacts of inorganic nitrogen application on mineralization of 14C-labelled maize and glucose, and on priming effect in saline alkaline soil. Soil Biology and Biochemistry 37: 681-691.
Confalonieri, R., Acutis, M., Bellocchi, G. and Donatelli, M. 2009. Multi-metric evaluation of the models WARM, CropSyst, and WOFOST for rice. Ecological Modelling 220: 395-1410.
Dalla Marta, A., Mancini, M., Ferrise, R., Bindi, M. and Orlandini, S., 2010. Energy crops for biofuel production: Analysis of the potential in Tuscany. Biomass and Bioenergy 34: 1041-1052.
Dambreville, C., Morvan, T. and Germon, J., 2008. N2O emission in maize-crops fertilized with pig slurry, matured pig manure or ammonium nitrate in Brittany. Agriculture, Ecosystems and Environment 123: 201-210.
Daynard, T.B. and Kannenberg, L.W., 1976. Relationships between length of the actual and effective grain filling periods and the grain yield of corn. Canadian Journal of Plant Science 56: 237-242.
Del Grosso, S.J., Halvorson, A.D. and Parton, W.J., 2008. Testing DAYCENT model simulations of corn yields and nitrous oxide emissions in irrigated tillage systems in Colorado. Journal of Environmental Quality 37: 1383-1390.
Denman, K.L., Brasseur, G., Chidthaisong, A., Ciais, P., Cox, P.M., Dickinson, R.E., Hauglustaine, D., Heinze, C., Holland, E., Jacob, D., Lohmann, U., Ramachandran, S., Da Silva Dias, P.L., Wofsy, S.C. and Zhang, X., 2007. Couplings between changes in the climate system and biogeochemistry. In: Solomon, S., Qin, D., Manning, M., Chen, Z., Marquis, M., Averyt, K.B., Tignor, M. and Miller H.L. (eds.) Climate Change 2007: the physical science basis. Contribution of working group I to the fourth assessment report of the intergovernmental panel on climate change. Cambridge University Press, Cambridge, UK, pp. 499-587.
Donatelli, M., Spallacci, P., Marchetti, R. and Papini, R., 1996b. Evaluation of CropSyst simulations of growth of maize and of water balance and soil nitrate content following organic and mineral fertilization applied to maize. In: Proceedings of the fourth European society for agronomy congress, 7-11 July 1996, Veldhoven/Wageningen, the Netherlands, pp. 342-343.
Donatelli, M., Stockle, C.O., Ceotto, E. and Rinaldi, M., 1996a. CropSyst validation for cropping systems at two locations of Northern and Southern Italy. European Journal of Agronomy 6: 35-45.
Duvick, D.N., 2005. The contribution of breeding to yield advances in maize (*Zea mays* L.). Advances in Agronomy 86: 83-145.
Evans, L.T., 1997. Adapting and improving crops: the endless task. Philosophical Transactions of the Royal Society of London Series B-Biology Sciences 352: 901-906.
Farahbakhshazad, N., Dinnes, D.L., Li, C., Jaynes, D.B. and Salas, W., 2008. Modeling biogeochemical impacts of alternative management practices for a row-crop field in Iowa. Agriculture, Ecosystems and Environment 123: 30-48.

Fisher, A.C., Hanemann, W.M., Roberts, M.J. and Schlenker, W., 2012. The economic impacts of climate change: evidence from agricultural output and random fluctuations in weather. The American Economic Review102: 3749-3760.

Food and Agriculture Organization (FAO), 2009. Harmonized world soil database (version 1.1). IIASA, Luxemburg, Austria.

Food and Agriculture Organization (FAO), 2010. Climate-smart agriculture: policies, practices and financing for food security, adaptation and mitigation. FAO, Rome, Italy.

Freibauer, A., Rounsevell, M., Smith, P. and Verhagen, A., 2004. Carbon sequestration in the agricultural soils of Europe. Geoderma 122: 1-23.

Garcia, A.G., Guerra, L.C. and Hoogenboom, G., 2009. Impact of planting date and hybrid on early growth of sweet corn. Agronomy Journal 101: 193-200.

Garrison, M.V, Batchelora, W.D., Kanwara, R.S and Ritchie, J.T., 1999. Evaluation of the CERES-Maize water and nitrogen balances under tile-drained conditions. Agricultural Systems 62: 189-200.

Gattinger, A., Muller, A., Matthias, H., Skinner, C., Fliessbach, A., Buchmann, N., Mäder, P., Stolze, M., Smith, P., El-Hage Scialabba, N., and Niggli, U., 2012. Enhanced top soil carbon stocks under organic farming. Proceedings of the National Academy of Sciences of the USA 109: 18226-18231.

Giltrap, D.L., Li, C. and Saggar, S., 2010. DNDC: a process-based model of greenhouse gas fluxes from agricultural soils. Agriculture, Ecosystems and Environment 136: 292-300.

Gomiero, T., Paoletti, M. and Pimentel, D., 2008. Energy and environmental issues in organic and conventional agriculture. Critical Reviews in Plant Science 27: 239-254.

Grant, B., Smith, W.N., Desjardins, R., Lemke, R. and Li, C., 2004. Estimated N_2O and CO_2 emissions as influenced by agricultural practices in Canada. Climatic Change 65: 315-332.

Halvorson, A.D., Del Grosso, S.J. and Reule, C.A., 2008. Nitrogen, tillage, and crop rotation effects on nitrous oxide emissions from irrigated cropping systems. Journal of Environmental Quality 37: 1337-1344.

Hastings, A.F., Wattenbach, M., Eugster, W., Li, C., Buchmann, N. and Smith, P., 2010. Uncertainty propagation in soil greenhouse gas emission models: an experiment using the DNDC model and at the Oensingen cropland site. Agriculture, Ecosystems Environment 136: 97-110.

Hlavinka, P., Eitzinger, J., Smutny, V., Thaler, S., Zalud, Z., Rischbeck, P. and Kren, J., 2010. The performance of CERES-Barley and CERES-Wheat under various soil conditions and tillage practices in Central Europe. Bodenkultur 61: 5-17.

Intergovernmental Panel on Climate Change (IPCC), 2001. Climate change 2001: impacts, adaptation and vulnerability. Contribution of working group II to the third assessment report of the Intergovernmental panel on Climate Change, Cambridge University Press, Cambridge, UK.

Intergovernmental Panel on Climate Change (IPCC), 2007. The IPCC fourth assessment report. Cambridge University Press, Cambridge, UK.

Jamieson, P.D., Porter, J.R., Goudriaan, J., Ritchie, J.T., Van Keulen, H. and Stol, W., 1998. A comparison of the models AFRCWHEAT2, CERES-Wheat, Sirius, SUCROS2 and SWHEAT with measurements from wheat grown under drought. Field Crops Research 55: 23-44.

Janzen, H.H., 2004. Carbon cycling in earth systems – soil science perspective. Agriculture, Ecosystems and Environment 104: 399-417.

Jones, S.K., Rees, R.M., Skiba, U.M. and Ball, B.C., 2007. Influence of organic and mineral N fertilizer on N2O fluxes from a temperate grassland. Agriculture, Ecosystems and Environment 121: 74-83.

Kim, S. and Dale, E.B., 2008. Effects of nitrogen fertilizer application on greenhouse gas emissions and economics of corn production. Environmental Science Technology 42: 6028-6033.

Kroeze, C., Mosier, A.R. and Bouwman, A.F., 1999. Closing the global N_2O budget: a retrospective analysis 1500-1994. Global Biogeochemical Cycles 13: 1-8.

Kustermann, B. and Hülsbergen, K.J., 2008. Emission of climate relevant gases in organic and conventional cropping systems. In: Proceedings of the 16th IFOAM Organic World Congress, June 2008, Modena, Italy. Available at: http://orgprints.org/12813/1/12813.pdf.

Lal, R., 2004. Soil carbon sequestration impacts on global climate change and food security. Science 304: 1623-1627.

Lal, R., 2007. Carbon management in agricultural soils. Mitigation and adaption strategies for global change. Springer 12: 303-322.

Lal, R., 2008. Crop residues and soil carbon. Conservation agriculture carbon offset consultation, 28-30 October 2008. Food and Agriculture Organization of the United Nations and Conservation Technology Information Center, Beck Agricultural Center, West Lafayette, IN, USA.

Lamers, M., Ingwersen, J. and Streck, T., 2007. Modelling nitrous oxide emission from water-logged soils of a spruce forest ecosystem using the biogeochemical model Wetland-DNDC. Biogeochemistry 86: 287-299.

Leifeld, J. and Fuhrer, J., 2010. Organic farming and soil carbon sequestration: what do we really know about the benefits? Ambio 39: 585-599.

Li, C., Aber, J., Stange, F., Butterbach-Bahl, K. and Papen, H., 2000. A process-oriented model of N_2O and NO emissions from forest soils: 1. Model development. Journal of Geophysical Research 105: 4369-4384.

Li, C., Frolking, S. and Butterbach-Bahl, K., 2005. Carbon sequestration in arable soils is likely to increase nitrous oxide emissions, offsetting reductions in climate radiative forcing. Climatic Change 72: 321-338.

Li, C., Frolking, S. and Frolking, T.A., 1992a. A model of nitrous-oxide evolution from soil driven by rainfall events. 1. Model structure and sensitivity. Journal of Geophysical Research – Atmospheres 97: 9759-9776.

Li, C., Frolking, S. and Frolking, T.A., 1992b. A model of nitrous-oxide evolution from soil driven by rainfall events. 2. Model applications. Journal of Geophysical Research – Atmospheres 97: 9777-9783.

Li, C., Frolking, S., Crocker, G.J., Grace, P.R., Klir, J., Körchens, M. and Poulton, P.R., 1997. Simulating trends in soil organic carbon in long-term experiments using the DNDC model. Geoderma 81: 45-60.

Li, C., Salas, W., De Angelo, B. and Rose, S., 2006. Assessing alternatives for mitigating net greenhouse gas emissions and increasing yields from rice production in China over the next 20 years. Journal of Environmental Quality 35: 1554-1565.

Lugato, E., Zuliani, M., Alberti, G., Delle Vedove, G., Gioli, B., Miglietta, F. and Peressotti, A., 2010. Application of DNDC biogeochemistry model to estimate greenhouse gas emissions from Italian agricultural areas at high spatial resolution. Agriculture, Ecosystems and Environment 139: 546-556.

Lynch, D.H., MacRae, R. and Martin, R.C., 2011. The carbon and global warming potential impacts of organic farming: does it have a significant role in an energy constrained world? Sustainability 3: 322-362

MacCarty, N., Ogle, D., Still, D., Bond, T. and Roden, C., 2008. A laboratory comparison of the global warming impact of five major types of biomass cooking stoves. Energy for Sustainable Development 12: 56-65.

Mack, M.C., Schuur, E.A.G., Bret-Harte, M.S., Shaver, G.R. and Chapin III, F.S., 2004. Ecosystem carbon storage e in arctic tundra reduced by long-term nutrient fertilization. Nature 431: 440-443.

Mackay, A.D. and Barber, S.A., 1986. Effect of nitrogen on root growth of two corn genotypes in the field. Agronomy Journal 78: 699-703.

Mann, L., Tolbert, V. and Cushman, J., 2002. Potential environmental effects of corn (*Zea mays* L.) stover removal with emphasis on soil organic matter and erosion. Agriculture, Ecosystems and Environment 89: 149-166.

Marinari, S., Masciandaro, G., Ceccanti, B. and Grego, S., 2000. Influence of organic and mineral fertilizer on soil biological and physical properties. Bioresource Technology 72: 9-17.

Marriott, E.E. and Wander, M.M., 2006. Total and labile soil organic matter in organic and conventional farming systems. Soil Science Society of America Journal 70: 950-959.

Morari, F., Berti, A., Borin, M. and Giardini, L., 2001. CropSyst model in simulating cropping systems with different input levels. Technology transfer. In: Proceedings of the 9th International Conference on the FAO ESCORENA. Network on recycling of agricultural, municipal and industrial residues in agriculture. 6-9 September 2000, Gargano, Italy, pp. 257-262.

Organization for Economic Cooperation and Development (OECD), 2010. Climate change and agriculture: impacts, adaptation, mitigation. OECD, Paris, France.

Panahi, M., Naseri, R. and Soleimani, R. 2010. Efficiency of some sweet corn hybrids at two sowing dates in central Iran. Middle Eastern Journal of Scientific Research 6: 51-55.

Pathak, H., Li, C. and Wassmann, R., 2005. Greenhouse gas emissions from Indian rice fields: calibration and up-scaling using the DNDC model. Biogeosciences 2: 113-123.

Paustian, K., Babcock, B.A., Hatfield, J., Lal, R., McCarl, B.A., McLaughlin, S., Mosier, A., Rice, C., Robertson, G.P., Rosenberg, N.J., Rosenzweig, C., Schlesinger, W.H. and Zilberman, D., 2004. Agricultural mitigation of greenhouse gases: science and policy options. Council on Agricultural Science and Technology (CAST) Report, 120 pp.

Paustian, K., Six, J., Elliott, E.T. and Hunt, H.W., 2000. Management options for reducing CO_2 emissions from agricultural soils. Biogeochemistry 48:147-163.

Pikul, J.L., Schumacher, T.E. and Vigil, M., 2001. Nitrogen use and carbon sequestered by corn in the northern Corn Belt, U.S. The Scientific World 1: 707-713.

Poirier, V., Angers, D.A., Rochette, P., Chantigny, M.H., Ziadi, N., Tremblay, G. and Fortin, J., 2009. Interactive effects of tillage and mineral fertilization on soil carbon profiles. Soil Science Society of America Journal 73: 255-261.

R Development Core Team, 2006. R: a language and environment for statistical computing. R Foundation for Statistical Computing, Vienna, Austria.

Reicosky, D.C., Evans, S.D., Cambardella, C.A., Allmaras, R.R., Wilts, A.R. and Huggins, D.R., 2002. Continuous corn with moldboard tillage: residue and fertility effects on soil carbon. Journal of Soil and Water Conservation 57: 277-284.

Rizzardi, M.A., Witech, D. and Deggerone, I., 1994. Grain yield and yield components of maize cultivars at two sowing dates. Ciencia Rural 24: 477-482.

Robertson, G.P. and Grace, P.R., 2004. Greenhouse gas fluxes in tropical and temperate agriculture: the need for a full-cost accounting of global warming potentials. Environment, Development and Sustainability 6: 51-63.

Roy, R.N., Finck, A., Blair, G.J. and Tandon, H.L., 2006. Plant nutrition for food security: a guide for integrated nutrient management. FAO Fertilizer and Plant Nutrition Bulletin 16. Food and Agriculture Organization of the United Nations, Rome, Italy.

Russell, A.E., Laird, D.A., Parkin, T.B. and Mallarino, A.P., 2005. Impact of nitrogen fertilization and cropping system on carbon sequestration in Midwestern Mollisols. Soil Science Society of America Journal 69: 413-422.

Sainju, U.M., Senwo, Z.N., Nyakatawa, E.Z., Tazisong, I.A. and Reddy, K.C., 2008. Tillage, cropping systems, and nitrogen fertilizer source effects on soil carbon sequestration and fractions. Journal of Environmental Quality 37: 880-888.

Saseendran, S.A., Ma, L., Nielsen, D.C., Vigil, M.F. and Ahuja, L.R., 2005. Simulating planting date effects on corn production using RZWQM and CERES-Maize Models. Agronomy Journal 97: 58-71.

Scott, A., Ball, B.C., Crichton, I.J. and Aitken, M.N., 2000. Nitrous oxide and carbon dioxide emissions from grassland amended with sewage sludge. Soil Use and Management 16: 36-41.

Sehy, U., Ruser, R. and Munch, J.C., 2003. Nitrous oxide fluxes from maize fields: relationship to yield, site-specific fertilization, and soil conditions. Agriculture, Ecosystems and Environment 99: 97-111.

Smith, P. and Olesen, J.E., 2010. Synergies between the mitigation of, and adaptation to, climate change in agriculture. Journal of Agricultural Science 148: 543-552.

Smith, P., 2004a. Carbon sequestration in croplands: the potential in Europe and the global context. European Journal of Agronomy 20: 229-236.

Smith, P., 2004b. Engineered biological sinks on land. In: Field, C.B. and Raupach, M.R. (eds.) The global carbon cycle. Integrating humans, climate, and the natural world. SCOPE 62, Island Press, Washington, DC, USA, pp. 479-491.

Smith, P., 2004c. How long before a change in soil organic carbon can be detected? Global Change Biology 10: 1878-1883.

Smith, P., Martino, D., Cai, Z., Gwary, D., Janzen, H., Kumar, P., McCarl, B., Ogle, S., O'Mara, F., Rice, C., Scholes, B., Sirotenko, O., Howden, M., McAllister, T., Pan, G., Romanenkov, V., Rose, S., Schneider, U., Towprayoon, S. and Wattenbach, M., 2007a. In: Metz, B., Davidson, O.R., Bosch, P.R., Dave, R. and Meyer, L.A. (eds.) Contribution of the working group on climate change 2007: mitigation of climate change. Working Group III contribution to the Fourth Assessment Report of the IPCC. Cambridge University Press, Cambridge, UK, pp. 497-540.

Smith, P., Martino, D., Cai, Z., Gwary, D., Janzen, H., Kumar, P., McCarl, B., Ogle, S., O'Mara,, F., Rice, C., Scholes, B., Sirotenko, O., Howden, M., McAllister, T., Pan, G., Romanenkov, V., Schneider, U., Towprayoon, S. and Wattenbach, M., 2007b. Policy and technological constraints to implementation of greenhouse gas mitigation options in agriculture. Agriculture, Ecosystems and Environment 118: 6-28.

Smith, P., Martino, D., Cai, Z., Gwary, D., Janzen, H., Kumar, P., McCarl, B., Ogle, S., O'Mara, F., Rice, C., Scholes, B., Sirotenko, O., Howden, M., McAllister, T., Pan, G., Romanenkov, V., Schneider, U., Towprayoon, S., Wattenbach, M. and Smith, J., 2008. Greenhouse gas mitigation in agriculture. Philosophical Transactions of the Royal Society B 363: 789-813.

Smith, W.N., Desjardins, R.L., Grant, B., Li, C., Lemke, R., Rochette, P., Corre, M. D. and Pennock, D., 2002. Testing the DNDC model using N_2O emissions at two experimental sites in Canada. Canadian Journal of Soil Science 82: 365-374.

Stockle, C.O., Martin, S. and Campbell, G.S., 1994. CropSyst, a cropping systems model: water/nitrogen budgets and crop yield. Agricultural Systems 46: 335-359.

Su, Y.Z., Wang, F., Suo, D.R., Zhang, Z.H. and Du, M.W., 2006. Long-term effect of fertilizer and manure application on soil-carbon sequestration and soil fertility under the wheat-wheat-maize cropping system in northwest China. Nutrient Cycling in Agroecosystems 75: 285-295.

Thelen, K.D., Fronning, B.E., Kravchenko, A., Min, D.H. and Robertson, G.P., 2010. Integrating livestock manure with a corn-soybean bioenergy cropping system improves short-term carbon sequestration rates and net global warming potential. Biomass and Bioenergy 34: 960-964.

Thomson, J.A., Giannopoulos, G., Pretty, J., Bags, E.M. and Richardson, J.D., 2012. Biological sources and sinks of nitrous oxide and strategies to mitigate emissions. Philosophical Transactions of the Royal Society B 367: 1157-1168.

Thornthwaite, C.W., 1948. An approach toward a rational classification of climate. Geographical Review 38: 55-94.

Todorovic, M., Albrizio, R., Zivotic, L., Abisaab, M., Stöckle, C. and Steduto, P., 2009. Assessment of AquaCrop, CropSyst, and WOFOST models in the simulation of sunflower growth under different water regimes. Agronomy Journal 101: 509-521.

Tonitto, C., David, M.B., Drinkwater, E.L. and Li, C., 2007. Application of the DNDC model to tile-drained Illinois agro-ecosystems: model calibration, validation, and uncertainty analysis. Nutrient Cycling in Agroecosystems 78: 51-63.

United States Environmental Protection Agency (US-EPA), 2006. Global anthropogenic non-CO_2 greenhouse gas emissions: 1990-2020. EPA 430-R-06-003, June 2006. Washington, DC, USA.

Wang, J., Cardenas, L.M., Misselbrook, T.H., Cuttle, S., Thorman, R.E. and Li, C., 2012. Modelling nitrous oxide emissions from grazed grassland systems. Environmental Pollution 162: 223-233

Wolf, J. and Van Diepen, C.A., 1995. Effects of climate change on grain maize yield potential in the European community. Climatic Change 29: 299-331.

Yang, X.M. and Wander, M.M., 1999. Tillage effects on soil organic carbon distribution and storage in a silt loam soil in Illinois. Soil and Tillage Research 52: 1-9.

Zebarth, B.J., Rochette, P. and Burton, D.L., 2008. N_2O emissions from spring barley production as influenced by fertilizer nitrogen rate. Canadian Journal of Soil Science 88:197-205.

Chapter 12.

Comparative assessment of soil bulk density by computerized tomography methods for carbon stock quantification

A. Segnini[1], C.M. P. Vaz[1], A. Posadas[1,2], M. D. Guastal[1], P.R. O. Lasso[1], A.C.C. Bernardi[3] and D.M.B P. Milori[1]

[1]Brazilian Agricultural Research Corporation (Embrapa), Embrapa Instrumentation, Rua XV de Novembro 1452, P.O. Box 74, 13560-970 São Carlos, SP, Brazil; [2]International Potato Center, Av. La Molina 1895, P.O. Box 1558, Lima 12, Peru; [3]Embrapa, Southeast Cattle Research Center, Rodovia Washington Luiz, km 234, P.O. Box 339, 13560-970 São Carlos, SP, Brazil; alinesegnini@hotmail.com

Abstract

Soil carbon (C) concentration and soil bulk density are essential parameters used to quantify soil C stocks and the potential of soil C sequestration. Bulk density was traditionally assessed by gravimetric methods but gamma ray computed tomography can be used to quantify variations in bulk density within the soil profile. Tomographic methods measure soil physical properties without disturbing the soil, and X-ray microtomography provides high resolution cross-section images of an object to create a virtual 3 dimensional-model of the object. As such, it was assumed that this technique may also be applied to evaluate soil bulk density. The objective of this study was to compare computerized tomographic results to estimate the bulk density of soils of different textures and under different management conditions. Results were compared with bulk density values determined by classical methods as volumetric ring or paraffin sealed clod methods. Comparisons among linear attenuation coefficients were undertaken to correlate them linearly with the soil bulk density parameter for each soil. However, some factors such as beam hardening effect and the polychromatic nature of X-rays microtomography make it difficult to directly quantify soil bulk density.

Keywords: gamma-ray tomography, X-ray microtomography, soil bulk density, gravimetric methods.

12.1 Introduction

An increase in soil organic carbon (C) content caused by a change in land management practices and consequently mitigation of climate change has implication for soil C sequestration assessments (Lal, 2004). Focusing on these studies, estimation of soil organic C (SOC) stocks requires bulk density measurements in addition to quantifying the concentration of C in soil. Variability in bulk density contributes to C stock uncertainty, influencing the range in SOC stock with time and space (Holmes *et al.*, 2012). There are many methodologies for assessing soil bulk density such as volumetric rings and clods; gamma-ray attenuation methods (Pires *et al.*, 2009) and gamma-ray computed tomography (Pires and Bacchi, 2010) among other nuclear methods (Timm *et al.*, 2005).

Computed tomography is a non-destructive and non-invasive technique used under *in situ* conditions. This technique has many multidisciplinary applications including medicine, but it may also be a useful tool in soil science because images obtained with X- or gamma-ray computed tomography can be easily converted to a map of soil density or/and water content (Crestana and Vaz, 1998). X-ray and gamma-ray computed tomography is based on the principle of eletromagnetic radiation attenuation by matter. When a beam of rays pass through a sample its number of photons in the direction of the beam is attenuated following the Beer-Lambert law (Macedo *et al.*, 1998). Soil bulk density is a physical attribute that can be evaluated by these methods, using the soil mass attenuation coefficient described by Costa *et al.* (2013). The main applications include studies on soil compaction, surface soil sealing, effects of wetting-drying cycles on soil structure, soil porosity and pore size distribution, water movement through soils and effects of agricultural management systems on soil structure. The advantage of this method is that most conventional image analysis techniques used in soil physics do not allow the investigation of undisturbed samples as they alter their structure or do not allow a second measurement at the same position.

Gamma-ray computed tomography can determine soil density with higher resolution, enabling the imaging of the core of undisturbed soil samples with a spatial resolution in the range of millimeters (Crestana and Vaz, 1998). To date, most studies using computed tomography to evaluate soil bulk density have generated images of soil samples from different management systems in order to evaluate changes in soil compaction as a result of tillage (Fante Junior *et al.*, 2002; Pedrotti *et al.*, 2005). Pires (2011) investigated the sensitivity of the gamma-ray computed tomography equipped with a ^{137}Cs source for assessing the structural deformation of a Brazilian clay soil compacted by agricultural machinery by characterizing soil bulk density and porosity. Core samples with a volume of 55 cm^3 were collected at the surface (0-10 cm), and showed significant changes in soil structure after two passages by tractor wheels, with extensive decreases in soil porosity. Computed tomography data allowed a continuous analysis of changes in soil bulk density and soil porosity distribution for 1.25 and 1.20 mm layers, which cannot be provided with millimetric resolution by traditional methods commonly used in soil physics studies.

Similar to gamma-ray tomography, X-ray microtomography can also be used to evaluate soil structure to help advance this challenging area of research (Vaz *et al.*, 2011). The aim of the present research was to apply X-ray computed microtomography to assess soil bulk density of samples from soils of different textures and under various management systems; and to compare these results to previous assessments based on gamma-ray tomography, conducted by the Brazilian Agricultural Research Corporation (Embrapa) and other soil research groups. Such comparisons are necessary because of the non-monoenergetic nature of X-ray microtomography. There are many opportunities for using this technique in soil studies including the evaluation of macro-and micro-pores of soils under various degrees of compaction caused by the trampling of livestock; studies on spatial variability and distribution of aggregates; characterizing water infiltration; and evaluating soil texture, density and particle size distribution (Lasso *et al.*, 2008). Parameters related to soil structure may have a direct relationship with the dynamics of C and its long-term sequestration in the soil (Torn *et al.*, 1997). Thus, the objective of this study was to assess the potential of computed tomography for estimating bulk density of soils with different textures and depths. Values from tomography were compared with those obtained by classical methods such as volumetric ring or clod methods.

12.2 Gamma-ray computed tomography to assess soil bulk density

Gamma-ray computed tomography was previously used to assess soil bulk density (Crestana and Vaz, 1998; Fante Junior *et al.,* 2002; Pedrotti *et al.*, 2005; Petrovic *et al.,* 1982; Pires *et al.*, 2010). Fante Junior *et al.* (2002) compared soil density with depth using the paraffin sealed clod method with computed tomography on compacted soil samples and found a good correlation between both techniques. Pedrotti *et al.* (2005) used bulk density data, obtained through volumetric ring and gamma-ray computed tomography, to evaluate soil compaction of a Brazilian Planosol under different cropping systems, and found a significant linear correlation coefficient (r^2=0.95). In another study, Pedrotti *et al.* (2005) used a computerized minitomographer with a radius-detector set, and a gamma-ray source of 59.6 keV (241 Am) in addition to a crystal scintillator detector of NaI (Tl) type, an electronic amplification and pulse counter, an electronic processing system and a computer to control and store data, and a lead circular collimators of 2.0 cm length and 0.40 cm diameter. Calibration was obtained through linear regression between the linear attenuation coefficient presented by the image reconstruction program with various tomography results from materials such as water, gasoline, soil (A and B horizons), aluminum, and the mass attenuation coefficient from each one. To obtain the attenuation mass coefficient (μm, cm^2/g), the number of incident photons ($I0$) and emergent photons (I) were measured, and mean measurements and the thickness at which the photons moved through the sample (cm (x)), were substituted into the Beer ratio. Once the equipment was properly operating, the tomographic images were obtained, and then with a suitable algorithm of reconstruction, bi-dimensional maps of the tomographic unities from the mean profiles were obtained. After the tomography analysis, the samples were weighed; the paraffin layer from the sample's edge was removed, and oven-dried at 105 °C for 48 h.

This drying process was developed in order to reduce the amount of water in the soil. During the drying procedure, soil bulk density was assumed to be constant, and the humidity distribution in the soil sample was considered homogeneous, based on the presence of caulinite as the main clay mineral. Using the tomography units (UT), the soil bulk density values were calculated from the average data of the calibration equation. Measures of mass attenuation coefficient of the soil and of the water (μm) were used to deduce the equations adjusted to the Planosol data. From the UT (straight line of the calibration of the and gamma-rays) and the values of the mass attenuation coefficient to the different materials, in the Beer's equation, the following formula was obtained to calculate the soil bulk density of the A and B horizons of the Planosol:

$$Dg = \frac{(UT/986.16) - (0.200\theta)}{0.267} \quad \text{for A horizons; and}$$

$$Dg = \frac{(UT/986.16) - (0.200\theta)}{0.297} \quad \text{for B horizons}$$

where UT is the mean value in each line (average of 5 pixels), obtained from the tomographic map provided by each sample; and θ is the water content in the soil sample (cm^3).

Costa *et al.* (2013) analysed the effect of soil mass attenuation coefficient variation, as a function of the sample thickness increase and collimator sizes on soil bulk density measurements. Tomographic images of core samples of two soils with different textures were investigated in the study by Costa *et al.* (2013). The maps were generated using distinct values considering their variation according to the sample thickness and the collimator size. Results from their study suggested that alterations in measured values of the soil mass attenuation coefficient was a function of the geometry adopted (collimator size and sample thickness) which could affect the determination of the soil physical properties such as bulk density.

12.3 Undisturbed samples for X-ray microtomography analyses: a case study

Microtomography assessment brings a new focus and alternatively research on new features of soil science (Peth *et al.*, 2008). X-ray microtomography has the advantage of providing non-destructive, high-resolution, three dimensional images of soil, allowing for the examination of samples without affecting its characteristics (Adams *et al.*, 2011). Alternatively, invasive techniques can alter the structure of the sample, such as aggregate size, pore size and distribution, water content, and solutes.

In our study, samples from four Brazilian soils representing four different soil textural classes were used including a Rhodic Hapludox or Ferralsol (>60% clay); a Typic Hapludox or Ferralsol (~35% clay); a Rhodic Hapludox (30-40% clay); and a Quartzipsamment or Leptosol (15% clay). Undisturbed 20 cm cubes of soil were extracted from different depths from each of these soil types, wrapped and transported to Embrapa in São Carlos, Brazil (Figure 12.1). The soil cubes were air-dried at room temperature for 30 days and subsequently cut in undisturbed cylindrical subsamples of 2 cm diameter × 3 cm height. Samples were shaped with a putty knife for CT scanning (to prevent unwanted movement during scanning and adhesive tape was

Figure 12.1. Undisturbed 20 cm cubes of soils were extracted from different depths and prepared for X-ray microtomography analysis.

used to mount the samples on the scanners rotation stage). The samples were analysed by X-ray microtomography using a SkyScan model 1172. Results were compared with bulk density values determined by the gravimetric method dried at 105 °C for 48 h using undisturbed soil samples (~100 cm^3).

12.4 Imaging acquisition and resolution from X-ray micro-tomography

In the process of acquiring the images the following parameters were used: filter aluminium/copper spatial resolution (voxel size) of 3-7 microns, rotation step of 0.4°, 180° rotation and 12 frames for averaging process. The process of image reconstruction of the computed tomography scan was performed using the SkyScan software NRecon with a beam hardening correction (Figure 12.2). The images produced were analysed by CTAn software which built a histogram of the linear attenuation coefficient of the X-rays for each sample, as well as measures of porosity. From the set of tomographic images of each sample, a region of interest (ROI) in the centre of the circular images was selected to avoid edge effects in soil sampling.

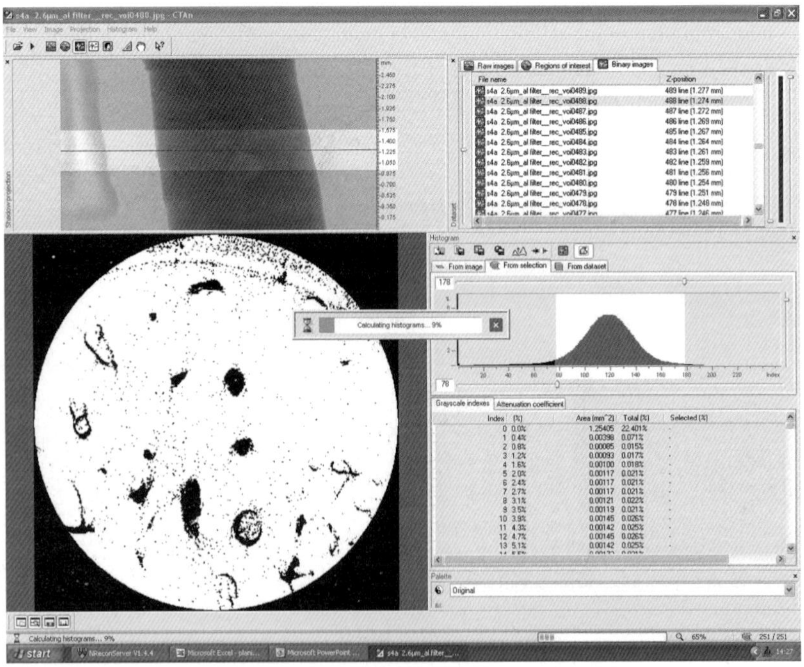

Figure 12.2. Skyscan CTan analyser software used to scan, reconstruct and analyse images.

Porosity was obtained by the segmentation process between the air and solid phases corresponding to the average value between the major peaks of air and solid phase histogram linear attenuation coefficient of the overall image. This process produces a binary image of the 3D image selected from the ROI. The software quantified porosity from the relationship between the volumes of the air phase and the total volume of interest (VOI) (Lasso *et al.*, 2008). As these coefficients were proportional to the density at each point of the sample, the average should be correlated with the overall density of the sample. The equipment allowed for the analysis of micro samples with a maximum diameter of 7 cm and a height of 10 cm, a resolution of a few µm, and samples with less than 1 cm in diameter, with a maximum resolution of the equipment, from about 1 µm. Different parameters for displaying images were investigated in order to observe the geometry of the pore system, inter-aggregate distribution, soil texture and structure. In the samples, the pixel size ranged between 4 and 9 µm. In the reconstruction of three-dimensional models, the image pixel was transformed mathematically into a voxel.

Figure 12.3 presents tomographic images of analysed soil samples with different textures. In all analyses, copper/aluminium filters were used. According Lasso *et al.* (2008), the best definitions of tomographic images are obtained with the use of these filters, which also resulted in a better separation of the peaks of the attenuation coefficients in the histograms (Figure 12.4). These filters were intended to filter low-energy radiation, which does not contribute to the overall analysis of dense samples. This prevented secondary rays from penetrating the sample and reach the detector. Tests performed with aluminium filters and lower voltages did not provide better results. As such, future research is needed to improve this correlation, for instance using other filters to provide more monoenergetic radiation.

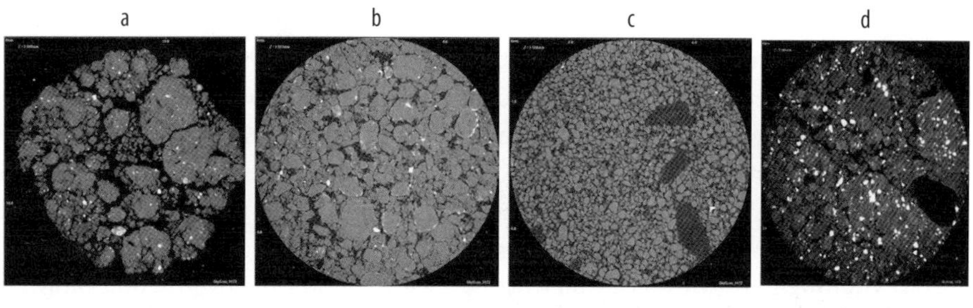

Figure 12.3. Reconstructed grayscale cross-sectional images of undisturbed soils. (a) Ferralsol with >60% clay; (b) Ferralsol with ~30% clay; (c) Ferralsol with 30-40% clay; and (d) a Leptosol with 15% clay, scanned using 2 cm sample sizes and a 3 to 7 µm resolution. An aluminium/copper filter was deployed during scanning.

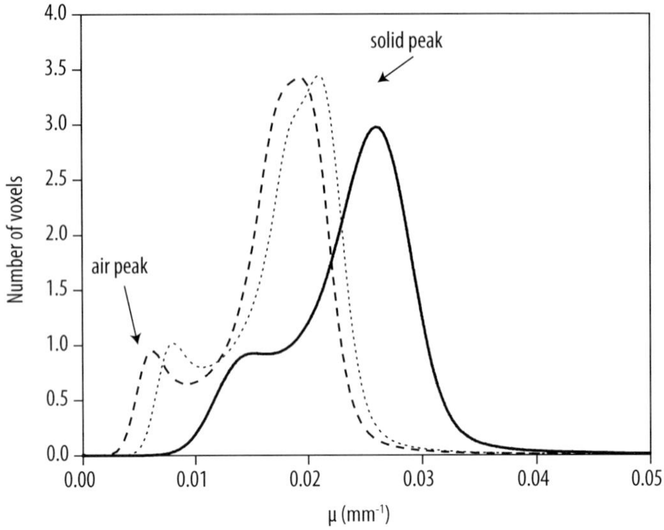

Figure 12.4. Linear attenuation coefficient histograms for reconstructed grayscale cross-sectional images (clay soil) scanned at a resolution 6 μm with an aluminium/copper filter. Different histograms are field replicates.

Histograms were used to obtain a separation of solid matter from air or material porosity (different peaks). These results demonstrated the potential of using this technique for the study of soil density (Figure 12.4).

The first attempt to compare linear attenuation coefficients obtained by the proposed technique and soil bulk densities values obtained by volumetric techniques (Figure 12.5).

Optimization of some parameters to evaluate soil density using X-ray microtomography was also investigated in our study. No correlation was obtained relating attenuation coefficients obtained by X-ray microtomography and soil bulk density obtained by clod method. According to Macedo and Crestana (1999) concepts such as density and moisture pertaining to the macro-scale cannot be extended to the micro-scale. At the macro-scale, the dominant features are soil aggregates and water in macropores. In contrast, dominant features at the micro-scale are micropores and hygroscopic moisture. It is therefore a challenge how to relate these two scales (upscaling or downscaling). This challenge is not new, but becomes increasingly evident, when techniques and imaging equipment with high spatial resolution such as X-ray microtomography are used to investigate soil characteristics.

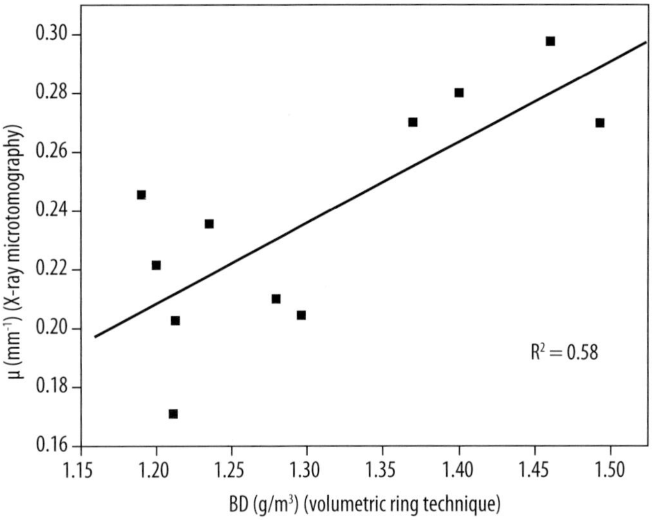

Figure 12.5. Relationship between the mean values of bulk density (BD, g/cm^3), determined by the volumetric ring technique and the linear attenuation coefficient (μ, mm^{-1}) determined by X-ray microtomography.

12.5 Conclusions

Computed tomography methods can be used to measure bulk density variations in the soil profile. The main advantage of using these advanced tools is to measure soil physical properties in a non-destructive way; in addition to obtaining other outputs such as a high resolution three-dimensional virtual model of an object. Previous work measuring soil bulk density provided interesting results showing a high correlation between results produced by gamma-ray tomography compared to classical methods. Some factors such as beam hardening effect and the polychromatic nature of X-rays microtomography make it difficult to directly quantify soil bulk density using the modern techniques. Further analyses should be performed in order to obtain relevant parameters which can be related to soil structure and be useful in evaluating soil C sequestration.

Acknowledgements

We are grateful for financial support from the State of São Paulo Research Foundation (FAPESP) (A. Segnini fellowship10/09211-4); the CNPq (158389/2011-5); CGIAR Research Program on Climate Change, Agriculture and Food Security (WLE) (A. Posadas); as well as to Embrapa Instrumentation. We also thank M.D. Guastali for soil sampling and V. Mares for editing the manuscript.

References

Adams, M., Crawford, J., Field, D., Henakaarchchi, N., Jenkins, M., McBratney, A., Courcelles, V.R., Singh, K., Stockmann, U. and Wheeler, I., 2011. Managing the soil-plant system to mitigate atmospheric CO_2. Discussion paper for the Soil Carbon Sequestration Summit, 31 January-2 February, 2011. The University of Sydney, Faculty of Agriculture, Food and Natural Resources, The United States Studies Centre at the University of Sydney, Australia. Available at: http://ussc.edu.au/s/media/docs/other/Science_Review_Paper.pdf.

Costa, J.C., Borges, J.A.R. and Pires, L.F., 2013. Soil bulk density evaluated by gamma-ray attenuation: analysis of system geometry. Soil and Tillage Research 129: 23-31.

Crestana, S. and Vaz, C.M.P., 1998. Non-invasive instrumentation opportunities for characterizing soil porous systems. Soil and Tillage Research 47: 19-26.

Fante Júnior, L., Oliveira, J.C.M., Bassoi, L.H., Vaz, C.M.P., Macedo, A., Bacchi, O.O.S., Reichardt, K. and Cavalcanti, A.C., 2002. Tomografia computadorizada na avaliação da densidade de um solo do semi-árido brasileiro. Revista Brasileira de Ciência do Solo 26: 835-842.

Holmes, K.W., Wherrett, A., Keating, A. and Murphy, D.V., 2012. Meeting bulk density sampling requirements efficiently to estimate soil carbon stocks. Soil Research 49: 680-695.

Lal, R., 2004. Soil carbon sequestration to mitigate climate change. Geoderma.123: 1-22.

Lasso, P.R.O., Vaz, C.M.P. and Naime, J.M., 2008. Recomendações para utilização do microtomógrafo de raios-X Sky Scan 1172 em ciências agrárias e biológicas. Embrapa Instrumentação Agropecuária, São Carlos, SP, Brazil.

Macedo, A. and Crestana, S., 1999. Macroporosity and soil particle density evaluation using X-ray microtomography. Revista Brasileira de Ciência do Solo 23: 763-771.

Macedo, A., Crestana, S. and Vaz, C.M.P., 1998. X-ray microtomography to investigate thin layers of soil clod. Soil and Tillage Research 49: 249-253.

Pedrotti, A., Pauletto, E.A., Crestana, S., Holanda, F.S.R., Cruvinel, P.E. and Vaz, C.M.P., 2005. Evaluation of bulk density of Albaqualf soil under different tillage systems using the volumetric ring and computadorized tomography methods. Soil and Tillage Research 80: 115-123.

Peth, S., Horn, R., Beckmann, F., Donath, T., Fischer, J. and Smucker, A.J.M., 2008. Three-dimensional quantification of intra-aggregate pore-space features using synchrotron-radiation-based microtomography. Soil Science Society of America Journal 72: 897-907.

Petrovic, A.M., Siebert, J.E. and Rieke, P.E., 1982. Soil bulk density analysis in three dimensions by computed tomographic scanning. Soil Science Society of America Journal 46: 445-450.

Pires, L. F., 2011. Gamma-ray computed tomography to evaluate changes in the structure of a clayey soil due to agricultural traffic. Acta Scientiarum Agronomy 33: 411-416.

Pires, L.F. and Bacchi, O.O.S., 2010. Mudanças na estrutura do solo avaliada com uso de tomografia computadorizada. Pesquisa Agropecuária Brasileira 45: 391-400.

Pires, L.F., Borges, J.A.R., Bacchi, O.O.S. and Reichardt, K., 2010. Twenty-five years of computed tomography in soil physics: a literature review of the Brazilian contribution. Soil and Tillage Research 110: 197-210.

Pires, L.F., Rosa, J.A., Pereira, A.B., Arthur, R.C.J. and Bacchi, O.O.S., 2009. Gamma-ray attenuation method as an efficient tool to investigate soil bulk density spatial variability. Annals of Nuclear Energy 36: 1734-1739.

Timm, L.C., Pires, L.F., Reichardt, K., Roveratti, R., Oliveira, J.C.M. and Bacchia, O.O.S., 2005. Soil bulk density evaluation by conventional and nuclear methods. Australian Journal of Soil Research 43: 97-103.

Torn, M.S., Trumbore, S.E., Chadwick, O.A., Vitousek, P.M. and Hendricks, D.M., 1997. Mineral control of soil organic carbon storage and turnover. Nature 389: 170-173.

Vaz, C.M.P., De Maria, I.C., Lasso, P.O. and Tuller, M., 2011. Evaluation of an advanced benchtop micro-computed tomography system for quantifying porosities and pore-size distributions of two Brazilian oxisols. Soil Science Society of America Journal 75: 832-841.

Chapter 13.

Evaluating the long-term effects of pre-conditioned biochar on soil organic carbon in two southern Ontario soils using the century model

M. Dil and M. Oelbermann

Department of Environment and Resource Studies, University of Waterloo, 200 University Avenue West, Waterloo, ON N2L 3G1, Canada; moelbermann@uwaterloo.ca

Abstract

Conventional agroecosystem management practices have accelerated the loss of soil organic matter. Adding organic matter from a relatively non-degradable source, such as biochar, may increase the long-term sequestration and stabilization of soil organic carbon (SOC). However, biochar contains a limited quantity of available mineral nutrients, requiring its addition in large quantities (>10 t/ha/y) and in association with nitrogen (N) fertilizer. Alternatively, biochar could be pre-conditioned by soaking it in a urea ammonium nitrate (UAN-enriched biochar), which may require an input of only 1 t/ha/y. The objective of this study was to evaluate the long-term (150 years) impact of the addition of UAN-enriched biochar on SOC, and on soil C fractions, compared to other commonly used agroecosystem management practices, in a coarse and medium textured soil using the Century Soil Organic Matter Model. Results showed that after 150 years, treatments with biochar had an increase in SOC in both soil types. However, the other agroecosystem management practices including continuous maize, no till and maize-soybean rotation decreased in SOC. Manure increased SOC in the coarse textured soil but decreased in the medium textured soil. An evaluation of the active, slow and passive C fractions revealed significant differences between the various agroecosystem management practices for both soil textures. Results showed a significantly greater ($P<0.05$) accumulation of C in the slow and passive fractions in the biochar treatments. There was also a significant difference ($P<0.05$) between the two soil types with respect to each C fraction and agroecosystem management practice. Results from this study suggested that the application of biochar pre-soaked in UAN, compared to all other management practices, may lead to an increase and the long-term stabilization of SOC.

Keywords: active soil organic carbon fraction, black carbon, carbon sequestration, passive soil organic carbon fraction, soil texture, temperate soils

DOI 10.3920/978-90-8686-788-2_13, © Wageningen Academic Publishers 2014

13.1 Introduction

One of the most pressing global issues currently facing humanity is climate change and its potential impact on agroecosystem productivity and food security (Lal, 2010). Land-use change and the adoption of conventional agroecosystem management practices, based on extensive tillage, combined with the removal or burning of crop residues, has accelerated soil degradation (Verhulst *et al.*, 2010) and led to the emission of CO_2 to the atmosphere (Oelbermann *et al.,* 2004). In order to maintain crop productivity, increasingly greater applications of nitrogen (N) fertilizers are required (Verhulst *et al.*, 2010), augmenting the emission of N_2O, a greenhouse gas (GHG) with a global warming potential 296 times greater than that of CO_2.

The depletion of soil organic carbon (SOC) parallels the adoption of conventional agroecosystem management practices, leading to a loss of ecosystem stability and resilience (Stavi and Lal, 2012). Stockmann *et al.* (2013) estimated that to date, 30 to 75% of the original SOC level has been lost from agricultural soils. This decrease is due to a lower quantity and quality of organic matter input, and enhanced soil organic matter (SOM) oxidation. Aertsens *et al.* (2013) proposed that the most prominent mitigation option for C sequestration is the implementation of conservation agriculture, including conservation tillage and no-till, cover cropping, residue management, and crop rotation. However, Stavi and Lal (2012) recommended that although the agronomic and environmental advantages of conservation agriculture are well recognized, most of these practices may have a limited potential for the long-term sequestration and stabilization of C in soil. For example, the adoption of no-till may not directly lead to an increase in soil C sequestration because a change in cultivation practices may cause in a change in crop productivity (Aertsens *et al.,* 2013). As such the net effect on soil C sequestration may be considerably less (Chatskikh *et al.,* 2008). Similarly, Govaerts *et al.* (2009) found that crop rotation changed the quantity and quality of organic matter input, affecting the long-term storage of SOC. Additionally, West and Post (2002) also observed that a more complex crop rotation cycle did not increase SOC.

The addition of organic matter from a relatively non-degradable source, such as biochar, may be an agroecosystem management practice that could increase the long-term sequestration and stabilization of SOC (Lee *et al.,* 2010) and lead to the mitigation of GHG emissions (Karhu *et al.,* 2011). Barrow (2012) also recommended that biochar may play an important role in rehabilitating degraded land and bring marginal soils into production, further expanding the global land base for potential C sequestration. Stavi and Lal (2012) estimated that the global capacity for sequestering C in soil amended with biochar may range from 7 to 110 Pg.

Although biochar is recalcitrant in nature, it does not remain unchanged after its application to the soil (Atkinson *et al.,* 2010). The incorporation of biochar can influence soil physical characteristics including structure, texture, porosity, particle size distribution and density (Atkinson *et al.,* 2010). This may result in an altered oxygen content, capacity to store water, and microbial and nutritional status of the soil in the plant rooting zone (Atkinson *et al.,* 2010). Due to biochar's large surface area, it has many micropores which may provide habitat for soil microorganisms and enhance arbuscular mycorrhizal fungal activity (Barrow, 2012). However, biochar contains a limited extent of available mineral nutrients for plant uptake, which requires it to be added in large quantities (>10 t/ha) and in association with nitrogen (N) fertilizer (Lee *et al.,* 2010). For example, when biochar was combined with mineral fertilizer, grain yields doubled on a Brazilian Ferralsol compared to only applying N fertilizer (Steiner *et al.,* 2007). Adding large quantities of biochar (≥10 t/ha) in addition to N fertilizer or manure may not be a cost-effective option for most agricultural producers in the temperate zone (Atkinson *et al.,* 2010). As a result, Resource Transformations International Ltd. in Waterloo, Ontario, Canada developed a method where biochar was pre-conditioned by soaking it in urea ammonium nitrate (UAN); reducing its application rate to 1 t/ha (J. Piskorz, pers. comm. 2010). This is because UAN is a concentrated N solution, composed of nitrate (NO_3^-), ammonium (NH_4^+), and urea [$CO(NH_2)_2$], which may promote the mineralization of C-rich materials such as biochar. While the potential agricultural and environmental benefits of adding biochar to soil have been identified in tropical regions, its use in temperate regions has not been evaluated sufficiently (Atkinson *et al.,* 2010); and information from long-term field studies is currently unavailable (Sohi *et al.,* 2010).

The use of empirical models to predict the long-term outcome of biochar addition to temperate soil may play a key role in identifying the potential of this agroecosystem management practice to sequester C. Process-based models like the Century Soil Organic Matter Model may be useful to predict changes in SOC as a result of different agroecosystem management practices when field data are not available (West and Six, 2007). This is because models are a powerful tool to predict changes in SOC over the long-term (Paustian *et al.,* 1992). Although the Century model has been successfully applied to evaluate the impact of various agroecosystem management practices on SOC (Lugato and Berti, 2008; Oelbermann and Voroney, 2011; Parton *et al.,* 1988), it has not yet been used to simulate the effects of biochar on SOC dynamics. The objective of this study was to evaluate the long-term (150 years) impact of the addition of UAN-enriched biochar on SOC, and its influence on soil active, slow and passive fractions compared to other conventional and conservation agriculture management practices using a coarse and a medium textured soil.

13.2 Materials and methods

The Century simulation was conducted using a coarse (Agriculture and Agri-Food Canada Research Station, Delhi, Ontario, Canada) and a medium (University of Guelph Research Station, Elora, Ontario, Canada) textured soil (Table 13.1). The criteria for selecting these two sites was based on their differences in soil texture in addition to well-documented information on past and current land management practices, climate data and soil characteristics.

The Century model is a site-specific model based on simplified soil-plant climate interactions (Smith *et al.*, 2009). The model evaluates agroecosystem management practices such as tillage, crop rotation, and fertilizer, manure or organic matter addition.

Table 13.1. Site characteristics (0-20 cm) and management practices of a coarse- and medium-textured soil in southern Ontario, Canada. Data adapted from Wanniarachichi *et al.* (1999) and Dil (2011).

	Coarse textured soil	Medium textured soil
Location	Delhi, Ontario	Elora, Ontario
Latitude/longitude	42°51'N/80°29'W	43°52'N/80°21'W
Altitude (m)	240	376
Mean precipitation (mm/y)	697	778
Mean temperature (°C)	7.8	6.0
Management practices	continuous maize	clover
Soil classification	Brunisolic Grey Brown Luvisol	Gleyed Melanic Brunisol
Sand (%)	85	21
Silt (%)	11	55
Clay (%)	4	24
Bulk density (g/cm^3)	1.4	1.3
pH	6.9	7.1
Soil organic carbon (%)	0.7	2.2
Soil total N (%)	0.06	0.19
C/N	12	12
NH_4^+ (mg/kg)	2.3	4.4
NO_3^- (mg/kg)	10.3	24.1
SMB-C (μg C_{mic}/g)[1]	147	440
SMB-N (μg N_{mic}/g)	53	33

[1] SMB = soil microbial biomass.

Century simulates above- and below-ground biomass production of crops, grasslands, savannas and forests, using a plant production model (Parton *et al.,* 1988). Changes in the level of SOC, determined using a soil organic matter (SOM) submodel, are based on plant residues decomposed by microbes and the resulting microbial products which are the substrate for humus formation (Parton *et al.,* 1988). The extent to which the decomposed compounds are stabilized is a function of soil texture (Parton *et al.,* 1988).

The Century model (version 4.0) was used to simulate temporal changes in SOC, using conventional and conservation agroecosystem management practices, including the addition of biochar, to a 20 cm depth in two temperate soils of different soil texture. In Century, monthly average maximum and minimum air temperature and monthly total precipitation data was obtained from on-site meteorological stations (Table 13.2). To initiate the model prior to simulating conventional and conservation agroecosystem management practices, the Century forest submodel was used to estimate equilibrium SOM levels and plant productivity of native forest (mixed deciduous-coniferous forest) under temperate conditions, and typical for southern Ontario, Canada, over ~10,000 years. The initialization of SOM steady state levels and subsequent deforestation events were used to bring the model to its first year of simulation using different agroecosystem management practices that began in 1915 with 55 years of rye-tobacco rotation, followed by continuous maize in the coarse textured soil (Delhi); and grass/hay for 52 years followed by continuous maize starting in 1967 in the medium textured soil at Elora (Table 13.3). These different agroecosystem management practices were simulated in Delhi and Elora until the year 2000. In 2001, conventional and conservation, including biochar addition, agroecosystem management practices were initiated and simulated for 150 years until 2150 (Table 13.3). At both sites it was assumed that there was a direct conversion of native forest vegetation to agricultural production in 1915.

Initial site-specific soil parameters (Table 13.1) including soil texture and pH were obtained from Wanniarachichi *et al.* (1999). Bulk density, SOC, soil total N, C/N were obtained from previous studies (Dil, 2011), and from data on file (NH_4^+, NO_3^-, soil microbial biomass (SMB)-C and SMB-N). The proportion of SOM initial values for the active, microbial, slow and passive fractions were 2, 71, 1547 and 762 g/m^2, respectively in the coarse textured (Delhi) soil, and 2, 179, 3,887, and 1,914 g/m^2, respectively in the medium textured (Elora) soil. Values for the maximum decomposition rate of SOM with active, slow and intermediate turnover were 4, 0.0013, and 0.05, respectively (Oelbermann and Voroney, 2011). Additional files were created in OMAD.100 representing biochar input (Table 13.3). Chemical parameters for the biochar were obtained from ABRI-Tech Inc.(Ottawa, Canada). The biochar consisted of a mix of maple (*Acer* spp.), oak (*Quercus* spp.) and birch (*Betula* spp.) woody biomass

Table 13.2. Climate data used for the Century simulation in Delhi, southern Ontario, Canada based on a 24-year average, and in Elora, southern Ontario based on a 33-year mean.

Month	Delhi (coarse-textured soil)			Elora (medium-textured soil)		
	Temperature (°C)		Precipitation (cm)	Temperature (°C)		Precipitation (cm)
	Min	Max		Min	Max	
January	-9.6	-2.4	6.9	-12.2	-4.1	5.8
February	-9.4	-1.3	5.7	-11.7	-2.9	4.9
March	-4.4	-3.9	8.4	-6.9	1.5	7.4
April	1.2	12.1	9.4	0.0	10.2	7.0
May	6.5	19.0	7.3	5.3	17.4	7.8
June	12.0	24.6	7.1	11.0	23.2	8.7
July	14.4	26.9	7.1	12.7	25.5	7.3
August	13.7	25.7	9.3	11.6	24.6	7.2
September	10.1	21.5	8.0	8.2	20.2	7.1
October	4.6	15.1	7.5	3.1	13.8	6.6
November	-0.1	7.4	8.2	-1.8	5.5	6.6
December	-6.3	0.6	8.5	-8.7	-1.6	7.2

and was pyrolized at 450 °C, with a nutrient content of 61.7% C, 0.24% N, 0.0024% phosphorus (P), and 0.17% potassium (K), 95% lignin and a C/N ratio of 257. In the Century simulation, the addition of biochar at 1 and 2 t/ha/y was equivalent to 61.7 g C/m^2 and 123.4 g $C/m^2/y$ respectively.

An additional file in FERT.100 was created that corresponded with N fertilizer input of 18.7 g $N/m^2/y$ for the various agroecosystem management practices (Table 13.3). This quantity of N fertilizer addition was typical for the production of maize (*Zea mays* L.) in southern Ontario. It was not possible to create a file in OMAD.100 that would have the equivalent of 100 or 200 $g/m^2/y$ biochar addition pre-conditioned with UAN. As such, for the biochar simulations, the FERT.100 file was used for the input of 18.7 g $N/m^2/y$ in addition to the biochar from the OMAD.100 file. Crop parameters used were those provided by Century. Default values set by the Century model were used for atmospheric N deposition and N_2-fixation rates. It is worth pointing out that the simulations used in this study were not undertaken to serve as an independent validation of the model. We were not able to evaluate model performance, where

Table 13.3. Scheduling of agricultural management practices in Century at Delhi (coarse textured soil) and Elora (medium textured soil) in southern Ontario, Canada.

Date	Management practices[1,2]
10,000 BC-1914	• Mixed deciduous/coniferous forest (MIX) ~10,000 years to estimate equilibrium soil organic matter levels and plant productivity and clear cut (CC) harvesting
1915-2000	• Delhi and Elora agroecosystem management practices Delhi: 1915 to 1970 rye-tobacco rotation; plowing (P), grain only (G) harvest; fertilizer 18.7 g $N/m^2/y$. 1971-2000 continuous maize (CHI high harvest); plowing (P); grain only harvest (G); fertilizer 18.7 g $N/m^2/y$. Elora: 1915 to 1966 grass/hay [100% cool (GCD)]; hay (H) harvest. 1967-2000 continuous maize (CHI high harvest maize); plowing (P); grain only harvest (G); fertilizer 18.7 g $N/m^2/y$.
2001-2150	• Continuous maize (CHI high harvest maize) Plowing (P); fertilizer at 18.7 g $N/m^2/y$; grain only (G) harvest • $Char^+$ 200 $g/m^2/y$ with maize (CHI high harvest maize) Plowing (P); biochar presoaked in UAN added at 123.4 $g/m^2/y$ (C/N = 257, lignin = 95%); grain only (G) crop harvest • $Char^+$ 100 $g/m^2/y$ with maize (CHI high harvest maize) Plowing (P); biochar presoaked in UAN added at 61.7 $g/m^2/y$; grain only (G) crop harvest • $Char^+$ 100 $g/m^2 \times 2$ years with maize (CHI high harvest maize) Year 1: plowing (P); biochar presoaked in UAN added at 61.7 $g/m^2/y$; grain only (G) crop harvest Year 1: plowing (P); fertilizer at 18.7 g $N/m^2/y$; grain only (G) crop harvest • Manure with maize (CHI high harvest maize) Plowing (P); cow manure at 1,500 g $C/m^2/y$ (C/N=20, lignin = 20%); grain only (G) harvest • No till with maize (CHI high harvest maize) No till (NT); fertilizer at 18.7 g $N/m^2/y$; grain only (G) harvest • Maize (CHI high harvest maize) – soybean (SYBN) rotation Plowing (P); fertilizer at 18.7 g $N/m^2/y$; grain only (G) harvest

[1] UAN = urea ammonium nitrate. The quantity of UAN used in presoaking the biochar was based on a 18.7 g $N/m^2/y$.

[2] CC, G, GCD, P, NT, MIX, SBYN and CHI are input characters for the model.

measured data is compared with simulated data, because there are currently no long-term (>5 years) field trials investigating biochar as a soil amendment (Sohi *et al.*, 2010).

In order to evaluate differences in soil C fractions (active, slow and passive) between the various agroecosystem management practices and between sites, a mean value over a 10-year period (years 2140 to 2150) was used. This 10-year period was representative of steady state conditions where changes in the active, slow and passive C fractions were no longer detected by Century (West and Six, 2007). The data were examined for homogeneity of variance and found to have normal distributions. Soil organic C in the active, slow and passive fractions was analysed using a multiple-factor analysis of variance in SPSS (SPSS Science Inc., 1989). Significantly different main effects were further tested using the least significant differences multiple comparison test (Steel *et al.*, 1997). Significant simple effects were tested using the estimated marginal means function in SPSS. For all statistical analyses, the threshold probability level for determining significant differences was $P<0.05$.

13.3 Results

Soil organic carbon stocks

Prior to clearing the forest, SOC was 9,815 g C/m^2 in Delhi (Figure 13.1), and 16,587 g C/m^2 in Elora (Figure 13.2). In 1915, when the sites were cleared and cultivated to implement agriculture, there was a spike in SOC, followed by rapid decline at both sites until the year 2000. In 2001, the continuation of conventional management practices or the initiation of conservation practices, including biochar addition, either increased or decreased SOC by the year 2150 at both sites. Soil organic C, in 2150, in

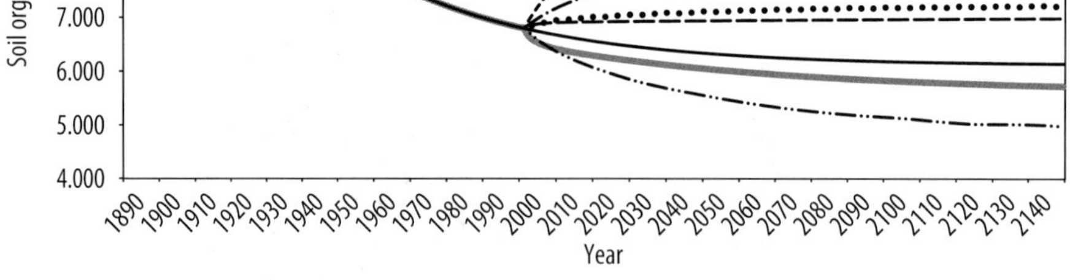

Figure 13.1. Soil organic carbon (g C/m^2) stocks simulated by the Century soil organic matter model using conventional and conservation agroecosystem management practices from 2001 to 2150 on a medium textured soil in Delhi, southern Ontario, Canada.

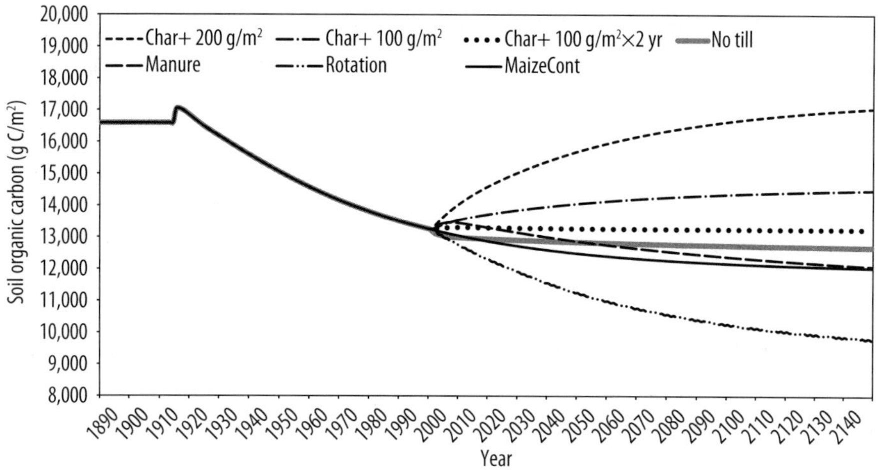

Figure 13.2. Soil organic carbon (g C/m^2) stocks simulated by the Century soil organic matter model using conventional and conservation agroecosystem management practices from 2001 to 2150 on a medium textured soil in Elora, southern Ontario, Canada.

Delhi followed a sequence of: Char$^+$ 200 g/m^2/y > Char$^+$ 100 g/m^2/y > Char$^+$ 100 g/m^{2}2 years > manure > continuous maize > no till > maize-soybean rotation (Figure 13.1). When comparing changes in SOC from 2001 to 2150, the various biochar amendments had the greatest increase followed by manure (Table 13.4). However, no till, continuous maize, and maize-soybean rotation showed a decline in SOC (Table 13.4). Differences in SOC between the various agricultural management practices in Delhi compared to continuous maize, in the year 2150, showed that the addition of biochar had the greatest increase followed by manure application (Table 13.4). However, when compared to continuous maize, no till and maize-soybean rotation showed a decline in SOC. In Elora, SOC followed a sequence of Char$^+$ 200 g/m^2/y > Char$^+$ 100 g/m^2/y > Char$^+$ 100 g/m^2×2 years > no till > manure > continuous maize > maize-soybean rotation (Figure 13.2). Comparing differences in SOC stocks between the years 2001 and 2150, an increase was observed only for Char$^+$ (Table 13.4). The remaining agroecosystem management practices showed a decline in SOC ranging from 12 g C/m^2 to 3,442 g C/m^2. However, when comparing the different management practices to continuous maize in the year 2150, only the maize-soybean rotation showed a decline in SOC (Table 13.4).

Table 13.4. Changes in soil organic carbon stocks (g/m^2), to a 20 cm depth, simulated by the Century soil organic matter model in a coarse-textured soil in Delhi and a medium-textured soil in Elora, southern Ontario, Canada using conventional and conservation agroecosystem management practices.

Site	Management practice	Soil organic carbon (g/m^2)		
		Year 2150	Change from year 2001	Change from continuous maize in year 2150
Delhi	Continuous maize	6,344	-456	-
	$Char^+$ 200 $g/m^2/y$	10,522	3,715	4,178
	$Char^+$ 100 $g/m^2/y$	8,319	1,515	1,975
	$Char^+$ 100 $g/m^2 \times 2$ years	7,208	404	864
	Manure	6,986	180	642
	No till	5,725	-1,077	-619
	Maize-soybean rotation	4,975	-1,825	-1,369
Elora	Continuous maize	12,042	-1,183	-
	$Char^+$ 200 $g/m^2/y$	17,048	3,658	5,006
	$Char^+$ 100 $g/m^2/y$	14,484	1,226	2,442
	$Char^+$ 100 $g/m^2 \times 2$ years	13,246	-12	1,204
	Manure	12,088	-1,194	46
	No till	12,680	-478	638
	Maize-soybean rotation	9,783	-3,442	-2,259

Soil carbon fractions

An evaluation of the active, slow and passive C fractions with Century showed significant differences between the various agroecosystem management practices at both sites (Table 13.5). In Delhi, the active C fraction followed a sequence of $Char^+$ 200 $g/m^2/y$ > $Char^+$ 100 $g/m^2/y$ > $Char^+$ 100 $g/m^2 \times 2$ years > manure ≈ continuous maize > maize-soybean rotation > no till. Only continuous maize and manure agroecosystem management practices were not significantly different. However, agroecosystem management practices were significantly different in the slow and passive C fractions, following a sequence of $Char^+$ 200 $g/m^2/y$ > $Char^+$ 100 $g/m^2/y$ > $Char^+$ 100 g/m^2 ×2 years > manure > continuous maize > no till > maize-soybean rotation (Table 13.5). In Elora, there was a significant difference between all agroecosystem management practices for each of the C fractions. For example, the active and passive C fractions followed a sequence of $Char^+$ 200 $g/m^2/y$ > $Char^+$ 100 $g/m^2/y$ > $Char^+$ $g/m^2 \times 2$ years > continuous maize > no till > maize-soybean rotation > manure. However, the slow

Table 13.5. Mean values of soil organic carbon (SOC) fractions (g/m^2) from the years 2140 to 2150 simulated by the Century soil organic matter model in a coarse-textured soil in Delhi and in a medium-textured soil in Elora, southern Ontario, Canada using conventional and conservation agroecosystem management practices. Standard errors are given in parentheses (n=10).

Site[1]	Management practices	Active SOC[2] (g/m^2)	Slow SOC (g/m^2)	Passive SOC (g/m^2)
Delhi	Continuous maize	222.03 (0.01)^D	4,411.65 (0.62)^E	1,161.39 (0.19)^E
	Char$^+$ 200 g/m^2/y	265.65 (0.03)^A	7,687.75 (3.09)^A	1,218.41 (0.68)^A
	Char$^+$ 100 g/m^2/y	244.36 (0.01)^B	6,078.19 (1.17)^B	1,190.75 (0.44)^B
	Char$^+$ 100 g/m^2 ×2 years	233.28 (0.05)^C	5,248.58 (0.23)^C	1,176.30 (0.31)^C
	Manure	225.87 (0.02)^D	5,114.18 (0.08)^D	1,170.95 (0.27)^D
	No till	131.08 (0.02)^F	4,395.26 (1.20)^F	1,079.43 (0.31)^G
	Maize-soybean rotation	168.77 (3.89)^E	3,455.90 (1.68)^G	1,104.77 (0.19)^F
Elora	Continuous maize	502.35 (0.04)^D	8,522.05 (1.62)^E	2,774.42 (0.10)^D
	Char$^+$ 200 g/m^2/y	609.35 (0.11)^A	13,523.09 (4.73)^A	2,888.68 (0.94)^A
	Char$^+$ 100 g/m^2/y	556.23 (0.03)^B	10,482.69 (1.43)^B	2,832.61 (0.42)^B
	Char$^+$100 g/m^2 ×2 years	529.32 (0.27)^C	9,499.15 (0.19)^C	2,803.92 (0.16)^C
	Manure	296.60 (0.56)^G	5,946.98 (13.97)^G	2,609.57 (1.73)^G
	No till	429.05 (0.13)^E	9,378.82 (0.60)^D	2,732.23 (0.29)^E
	Maize-soybean rotation	374.53 (1.09)^F	6,597.15 (5.60)^F	2,656.97 (1.05)^F

[1] Mean values are significantly different between sites (soil texture) for each management practice within each soil organic carbon (SOC) fraction. Interaction effects of site (soil texture)-by-management practices were significant for the soil active [F (6, 126)=11.949, P=0.001], slow [F (6, 126)=192.205, P=0.001], and passive [F (6, 126)=4.478, P=0.001] carbon fractions.

[2] Mean values followed by upper case letters, comparing differences between management practices and within each of the SOC fractions (active, slow and passive) and within each site (soil texture) are not significantly different at P<0.05.

C fraction followed a sequence of Char$^+$ 200 g/m^2/y > Char$^+$ 100 g/m^2/y > Char$^+$ 100 g/m^2 ×2 years > no till > continuous maize > maize-soybean rotation > manure (Table 13.5).

There was a significant difference between the two sites with respect to each C fraction and agroecosystem management practice (Table 13.5). Interaction effects of site (soil texture)-by-management practices were significant for the soil C active, slow and passive fractions (Table 13.5). When comparing differences for each C fraction and agroecosystem management practice between the year 2001 and 2150 in Delhi, biochar management practices accumulated C in the soil whereas the remaining practices lost SOC in one or two of the C fractions (Table 13.6). In Elora, only the Char$^+$ 100

Table 13.6. Mean values of changes in the different soil organic carbon (SOC) fractions (g/m²) between the years 2001 and 2150 in a coarse-textured soil in Delhi and a medium-textured soil in Elora located in southern Ontario, Canada, using conventional and conservation agroecosystem management practices.

Site	Management Practices	Active SOC (g/m²)	Slow SOC (g/m²)	Passive SOC (g/m²)
Delhi	Continuous maize	-7.13	-686.04	37.15
	Char⁺ 200 g/m²/y	36.50	2,591.08	96.87
	Char⁺ 100 g/m²/y	15.11	972.75	68.07
	Char⁺ 100 g/m² ×2 years	2.95	153.51	52.66
	Manure	-4.93	16.89	47.11
	No till	-100.58	-707.62	-47.13
	Maize-soybean rotation	-72.10	-1,647.47	-21.28
Elora	Continuous maize	-25.87	-1,155.50	-1.44
	Char⁺ 200 g/m²/y	79.52	117.69	3,560.93
	Char⁺ 100 g/m²/y [1]	27.40	817.36	59.21
	Char⁺ 100 g/m² ×2 years	1.15	-173.66	29.29
	Manure	-244.93	3,791.18	173.96
	No till	-71.44	-301.54	-44.38
	Maize-soybean rotation	-157.39	-3,099.29	-126.37

g/m²/y and 200 g/m²/y gained SOC in all three fractions whereas the remaining treatments lost SOC (Table 13.6). The accumulation of C within each fraction varied between agroecosystem management practices and between sites. For example, in Char⁺ 200 g/m²/y, the greatest C gain occurred in the slow fraction in Delhi but in Elora it occurred in the passive fraction.

13.4 Discussion

Results suggested that differences in SOC between Delhi and Elora, prior to forest clearing and upon initiation of the Century simulation, were likely due to natural variations in soil physical and chemical properties (West and Six, 2007). This may also related to the quantity and quality of organic matter entering the soil, and how this organic material is mineralized depending on ambient temperature and moisture (Dungait *et al.*, 2012) in addition to differences in soil texture (Lane *et al.*, 1998). Lane *et al.* (1998) noted that coarse-textured soils had a lower productivity and site fertility compared to soil with a finer texture due to their higher clay content. Bricklemyer *et al.* (2007) also suggested that soil with a finer texture, such as that found in Elora, may have a stronger influence on the long-term stabilization of C compared to a coarse textured soil. They also observed that clay content in the Century model affected the proportion of SOC transferred from the active fraction to the slow fraction (Bricklemeyer *et al.*, 2007). This is because clay particles protect SOC against microbial decay (Strong *et al.*, 1999). The initial spike in SOC in the year 1915, when the forest was cleared, was likely due to a sudden and large input of organic matter from the cleared forest. The slash left on site may have contributed an abrupt and higher input of organic matter to the soil resulting in a short-term increase in SOC (Henderson, 1995). Kirschbaum *et al.* (2008) found that most of this organic matter may be derived from coarse tree roots that remained on the site after clearing; making a substantial contribution to the overall input of C to the soil.

The implementation of agricultural practices in 1915 and the continued maize monocrop at both sites led to a rapid decline in SOC. Varvel (2006) also observed a decline in SOC as a result of continuous maize agroecosystem management practices. He found a 6.2% decrease in SOC (0-30 cm) after 8 years of continuous corn on a silty clay loam in Nebraska (Varvel, 2006). Soil organic C loss in the continuous maize management practice in our study, relative to native forest soil, was within the range of that reported by Lal (2004a).

Although crop rotation, manure application, and no till are conservation management practices (Verhulst *et al.*, 2010), our study showed that these practices resulted in a SOC decline over a 150 years. Similarly, West and Post (2002) observed a loss of 19 g $C/m^2/y$ when changing from continuous maize to a maize-soybean rotation. They suggested that continuous maize produces more residue biomass and therefore a greater C input compared to a maize-soybean rotation (West and Post, 2002). In the United States, Johnson *et al.* (2006) reported that the mean quantity of crop residue input from maize was 680 $g/m^2/y$ and 250 $g/m^2/y$ from soybeans. West and Six (2007) also found that crop rotation affected levels of SOC through the quality of residue input, which could influence the amount of C sequestered in the different fractions (Govaerts *et al.*, 2009). A global database of 67 long-term experiments reported that a change from conventional till to no till led to a smaller loss of SOC compared to a change from continuous maize to a maize-soybean rotation (West and Post, 2002); and was similar to our results. However, Govaerts *et al.* (2009) reported various results with respect to no till or conventional till and the effects on SOC. They found that out of 78 studies, 7 showed a decrease in SOC under no till compared to conventional till; 40 showed an increase in SOC under no till; and 31 showed no change. Govaerts *et al.* (2009) pointed out that the principal mechanisms directing the equilibrium between an increase, a decrease or no change in SOC between conventional till and no till are not yet understood.

The effect of farmyard manure on SOC has been assessed since 1852 as part of the Hoosfield continuous barley long-term experiment at Rothamsted Research, England (Jenkinson and Rayner, 1977). This long-term research showed that the addition of 3,500 g C/m^2/y farmyard manure increased the SOC stock by ~71% over a 120 year period (Jenkinson and Rayner, 1977). Although SOC increased at a much lower percentage (~3%) over 150 years in Delhi compared to the soil at Rothamsted Research, this difference may be due to a lower C input (1,500 g C/m^2/y) in our study. Comparatively, the amount of SOC decreased (~9%) over the 150 year simulation in the medium textured soil in Elora through the addition of manure. This suggested that even in a medium textured soil an input of 1,500 g C/m^2/y derived from manure was not sufficient to maintain the soil's C balance. Stewart *et al.* (2009) proposed that this may be due to a lower C stabilization efficiency in a soil with an already higher SOC content compared to soil with a lower intrinsic SOC content (e.g. Delhi). In both Delhi and Elora, manure application resulted in a higher accumulation of SOC compared to the continuous maize treatment. This coincides with the results from Lal (2004b), who found that the addition of manure led to a greater accumulation of SOC compared to the addition of the equivalent amount of nutrients in the form of mineral fertilizer.

Each of the different biochar management practices either led to an increase or no change in SOC at both sites. This was likely because biochar, as a result of incomplete combustion, is the only plant-derived organic matter that may have a higher intrinsic stability than other sources of organic matter (Rumpel *et al.*, 2012). Although biochar is stable over the long-term, it can be degraded microbially (Rethemeyer *et al.*, 2005). Enhanced microbial activity may lead to a small proportion of the biochar to become mineralized (Lehmann and Joseph, 2009), promoting the decay of more labile compounds over the short-term (Harner *et al.*, 2004). Our results suggested that an increase in SOC, as a result of biochar addition, may also be dependent on soil texture, where the coarse textured soil from Delhi showed the greatest increase in SOC. This is because C mineralization rates are typically higher in soils with a lower clay content compared to soil with greater clay content (Franzluebbers, 1999). Due to their lower clay content, coarse textured soils may have a lower physical and chemical protective capability (Carter *et al.*, 2003). Ultimately, this could result in a greater C stabilization efficiency (Stewart *et al.*, 2009) leading to a higher capacity to sequester C compared to soil with a finer texture (Carter *et al.*, 2003). Liang *et al.* (2010) also observed a greater specific surface area for microbial activity in sandy soil amended with biochar compared to a sandy soil that was not amended with biochar.

A decrease in each of the C fractions over 150 years, except those amended with biochar, may be due to an insufficient input of organic matter (Kimetu *et al.*, 2009).

Lugato *et al.* (2006) found that C input and N application rate had a direct effect on microbial growth and the rate of C mineralization. However, the addition of biochar may increase soil microbial biomass and the amount of recalcitrant C (Kautz *et al.,* 2004), leading to an increase of C in the active, slow and passive fractions as was observed in our study. Similarly, a greater soil microbial biomass C in treatments with UAN pre-conditioned biochar was observed in a separate study by the same authors as this chapter (data on file). This suggested that input of a recalcitrant source of C may lead to its long-term stabilization, most predominantly in the slow and passive fractions. Our results also suggested that biochar pre-conditioned with UAN could be added at a much lower rate (1 to 2 t/ha/y), and still enhance SOC and its long-term stabilization than suggested in other studies (Lee *et al.,* 2010; Steiner *et al.,* 2007). Input of a larger fraction of stable C could make the soil more resilient to changes in temperature as a result of global warming (Lehmann and Sohi, 2008).

A variation in the amount of C stabilized between the two sites may also be due to differences in soil texture. In the Century model, soil texture influences the rate of organic matter decomposition in the active C fraction and its rate of stabilization in the slow and passive C fractions. The higher clay content in Elora (24%) may have contributed to a greater proportion of C stored in the slow and passive fractions compared to that at Delhi with a lower (4%) clay content. Bricklemyer *et al.* (2007) observed that as clay content increased, SOC estimates from Century also increased at five different sites in Montana. This is because soil with a finer texture has a lower bulk density, which is associated with a higher SOM content (Sohi *et al.,* 2010).

Differences in the quantity of C allocated to the active, slow and passive fractions showed that conventional and conservation agroecosystem management practices may impact C stabilization differently. Paustian *et al.* (1992) suggested that this may be due to the quantity and recalcitrance of the organic matter added to the soil. For example, Parfitt *et al.* (1997) observed a decrease in the various C fractions after conversion of forest to pasture or forest to maize on a Nitisol in the United States. These results were similar to our study, where a decrease in all C fractions in all management practices, except for the manure and biochar practices, was observed. We found the greatest proportion of C was allocated to the slow fraction followed by the passive and active fractions. Similar results were reported by Oelbermann and Voroney (2011) in a temperate soybean-maize-wheat (*Triticum aestivum* L.) rotation under no-till. However, the recalcitrant nature of biochar likely led to a greater stabilization in all three C fractions. For example, Vignesh *et al.* (2012) found an increase in the slow and passive C fractions as a result of biochar addition.

Results from our study showed the potential of using a process-oriented model, validated under local conditions, as a tool to predict changes and the long-term stabilization of SOC under various agroecosystem management practices, including biochar addition. It should be noted that Century assumes soil bulk density remains constant throughout the simulation period (Pennock and Frick, 2001). In the field, however, soil bulk density will change over time when the quantity and quality of organic matter input changes. For example, biochar has a low bulk density (0.3 g/cm^3) compared to 1.4 g/cm^3 for the coarse textured soil (Delhi) and 1.3 g/cm^3 for the medium textured soil (Elora). As such, the addition of biochar may change soil physical characteristics with time resulting in an increased soil volume (Lehmann, 2009) which will influence the distribution and overall amount of C sequestered.

13.5 Conclusions

A 150 year simulation was generated using the Century Soil Organic Matter Model to evaluate the effect of conventional and conservation agroecosystem management practices, including the addition of biochar, on SOC. The goal of this simulation was to predict the long-term impact of biochar pre-conditioned with UAN and its effect on SOC stocks and C stabilization in soil with contrasting textures. Results showed that the application of biochar, compared to all other management practices, led to an increase and the long-term stabilization of SOC. Our results also indicated that biochar may be added at a much lower quantity (1 to 2 t/ha/y), compared to the typical amount (>10 t/ha/y) suggested by previous field studies, and still lead to an increase in SOC and its long-term stabilization in temperate biomes. This study also showed that the quantity of SOC and the fraction in which it was stabilized was influenced by soil texture. As such agroecosystem management practices, depending upon soil texture, likely affect the long-term balance and stabilization of C in soil. Given that the production of biochar and its incorporation into temperate soils is a relatively novel concept, it is pertinent to conduct field research using different soil textures and climate. Ultimately, these data could be used in the calibration of models and help predict the long-term changes SOC as a result of biochar addition.

Acknowledgements

We thank the Natural Sciences and Engineering Research Council of Canada (NSERC), and the Canadian Foundation for Innovation (CFI) for providing financial assistance and research infrastructure. We also thank NSERC-Industrial for financial assistance to M. Dil.

References

Aertsens, J., De Nocker, L. and Gobin, A., 2013. Valuing the carbon sequestration potential for European agriculture. Land Use Policy 31: 584-594.

Atkinson, C.J., Fitzgerald, J.D. and Hipps, N.A., 2010. Potential mechanisms for achieving agricultural benefits from biochar application to temperate soils: a review. Plant and Soil 337: 1-18.

Barrow, C.J., 2012. Biochar: potential for countering land degradation and for improving agriculture. Applied Geography 34: 21-28.

Bricklemeyer, R.S., Miller, P.R., Turk, P.J., Paustian, K., Keck, T. and Nielsen, G.A. 2007. Sensitivity of the century model to scale-related soil texture variability. Soil Science Society of America Journal 71: 784-792.

Carter, M.R., Anger, D.A., Gregorich, E.G. and Bolinder, M.A., 2003. Characterizing organic matter retention for surface soils in eastern Canada using density and particle size fractions. Canadian Journal of Soil Science 83: 11-23.

Chan, K.Y., Van Zwieten, L., Meszaros, I., Downie, A. and Joseph, S., 2007. Agronomic values of greenwaste biochar as a soil amendment. Australian Journal of Soil Research 45: 629-634.

Chatskikih, D., Olesen, J.E., Hansen, E.M., Elsgaard, L. and Petersen, B.M., 2008. Effects of reduced tillage on net greenhouse gas fluxes from loamy sand soil under winter crops in Denmark. Agriculture, Ecosystems and Environment 128: 117-136.

Dil, M., 2011. The effects of nitrogen-enriched biochar on maize (*Zea mays*) productivity and soil organic carbon. MSc Thesis, Department of Environment and Resource Studies, University of Waterloo, Waterloo, Canada.

Dungait, J.A., Hopkins, D.W., Gregory, A.S. and Whitmore, A.P. 2012. Soil organic matter turnover is governed by accessibility not recalcitrance. Global Change Biology 18: 1781-1796.

Franzluebbers, A.J., 1999. Potential C and N mineralization and microbial biomass from intact and increasingly disturbed sols of varying texture. Soil Biology and Biochemistry 31: 1083-1090.

Govaerts, B., Verhulst, N., Sayre, K.D., Dixon, J. and Dendooven, L., 2009. Conservation agriculture and soil carbon sequestration: between myth and farmer reality. Critical Reviews in Plant Sciences 28: 97-122.

Harner, U., Marschner, B. Bordowski, S. and Amelung, W. 2004. Interactive priming of black carbon and glucose mineralisation. Organic Geochemistry 35: 823-830.

Hassink, J., Bouwnan, L.A., Zwart, K.B., Bloem, J. and Brussaard, L., 1993. Relationships between soil texture, physical protection of organic matter, soil biota, and C and N mineralization in grassland soils. Geoderma 57: 105-128.

Henderson, G.S., 1995. Soil organic matter: a link between forest management and productivity. In: McFee, W.W. and Kelly, J.M. (eds.) Carbon forms and functions in forest soils. Soil Science Society of America Inc., Madison, WI, USA, pp. 419-436.

Jenkinson, D.S. and Rayner, J.H., 1977. The turnover of soil organic matter in some of the Rothamsted classical experiments. Soil Science 123: 298-305.

Johnson, J.M.F., Allmaras, R.R. and Reicosky, D.C. 2006. Estimating source carbon from crop residues, roots and rhizodeposits using the national grain-yield database. Agronomy Journal 98: 622-636.

Karhu, K., Mattila, T., Bergström, I. and Regina, K., 2011. Biochar addition to agricultural soil increased CH_4 uptake and water holding capacity –results from a short-term pilot field study. Agriculture, Ecosystems and Environment 140: 309-313.

Kautz, T., Wirth, S. and Ellmer, F., 2004. Microbial activity in a sandy arable soil is governed by the fertilization regime. European journal of Soil Biology 40: 87-94.

Kimetu, J.M., Lehmann, J., Kinyangi, J.M., Cheng, C.H., Thies, J., Mugendi, D.N. and Pell, A., 2009. Soil organic C stabilization and thresholds in C saturation. Soil Biology and Biochemistry 41: 2100-2104.

Kirschbaum, M.U.F., Harms, B., Mathers, N.J. and Dalal, R.C., 2008. Soil carbon and nigrogen changes after clearing mulga (*Acacia aneura*) vegetation in Queensland, Australia: observations, simulations and scenario analysis. Soil Biology and Biochemistry 40: 392-405.

Lal, R., 2010. Managing soils to address global issues of the twenty-first century. In: Lal, R. and Stewart, B.A. (eds) Food security and soil quality. Taylor and Francis, Boca Raton, FL, USA, pp. 5-22.

Lal, R., 2004a. Soil carbon sequestration impacts on global climate change and food security. Science 304: 1623-1627.

Lal, R., 2004b. Soil carbon sequestration to mitigate climate change. Geoderma 123: 1-22.

Lane, D.R., Coffin, D.P. and Lauenroth, W.K., 1998. Effects of soil texture and precipitation on above-ground net primary productivity and vegetation structure across the central grassland region of the United States. Journal of Vegetation Science 9: 239-250.

Lee, J.W., Hawkins, B., Day, D.M. and Reicosky, D.C. 2010. Sustainability: the capacity of smokeless biomass pyrolysis for energy production, global carbon capture and sequestration. Energy and Environmental Science 3: 1965-1705.

Lehmann, J. and Joseph, S., 2009. Biochar for environmental management: an introduction. In: Lehmann, J. and Joseph, S. (eds.) Biochar for environmental management. Earthscan Publishers, London, UK, pp. 1-12.

Lehmann, J. and Sohi, S.P., 2008. Comment on 'fire-derived charcoal causes loss of forest humus'. Science 321: 1295.

Liang, B., Lehmann, J., Sohi, S.P., Thies, J.E., O'Neill, B., Trujillo, L., Gaunt, J., Solomon, D., Grossman, J., Neves, E.G. and Luizão, F.J., 2010. Black carbon affects the cycling of non-black carbon in soil. Organic Geochemistry 41: 206-213.

Lugato, E. and Berti, A., 2008. Potential carbon sequestration in a cultivated soil under different climate change scenarios: a modelling approach for evaluating promising management practices in north-east Italy. Agriculture, Ecosystems and Environment 128: 97-103.

Lugato, E., Berti, A. and Giardini, L., 2006. Soil organic carbon (SOC) dynamics with and without residue incorporation in relation to different nitrogen fertilisation rates. Geoderma 135: 315-321.

Oelbermann, M. and Voroney, R.P., 2011. An evaluation of the century model to predict soil organic carbon: examples from Costa Rica and Canada. Agroforestry Systems 82: 37-50.

Oelbermann, M., Voroney, R.P. and Gordon, A.M. 2004. Carbon sequestration in tropical and temperate agroforestry systems: a review with examples from Costa Rica and Canada. Agriculture, Ecosystems and Environment 104: 359-377.

Parfitt, R.L., Theng, B.K.G., Whitton, J.S. and Shepherd, T.G., 1997. Effects of clay minerals and land use on organic matter pools. Geoderma 75: 1-12.

Parton, W.J., Stewart, J.W.B. and Cole, C.V., 1988. Dynamics of C, N, P and S in grassland soils: a model. Biogeochemistry 5: 109-131.

Paustian, K., Parton, W.J. and Persson, J., 1992. Modeling soil organic matter in organic-amended and nitrogen-fertilized long-term plots. Soil Science Society of America Journal 56: 476-488.

Pennock, D.J. and Frick, A.H., 2001. The role of field studies in landscape-scale applications of process models: an example of soil redistribution and soil organic carbon modelling using CENTURY. Soil tillage Research 58: 183-191.

Rethemeyer, J., Kramer, C. and Gleixner, G., 2005. Radiocarbon analysis of functional-defined and molecular organic matter fractions from agricultural soil profiles. Geoderma 128: 94-105.

Rumpel, C., Chabbi, A. and Marschner, B. 2012. Carbon storage and sequestration in subsoil horizons: knowledge, gaps and potentials. In: Lal, R., Lorenz, K., Hüttl, R.F., Schneider, B.U. and Von Braun, J. (eds.) Recarbonization of the biosphere: ecosystems and the global carbon cycle. Springer, Heidelberg, Germany, pp. 445-463.

Smith, J.L., Bell, J.M., Bolton Jr., H. and Bailey, V.L. 2009. The initial rate of C substrate utilization and longer-term soil C storage. Biology and Fertility of Soils 44: 315-320.

Sohi, S.P., Krull, E., Lopez-Capel, E. and Bol, R., 2010. A review of biochar and its use and function in soil. Advances in Agronomy 105: 47-82.

Stavi, I. and Lal, R., 2012. Agroforesry and biochar to offset climate change: a review. Agronomy for Sustainable Development 33: 81-96.

Steel, G.C., Torrie, J.H. and Kikey, D.A., 1997. Principles and procedures of statistics: a biometrical approach. McGraw-Hill, New York, NY, USA.

Steiner, C., Teixeira, W.G., Lehmann, J., Nehls, T., Vasconcelos de Macêdo, J.L., Blum, W.E.H. and Zech, W., 2007. Long-term effects of manure, charcoal and mineral fertilization on crop production and fertility on a highly weathered central Amazonian upland soil. Plant and Soil 291: 275-290.

Stewart, C.E., Paustian, K., Conant, R.T., Plante, A.F. and Six, J. 2009. Soil carbon saturation: implications for measureable carbon pool dynamics in long-term incubations. Soil Biology and Biochemistry 41: 357-366.

Stockmann, U., Adams, M.A., Crawford, J.W., Field, D.J., Henakaarchchi, N., Jenkins,M., Minasny, B., McBratney, A.B., De Remy de Courcelles, V., Singh, K., Wheeler, I., Abbott, L., Angers, S.A., Baldock, J., Bird, M., Brookes, P.C. Chenu, C., Jastrow, J.D., Lal, R., Lehmann, J., O'Donnell, A.G., Parton, W.J., Whitehead, D. and Zimmermann, M. 2013. The knowns, known unknowns and unknowns of sequestration of soil organic carbon. Agriculture, Ecosystems and Environment 164: 80-89.

Strong, D.T., Sale, P.W.G. and Helyar, K.R., 1999. The influence of the soil matrix on nitrogen mineralisation and nitrification: iv texture. Australian Journal of Soil Research 37: 329-344.

Varvel, G.E., 2006. Soil organic carbon changes in diversified rotations of the western corn belt. Soil Science Society of America Journal 70: 426-433.

Verhulst, N., Govaerts, B., Verachtert, E., Castellanos-Navarette, A., Mezzalama, M., Wall, P.C., Chocobar, A., Deckers, J. and Sayre, K.D., 2010. Conservation agriculture, improving soil quality for sustainable production systems? In: Lal, R. and Stewart, B.A. (eds) Food security and soil quality. Taylor and Francis, Boca Raton, FL, USA, pp.137-208.

Vignesh, N.S., Rajkishore, S.K., Punith Raj, T.S., Dhumgond, P., Sharanbhoopal, R. and Asha, L., 2012. Influence of amendments on carbon sequestration in cocoa (*Theobroma cacao*) cultivated soils. Research Journal of Agricultural Sciences 3: 10-19.

Wanniarachichi, S.D., Voroney, R.P., Vyn, T.J., Beyaerts, R.P. and MacKenzie, A.F. 1999. Tillage effects on the dynamics of total and corn-residue derived soil organic matter in two southern Ontario soils. Canadian Journal of Soil Science 79: 473-480.

West, T.O. and Post, W.M., 2002. Soil organic carbon sequestration rates by tillage and crop rotation: a global data analysis. Soil Science Society of America Journal 66: 1930-1946.

West, T.O. and Six, J., 2007. Considering the influence of sequestration duration and carbon saturation on estimates of soil carbon capacity. Climatic Change 80: 25-41.

Index

D

E

F

G

H

I

K

L

M